EVOLUTIONARY THEORIES AND RELIGIOUS TRADITIONS

SCIENCE AND CULTURE IN THE NINETEENTH CENTURY

BERNARD LIGHTMAN, EDITOR

EVOLUTIONARY THEORIES AND RELIGIOUS TRADITIONS

National, Transnational, and Global Perspectives, 1800–1920

Edited by Bernard Lightman and Sarah Qidwai

UNIVERSITY OF PITTSBURGH PRESS

Published by the University of Pittsburgh Press, Pittsburgh, Pa., 15260

Manufactured in the United States of America
Printed on acid-free paper
10 9 8 7 6 5 4 3 2 1

Cataloging-in-Publication data is available from the Library of Congress

ISBN 13: 978-0-8229-4792-9
ISBN 10: 0-8229-4792-7

Cover art: John Bentley, *The Hindu Solar System*. Wren Library, Bentley papers, Add. Ms.a.617, by permission of the Master and Fellows of Trinity College Cambridge.

Cover design: Melissa Dias-Mandoly

CONTENTS

ACKNOWLEDGMENTS

This edited volume is an integral part of the historical research undertaken for the Science and Religion Exploring the Spectrum project, centered at Birmingham University and led by Professor Fern Elsdon-Baker. It is the sister project to a volume on evolution and religion in global popular media in the twentieth century edited by Alex Hall and Will Mason-Wilkes. Our thanks to Alex, Will, Fern, and the entire team for their encouragement and advice.

An in-person workshop was planned for the summer of 2020 but was derailed due to the COVID-19 pandemic. The subsequent online workshop, with participants from thirteen different time zones, resulted in the rich and illustrative edited volume in its current form. We wish to thank this superb group of scholars who enthusiastically worked on their papers during the pandemic. We are also indebted to Abby Collier and the team at the University of Pittsburgh Press, the two anonymous reviewers, and all the individuals who provided feedback and suggestions in various stages of this project.

The editor and publisher would like to thank the Faculty of Liberal Arts and Professional Studies, York University, Toronto, Canada, for the financial support it provided to this work.

EVOLUTIONARY THEORIES AND RELIGIOUS TRADITIONS

INTRODUCTION

Bernard Lightman and Sarah Qidwai

The studies gathered in this collection showcase the rich and diverse responses to evolutionary ideas around the globe in the long nineteenth century. Evolutionary figures who championed a variety of theories appealed to different religious traditions. As a result, no single pattern, whether it be harmony, integration, or least of all conflict, is comprehensive enough to capture the complexity of the global cultural engagement with evolution and religion in this period. Consider these examples of very different relations between religious and evolutionary doctrines in Argentina, Türkiye, and Japan. Three years after the publication of the *Origin of Species* (1859), an Argentinian lawyer named José Manuel Estrada (1842–1894), who was also a defender of Catholic thought, complained about those intellectuals in Buenos Aires who had been touting Jean-Baptiste Lamarck's transformism. Estrada was a key figure during parliamentary discussions about secular education in the early 1880s. Although fifty years earlier Charles Darwin (1809–1882) traveled through Argentina while on the *Beagle* voyage, he was not the only evolutionist discussed in the controversies during the eighties. In fact, Darwin's theory was understood through the evolutionary concepts of Herbert Spencer. But Spencer's evolutionary theory diverged significantly from Darwin's, particularly when it came to its implications for religion. In some regions of the world, Darwin was read through the evolutionary perspective of other evolutionary figures in addition to Spencer. In fact, during the late nineteenth century, Muslim intellectuals in Ottoman Türkiye found Ernst Haeckel's writings important in their attempts to build a broad worldview. Inspired by their reading of Haeckel, they created a connection between materialism and evolution that was compatible with their faith. This allowed the introduction of ideas of previous Islamic intellectuals, such as the twelfth-century scholar, mystic, and philosopher Ibn 'Arabî (1165–1240). But neither Haeckel nor materialism appealed to Japanese intellectuals with a commitment to Buddhism. In the beginning of the twentieth century, members of Jinsei gakkai (the Society for Studies of Humanity), an academic society in Japan, organized a conference at the University of Tokyo to mark the momentous occasion of the fiftieth anniversary of the publication of the *Origin*. The lead organizer, Fujikawa Yū (1865–1940), a medical doctor and polymath, was the driving force behind a new publication that focused on the scientific understanding of humankind, the journal *Jinsei*. Fujikawa was an adherent of Jōdo Shinshū (the True School of Pure Land). His commitment to Buddhism was not apparent at the conference, nor did he see any conflict between his faith in Buddhism and evolutionary theory.

In this volume our goal is to explore how intellectuals and the public around the world viewed the relationship between science and religion from about 1800 to about 1920. We have attempted to provide a broad range of national and transnational perspectives. This goal has taken us to Britain, Spain, and the United States but also to New Zealand, Australia, India, Egypt, Argentina, Sri Lanka, China, Japan, and the Ottoman Empire. While various European countries and the United States of America receive some attention, our emphasis is on other parts of the world that are underexamined in modern histories of evolution and religion developed in the United Kingdom and North America. Some chapters deal with a single country or geographic region, while others take a transnational viewpoint. We have examined a variety of religious traditions, including Christianity, Hinduism, Buddhism, Confucianism, and Islam, as well as the intersection of multiple religious traditions. We have drawn attention not only to European scientists other than Darwin who played a significant role in the dissemination of evolutionary ideas but also to the central figures in national contexts who appropriated scientific theories for their own purposes. Many of these central figures were not scientists. They were missionaries, religious figures, journalists, cartoonists, indigenous leaders, novelists, popularizers of science, educators, and politicians. It is important to note that some religious figures were discussing and incorporating evolutionary theory into their thought well before Darwin came on the scene. Although many chapters focus on evolution across the life sciences, others examine the closely connected themes of social evolution and human origins. Disputes about the validity of evolutionary theories and their implications for religion are further situated within interconnected debates about colonialism, imperialism, racism, orientalism, secularization, materialism, and education. Even cultural historians who are not focused on science will therefore find many points of interest for their work. We have thought carefully about how conceptions of the relationship between science and religion have circulated globally and through what cultural media. We have selected the middle of the 1920s as a cutoff point, as this is when a new medium, radio, was introduced, altering the way views on science and religion were communicated.

Any study of the transnational history of evolution and religion in the long nineteenth century must draw on, and bring together, multiple areas of historical research. At minimum, this interdisciplinary study would include previous scholarship on global history, the global history of religion, the global history of science, the history of science and religion, and the history of evolutionary theory as these pertain to the period and places we examine. Sophisticated work in global history began at the beginning of the twenty-first century. Two works published in 2006 argued that globalization had a long history and explored how globalization had been written about from a historical perspective.[1] A more recent study attempted to explain why global and world history have been among the fastest-growing fields for several decades, while another work focuses specifically on the nineteenth century.[2]

As Michael Bergunder sums up, "the historicization of 'religion' remains an unresolved question in the academic discipline dedicated to its study."[3] Thus, for a volume that speaks to a global or transnational view of religion, it is imperative to acknowledge that in the nineteenth century, various worldviews started to define themselves as religious traditions, and any global history of evolution and religion will include a critical examination of this phenomenon.

In the last few decades, the "global" turn in the history of science has marked an important step for scholarship in the discipline.[4] One consequence is that scholars have begun to stress the significance of examining non-Western contexts as well as entanglements driven by imperial encounters. As a result, the call to decolonize scholarship in the history of science is stronger than ever.[5] In the introduction to the eighth volume of *The Cambridge History of Science*, which focuses on modern science in national, transnational, and global context, Hugh Richard Slotten offers a valuable synthesis of recent scholarship in the global history of science.[6] Slotten sums up a key theme in that volume: the recent emphasis on the "situatedness of science," which places "activities, practices, and knowledge in their proper local context."[7] The driving point of analysis is to view the history of modern science "not only in particular local, national, and regional contexts but also with respect to the flow or circulation of knowledge, tools, methods, people, and artifacts across national borders."[8] This orientation decenters the approach to the history of science based on Western accounts, treats non-Western regions as fundamental to the development of modern science, and pictures all regions and nations as interconnected on a global scale. Building on this historiographical vision in their chapter "Science and Imperialism since 1870," Pratik Chakrabarti and Michael Worboys argue that the story is complex and "the history of modern science is one of hybrid origin and articulation and that modern science is neither Western nor colonial, but 'global.'"[9] While Chakrabarti and Worboys's study takes a global approach, entanglements between imperialism and various sciences cast a looming shadow in most accounts.[10] In recent years, scholars have demonstrated that this relationship is imperative to understanding the development of nineteenth-century science. In *The Routledge Handbook of Science and Empire* edited by Andrew Goss, for example, the essayists examine how the symbiotic relationship between science and empire spawned complex systems, institutions, and networks that sustain each other.[11] This would include religious institutions across the globe. While these texts do not directly address the global history of evolution and religion, we grapple with similar issues across this volume.

Incorporating the global turn into the history of science and religion began a little over a decade ago. Two publications charted a way forward toward a more "global" approach to the field of science and religion, which had for two decades been consumed by the critique of the conflict thesis and the development of the complexity principle.[12] In their pioneering edited collection, *Science and Religion around the World*, John Hedley Brooke and Ronald L. Numbers shifted the discussion to include non-Western contexts and non-Abrahamic religions.[13] In addition to chapters on early and modern Judaism, Christianity, and Islam, the collection offered chapters on early Chinese religions, Indic religions, Buddhism, African religions, and unbelief. Acknowledging that the field had been too preoccupied with Christianity's historical relationship to science, they attempted to establish new directions that encompassed a "global" approach. But since there was already a well-developed scholarship on Judaism and Christianity, the chapters devoted to them provided a much fuller picture than the chapters dealing with the other religious traditions. In the sections that covered non-Western topics, the weakness of the previous historiography was all too apparent. Nonetheless, these chapters charted a way forward. Around the same time, in his chapter "A Global History of Science and Religion" in the edited collection *Science and Religion: New Historical Perspectives*, Sujit Sivasundaram posed a number of provocations for what a global history of this field could look like.[14] Sivasundaram argued that "global history" required a historical methodology based on "the analysis of broad patterns and connections across space, rather than a comprehensive history of all regions."[15] More importantly, he addressed the question of colonial encounters, stressing the fact that different intellectual traditions borrowed ideas from each other. These publications represent two independent attempts to open up the historical study of the relationship between science and religion to non-Western religions. Both push scholars to go beyond the Judeo-Christian traditions when referring to religion. Furthermore, they stress a move away from center-periphery frameworks overall. The new set of non-Western actors highlighted in recent research engaged with and, often, reconfigured imported and indigenous scientific ideas and practices and religious traditions, which they did not necessarily systematically distinguish from each other. It is important to emphasize that the accounts of science and belief in non-Western religions do not begin with colonial encounters. We have to examine the rich and local traditions flourishing well before the nineteenth century, accounting for how these systems and relationships change or are destroyed, absorbed, or even appropriated. Sivasundaram's piece was followed six years later by Yiftach Fehige's edited collection *Science and Religion: East and West* (2016), which also attempted to broaden the study of science and religion so as to include the Global South in a significant way. In the introduction, Fehige presents an overview of the development of science and religion as an academic field. But he points out the increasing awareness of the role of Eastern cultures in the rise of modern science. "The interactions between East and West," Fehige declares, "are just another aspect of the trans-cultural character of modern science and not an indication that modern science is constrained by language, geographical location, religious context or national identities."[16] Arguing for a less Eurocentric approach, this is a groundbreaking effort to move toward a transcultural history of science and religion. Numbers, Brooke, Sivasundaram, and Fehige laid the groundwork for a more sophisticated study of the transnational and global history of science and religion.

Given the underdeveloped state of the field, we are not yet at the point where a full global history of science and religion is possible in any meaningful sense. That is a far too ambitious project. In this volume, we have tried to move the field forward by focusing our attention on a particular period, the nineteenth century, on the scientific theme of evolution, and on national and transnational perspectives. We have also tried to explore a broad spectrum of religious traditions. The nineteenth century is a particularly important period for the history of the relationship between science and religion. The modern Western category of science was not established until the nineteenth century.[17] In that century science came to be linked to a purportedly unified set of practices known as "the scientific method," connected to a distinct group of individuals known as scientists, and cleansed of concepts that had previously been seen as defining features that we would now view as metaphysical and religious.[18] The very definition of what constituted science was under debate. As a result, the scientific disciplines were transformed and new disciplines arose. In the early decades of the western European nineteenth century, a new disciplinary landscape developed that included more specialized bodies of knowledge, referred to as physics, chemistry, geology, astronomy, and biology. The fluidity of science in the nineteenth century cannot be overestimated. This volatile atmosphere was destabilized even further in the middle of the century by the theory of evolution, which was not applied merely to biological issues. Specialists from a range of sciences, including astronomy and geology, convened around evolutionary themes. And in turn, evolutionary themes impacted the subsequent growth of each discipline. Moreover, evolution was seen as relevant "outside" the sciences and, some argued, provided the key to understanding every aspect of human culture, including religion. Thus, while historians of science and religion adopted the global approach to bring into the fold the study of religions other than Judaism and Christianity, focusing on a specific field—evolution, for example—will paint an even broader picture. It should also be remembered that the nineteenth century was not only a period in which science was transformed. This was also an era when imperial European powers scrambled to colonize other parts of the world, when industrialization was transforming the social and economic structure of both Europe and its colonies, and when traditional political structures were disrupted and altered. The nineteenth century, after all, was the era in which both liberalism and socialism were born. The transformation of science was closely tied to these other developments. Thus, the accounts are no longer about simple binaries of science and religion; they involve a whole array of social, economic, cultural, and political circumstances as well.

A small number of scholarly works have covered the relationship between evolution and modern religion in various geographical settings. They tend to focus on responses and reactions to Charles Darwin or Darwinism. In his *Dealing with Darwin* (2014), David Livingstone emphasizes the instability of scientific meaning across local spaces. Darwin's evolutionary theory was deployed for different political purposes in Charleston, South Carolina; Wellington, New Zealand; St. Petersburg, Russia; and Cape Town, South Africa. Place, cultural politics, and rhetorical style, Livingstone argues, matter in discussions about Darwin in religious communities. In this book Livingstone illustrated his geography of science-and-religion by focusing on Scottish Calvinists, one "spatially distributed but consciously self-identifying confessional family," and tracing how their confrontation in a range of different places with Darwin generated very different responses.[19] In some cases, Scottish Calvinists rejected Darwinism outright; others tolerated it; and still others welcomed it.

While still focused on a specific religion, by demonstrating diverse responses in different spaces Livingstone demonstrated the importance of examining local contexts. More recently, C. Mackenzie Brown's edited collection *Asian Religious Responses to Darwin* (2020) presents one of the more pluralistic studies on the topic. Emphasizing the importance of cultural, historical, and religious contexts, this volume centered on Asian religions. Brown stated that among all the concerns surrounding the Darwinism model, the lack of "cosmic teleology"—the conception of a universe devoid of transcendent direction, meaning, and purpose—was most common among Asian religious thinkers.[20] However, as with Livingstone, the responses were specifically to Darwin's theory.

This volume could have been organized in many different ways. We might have divided the chapters up by geographical region, which is largely the approach taken in Slotten, Numbers, and Livingstone's *Modern Science in National, Transnational, and Global Context*. But the main disadvantage of that approach is that the transnational features of the topic would have been obscured. We also could have chosen to structure the volume on the basis of scientific discipline, but that would have elided the way evolutionary themes were explored across biology, astronomy, geology, and the study of human antiquity. A third possibility was to focus each section on a different religious tradition, gathering all of the chapters on Christianity, followed by a section on Confucianism, Buddhism, and Hinduism, and then another one on Islam. Brown's edited collection *Asian Religious Responses to Darwin* is structured thus, around the responses of different religions traditions in Asia. But this would have prevented bringing out the parallels that existed between religious traditions in their responses to evolutionary theories. Finally, we could have simply ordered the chapters chronologically, which would have made more apparent any developments over time. However, we settled on a thematic approach that divided the chapters into three groups: "Empire and Colony," "Authority and Minority," and "Appropriation and Response." This structure will allow the reader to more easily see the important thematic links between the chapters, pointing to the transnational dimensions of the project. These themes cut across several chapters in different categories but provide new approaches to the study of evolution and religion. Most importantly, these themes allow the authors to cut across the "East and West" binary and challenge Eurocentric biases.

In the first section, "Empire and Colony," the chapters deal with imperialism and empire as a central category of analysis. The networks, publications, and scholarly communities discussed are seen as part of, or responses to, a web of empire in which there were connections between colonies as well as between colonies and Western metropoles. The chapters explore the complex connections between evolutionary thought, religion, and the growth of European empires. In his chapter, "British Orientalism on Histories of Religion and Astral Sciences in Northern India," Schaffer explores debates within the Asiatic Society revolving around the accuracy and age of ancient Sanskritic astral sciences. The implications for the place of India in the British empire were enormous. At stake was the credibility of Calcutta pandits, Brahmin astronomers, Parisian savants, British orientalists, and biblical scholars. Whereas Schaffer focuses on how issues of science and empire become entangled in discussions about the relationship between Christianity and Hinduism, Fernando and Qidwai draw attention to how the religious diversity of South Asia affected the response to evolution in their "Debating Evolution and Religion in Nineteenth-Century South Asia." Here, where heterogeneous religious worlds existed side by side in a colonial context, science could play a crucial role in their justifications for the superiority of their faith. Putting a Muslim, Sayyid Ahmad Khan, and a Buddhist, Anagarika Dharmapāla, in dialogue with each other, Fernando and Qidwai discuss the advantages and dangers of using "South Asia" as an analytic category in the study of the global history of evolution and religion. In his chapter, Barnes takes us to colonial Australia to examine the world of university education. "Evolution in Colonial Australia: Institutions, Religion, and Moral Formation" treats universities as scientific and religious sites, in which imperatives for the moral, religious, and intellectual formation of an intellectual elite shaped an agenda that excluded the hiring of Darwinians until the 1880s. Barnes, then, argues that midcentury Australian universities were conservative by nature and aimed to transmit received knowledge rather than questioning it or adopting the new discoveries about evolution coming out of Britain. The last chapter in this section, Guimont's "As Above, So Below: The Role of Astronomical Evolution, Imperialism, and Religion in the Long Nineteenth Century," investigates how notions of evolution, astronomy, and extraterrestrial life were informed by racial ideologies of empire. By tracking the secularization of religious pluralism in science fiction, astronomical research, and, finally, socialist theory, Guimont demonstrates the links between pluralism, imperialism, and radical political projects.

The second section, "Authority and Minority," concerns how power dynamics operate at a more focused and local level. Chapters in this section, for example, tackle the themes of how dominant institutions interact with minorities; or how missionaries relate to indigenous populations; or how intermediary figures, or go-betweens, play an important role in how the relationship between evolution and religion unfolds in local contexts. This section begins with Navarro's chapter "Evolution in Times of Political Transformations: Science, Religion, and the Media in Restoration Spain (1874–1898)," which focuses on the role of the Catholic Church and its enemies, largely republicans and liberals, in the debates surrounding evolution. Navarro points out that the debates in the period of turmoil just after the restoration of the monarchy in 1876 were as much about politics as about a scientific theory and its religious implications. Navarro pays special attention to how the controversy played out in the pages of periodicals, with their satire and cartoons, which allows him to analyze the ways in which evolution was used at a popular level. Silva also considers the interplay between the powerful Catholic Church and its opponents, but this time in Argentina, in his "Argentine Positivism on Evolution and Religion in the Late Nineteenth Century." In this case, the opponents were positivists, though Silva notes that one group of Comtists was more congenial to some Catholic thinkers. Here again the scientific and religious issues were linked closely to discussions about what kind of education was needed most by the country, one based on a scientific curriculum or one based on the humanities and the classics. Gold brings us back across the Atlantic to Egypt in her "Prehistoric and Primeval Pasts: Antique Chronology and Civilizational Progress in Semicolonial Egypt." She delves into how the study of human antiquity in Britain and Egypt offered both scientific and religious explanations for human origins. Approaching the topic through the eyes of Joseph Hekekyan, a Europeanized, Catholic, Turkish-Armenian, and Cairo-based civil engineer, Gold shows that nineteenth-century Egyptian treatises on primeval chronology were crucial landscapes for debating the topic of human origins as well as for establishing scientific authority. In the final chapter in this section, titled "Evolution, Religion, and Racial Politics in New Zealand, 1814–1930," Stenhouse tackles the topic of how colonial politicians, men of science, and settlers read Darwin as sanctioning the extinction of the indigenous population and how some missionaries and Māori leaders contested that reading. By placing science, religion, politics, and race within this analytic framework, Stenhouse shows that the racial science and politics of the post-*Origin* decades in New Zealand fed into a larger secularizing process solidifying the cultural authority of laymen, often with ties to Anglicanism.

In the third and final section, "Appropriation and Response," the chapters focus on the appropriation of evolutionary ideas or responses to key evolutionary publications in this period. Emphasizing appropriation and response counteracts the outdated dissemination and passive reception models of Darwin and evolution. Many of the chapters take a transnational approach to the topic and open up understudied geographic and religious contexts for examination. In their collaborative piece titled "Evolution and Religion in Transnational Contexts: Britain, Japan, and China, 1859–1920," Fan and Lightman argue that new global spaces developed in which British, Japanese, and Chinese intellectuals actively grappled with the ramifications arising for various world religions as a result of evolutionary theory. British figures such as Thomas Henry Huxley and Spencer engaged extensively with Eastern religions in working out the broader significance of evolutionary theory, while Japanese and Chinese intellectuals turned to these two evolutionists to understand Buddhism, Confucianism, and Daoism. Though disruptive to traditional religious belief systems, evolution provided these global intellectuals with a new way to make sense of a rapidly changing world. Yoshida and Weldon's chapter, "The Jōdo Shinshū Embrace of Science in Late Meiji and Taishō Japan: Science, Secularism, and Buddhism in the Thought of Ishikawa Seishō and Fujikawa Yū," explores how two religiously devout Japanese scientists, Ishikawa Seishō and Fujikawa Yū, tied together a modern scientific worldview with their Buddhism. The authors assert that appropriation of evolutionary theory was more easily accomplished by them than by Christian intellectuals since Buddhism rejected hardened dogmas, embraced naturalism, and perceived the world as in a state of continuous transformation. As a result, Japanese Buddhists rarely became involved in aggressively anti-scientific movements. At the same time, they were concerned about the rise of secularism in the scientific community and promoted Jōdo Shinshū Buddhism as a way to counteract it. Wan's study, "Cosmic, Theistic Evolution and Kang Youwei: 'The Martin Luther of Confucianism,'" examines a Chinese intellectual who appropriated cosmic evolutionism in order to modernize Confucianism. Like Robert Chambers and Herbert Spencer, Kang endorsed a systematic view of nature in which the evolutionary process connected all phenomena, including scientific, religious, cultural, and political, in a united whole. But unlike Chambers and Spencer, he grafted evolutionary theory onto Chinese religious traditions rather than onto Christianity. Yalçınkaya's "Evolution and Constructions of Islam in the Ottoman World, 1870–1920," shifts the focus from Buddhism and Confucianism to the appropriation of evolutionary theory in Islam. He argues that debates on evolution in this context are better understood as part and parcel of a much broader discussion on materialism that led to the revival of esoteric traditions within Islamic thought. In this story, periodicals, not books, played a crucial role in how Muslim intellectuals engaged with evolutionary ideas. Unexpectedly, Ernst Haeckel and Ludwig Büchner are the important evolutionists, not Darwin. Haeckel takes center stage in Halverson's account of American biologists, evolution, and religion. Titled "'Science Is Justified by Works, Not by Faith': American Biologists and Ernst Haeckel's Evolutionary Science and Religion," this chapter demonstrates that while Haeckel may have been one of the most well-known evolutionists of his time, his concept of monistic religion gained little traction in the United States. American biologists viewed Haeckel's work as too speculative and radical but were, nevertheless, attracted to Herbert Spencer's synthetic philosophy as re-envisioned in somewhat religious terms by their fellow countrymen John Fiske and Edward Youmans.

We conclude with a coda, which discusses how our overall goal in the volume is to move the history of science and religion field forward by shifting the emphasis from complexity to a transnational approach. Throughout the volume we have pointed to how diverse evolutionary theories, not just the Darwinian one, were appropriated by global intellectuals; how we need to consider diverse religious traditions in order to move away from the emphasis on Christianity; and how it is essential to integrate scholarship on empire, colonialism, indigeneity, and race into our analysis of nineteenth-century science and religion. Recent works have called for critical approaches to the topic of science and religion as well.[21] Once we have a better grasp of the transnational dimensions of science and religion in this period, we may begin to work toward a big picture that operates at the global level.

PART I

Empire and Colony

1

British Orientalism on Histories of Religion and Astral Sciences in Northern India

Simon Schaffer

> Is not our wonder heightened, our interest in this singular people increased, when we learn further, that the outlines of the Copernican system of astronomy, together with some of the leading principles of the Newtonian philosophy, and the abstruse speculations of arithmetic and algebra, had been taught in Hindostan long before they were discovered in Europe?
>
> John Mitchell, Glasgow University Prize Essay, 1805

> With respect to Science, we certainly never supposed that the discoveries of Newton or of La Place had been anticipated by the Brahmans: nor that the existing boundaries of scientific knowledge were likely to be extended by our intercourse with India. We still think, however, that the *history* of science may derive important contributions from that source.
>
> Alexander Hamilton, *Edinburgh Review*, 1820

Following the East India Company's assumption of power in Bengal during the 1760s, some of the small number of British scholars and officials in the populous and far-reaching entrepôt of Calcutta became increasingly familiar with Indian astral sciences. The encounter nourished debates about sciences' evolution across a variety of confessional and colonial settings.[1] The aim of this chapter is to make sense of the different roles played by key networks of information and administration in assigning authority over rival histories of religion and sciences in the extremely heterogeneous cultures that colonial powers sought to interpret and thus somehow govern. The remarkable ranges of creeds and practices with which the company was to be concerned disoriented and transformed received accounts of how sciences had changed. As is shown in several other chapters of this book, the very definition of the sciences in this period was neither stable nor effortlessly powerful. Astral sciences provide a peculiarly rich case for the comparative concerns of the histories in this collection since these sciences were explicitly focused on global histories and geographies, often in conflicted settings of colonial strife. Celestial sciences and their associated theologies were directly involved in chronology and cosmography. Tabulated numbers and ingeniously divided instruments seemingly let these astral sciences maneuver between precise times and accurate distances, thus defining comparative positions and epochs. At the start of the long nineteenth century, European protagonists were wont to assert that distance from their metropoles was like passage into the past, as though exotic and remote were ancient and primordial. The statesman Edmund Burke (1729–1797) told the Scottish churchman and historian of empire William Robertson (1721–1793) that "the Great Map of Mankind is unrolled at once; and there is no State or Gradation of Barbarism, and no mode of refinement which we have not at the same instant under our View."[2] Contrasts between what has been labeled the West and the Rest must certainly not be assumed; rather, their construction and use must be charted and subject to critique. Asymmetric geographies of past and future, civilization and barbarism, are entirely hopeless in analyzing orientalism and status of religions and sciences in cultural encounters. But they offer topics for this analysis.

Many British orientalists held that in northern India, especially in territories coming under company control in Bengal and across the Indo-Gangetic Plain, ancient astral knowledge resembling modern European sciences had flourished and degenerated but might be recovered. They assumed such astronomy must be part of a religious worldview. To explain, indeed, to explain away, these astral sciences' fate, some told of ancient practices allegedly distorted or suppressed by indigenous religious institutions. This was certainly no contrast between secular empiricism and devout cosmology. Rather than concede the realities of vicious exploitation, Europeans' astrotheological antiquarianism might instead insinuate colonizers' return to territories once home to their privileged knowledges. An influential critique of historicism's imperial role claims "the modern European idea of history came to non-European peoples in the nineteenth century as somebody's way of saying 'not yet' to somebody else."[3] Orientalism's history of astral sciences was sometimes somebody's way of saying "no longer" about somebody else.

These issues mattered when, on May 2, 1803, British scholars and company officials met, as customary, at the Calcutta residence of the revenue official and judge John Herbert Harington (1765–1828).[4] Guests included members of the Asiatic Society, exclusive rally of British orientalism in the city, founded two decades before by the lawyer William Jones (1746–1794). At

Harington's house in 1803 were contributors to the society's journal, *Asiatick Researches*. Several held positions at the new College of Fort William to train company recruits in rudiments of indigenous law and languages. "Founded on the principles of the Christian religion," its charter written in consultation with the city's British chaplains, the college was designed to combat Jacobin and republican politics it was feared might spread among younger vulnerable company servants.[5] Harington served as its professor of British law in India. His party included the college's Sanskrit professor, Henry Thomas Colebrooke (1765–1837), the first European anywhere to hold such a chair and a company jurist some regarded as sole expert on Hindu law.[6] Many diners also shared astronomical interests. Colebrooke launched studies of Sanskrit because of his work on Indian astronomy, on which he planned a large treatise, and later became president of the new Astronomical Society of London.[7]

But at dinner on May 2, 1803, Colebrooke engaged in a "violent argument" with the Calcutta commercial agent and credit-broker John Bentley (1757–1824) about Bentley's Asiatic Society paper "On the Antiquity of the Suryá Siddhánta," a commentary on the age of Sanskrit astronomical writings.[8] Bentley held these were by no means as old as supposed by recent Paris and Edinburgh commentaries on Indian antiquity and religion. He alleged that apparently ancient Sanskrit treatises he had obtained that computed solar, planetary, and lunar positions were modern forgeries. According to Bentley's reviewers, "astronomical evidence is the only substantial ground adduced by the Hindus for attributing high antiquity to their nation, for historical records do not exist there." Orientalist claims that, lacking Indian historical records, astronomical chronologies were sole authorities had political consequences. Bentley reckoned Hindu priestcraft exploited astral sciences, "the fancy of their Bráhmens and poets in disguising and embellishing their history with allegory and fiction" having "blended the whole together in a mass of absurdity and contradiction." He claimed to have penetrated this deceit using astronomical principles. The expert calculator did not denounce their precision, judging Sanskritic astral sciences highly accurate. So, Bentley argued, Sanskrit texts provided celestial evidence to date them properly, thus unmasking the religious conspiracies behind them.[9]

Colebrooke the administrator explained what was at stake. He judged that Hindus' astronomy served religious purposes, so "they have been content with a very inaccurate practice of it." In unpublished notes he described their ancient instruments' crudity, while in print he argued that Bentley supposed "the age of an astronomical Hindu treatise can be determined *with precision*. I have always contended that their practical astronomy has been too loose and imperfect for that test unless as an *approximation*." An advocate of the antiquity and authority of Sanskritic tradition, Colebrooke was unimpressed by its exactitude. "Mr Bentley concludes forgery and imposture where I only infer carelessness and inaccuracy."[10] Bentley's charges were endorsed by authorities including the heads of the Greenwich and Paris Observatories, while he discussed these views with the unitarian philosopher and Bengali educator Ram Mohan Roy (1772–1833). The reported violence of the 1803 dinner table argument and the status and age of what the disputants labeled "Hindu astronomy" illuminate interests in historic relations between religious beliefs and sciences during major crises in orientalism.

Threats allegedly manifest in subversive politics' Indian ramifications were indicated by the militant program for the College at Fort William composed with the company's Christian ministers. Missionaries' role was the topic of fierce controversy until at least 1813, when, under the Charter Act that ended almost all the company's commercial monopoly and asserted state authority, it became legal to evangelize in company territories. Pamphlet wars were stoked by proselytization and by mutiny of company troops blamed on sepoys' fears of Christian conversion.[11] The term *Hinduism*, increasingly current from the later eighteenth century, seems first to have been used where apostasy and conversion were involved: by Calcutta merchant Thomas Motte describing the defection of an Orissa landlord from the Hindu faith and by evangelical company official Charles Grant (1746–1823) lobbying for missionary campaigns in Bengal.[12] A so-called pious clause proposed by Grant and his allies would insert into the company's charter provisions "for sending chaplains, schoolmasters, and missionaries, to promulgate useful knowledge and the true religion among the natives of India." The clause was rejected by Scottish lawyer Henry Dundas (1742–1811), metropolitan manager of Indian affairs, but only because "Hindoos, though they were a timid and innocent people, yet had been singularly bigoted in their religious prejudices," so were unlikely to convert.[13] Dundas nevertheless had no doubts about the need for conservative vigilance. He reckoned threats to order in India so great he was concerned that the new College at Fort William might itself incubate subversion: "Such an assemblage of literary and philosophical men would degenerate into a school of Jacobinism in India. I hate Jacobinism everywhere, but in India I should consider it as the Devil itself and to be guarded against with equal assiduity."[14]

INDIA WAS THE DEBATABLE GROUND

Apparently, such fear and loathing especially affected Indian astronomy. In his 1795 *History of Hindostan* the amply beneficed cleric Thomas Maurice (1754–1824), "official mythographer of the East India Company," seeking traces of Trinitarian doctrine within Hindu cosmology, made religion a sign of Hindus' original kinship with the British, followed by subsequent degeneration from primitive sanctity. He thundered that radical French writers perverted "the noble science of astronomy to the subversion and annihilation of everything hitherto considered sacred . . . in the vain hope of proving Christianity a system equally baseless and contracted." This was a hideous threat because British astronomy vouchsafed "a theology so inseparably connected with the national government."[15] Distorted histories of Sanskritic astronomy had allegedly provided weapons for atheist anarchy threatening company rule and Christian faith. "It is notorious, also, that India was the *debatable ground* on which the fury of the Jacobin hostility had reared her most triumphant banner; the imagined antiquity of that empire, and the uncounted ages during which the arts and sciences were asserted to have flourished among the Brahmins, being boldly announced by . . . the whole host of infidel writers."[16] In several imperial projects, traces of ancient sciences were used to debate European histories. Materials gathered during French occupation of Egypt from 1799, especially astronomical and zodiacal signs detected in Nile valley temples, were taken to challenge scriptural chronologies.[17] But for Maurice and his colleagues, Indian astral sciences seemed especially, disturbingly, salient.

Astral sciences in Indian societies included sophisticated methods, some from Hellenistic and Islamic formulations, for calendrical determination of celestial events. Horoscopes and divination were reliant on detailed tabulations of planetary positions derived from artfully defined epochs. Learned debates were long conducted at centers of scholarship and devotion about this set of practices, *jyotiḥśāstra*, in relation to theogonic cosmologies in the puranic traditions.[18] One specialist in astral sciences, the Bengali pandit Rāmacandra Sabhāpati, reportedly told Jones that mythical cosmologies, stocked with tales of a flat Earth, milky oceans, and Moon-eating dragons, must be rejected: "We know," Rāmacandra apparently insisted of lunar eclipses, "that the supposed head and tail of the dragon mean only the nodes, or points formed by the intersections of the ecliptic and the moon's orbit." Contrasting his astral sciences with puranic cosmology, the pandit allegedly argued that "they have imagined a system which exists only in their fancy, but we consider nothing as true without such evidence as cannot be questioned." Jones added, "We must never confound the system of the *Jyautishicas* or mathematical astronomers, with that of the *Pauránicas*, or poetical fabulists; for to such a confusion alone must we impute the many mistakes of Europeans on the subject of Indian science."[19]

The Asiatic Society's fellows reckoned authority hinged on formal codes sanctioned by antiquity. In legal, religious, and astronomical practices, it was held that older authoritative texts lay behind what seemed unruly, sometimes barely comprehensible, patterns of customary behavior in courts, bazaars, or temples.[20] Pandits were sometimes understood as ritualists cajoled to reveal traces of texts upon which these behaviors supposedly relied. Since indispensable, they were suspect.[21] If principles could not be had, British orientalists occasionally judged that Brahmins "deem'd themselves precluded by laws in their opinion of sacred and eternal obligation from any development of their secret doctrines."[22] At times it was held instead that they followed rules whose deeper laws they did not know or had forgotten. Thus Dundas was told in 1788, "It is certain that in Bengal there is a mean profession of people, who annually compile almanacks from ancient Tables and calculate eclipses with considerable exactness, but are altogether ignorant of the principles on which their calculations depend."[23]

Versions of British orientalism did not depend on primordial conflicts between religion and sciences. Rather, definitions of such enterprises were constantly in question. Provisional determination of what might count as calculation and what as creed was in part the complex result, never the directive cause, of these labors. For many British disciples of Newtonian cults, celestial mechanics rested on laws of governance vouchsafed by a wisely benevolent deity through the solar system's heavenly book. The doctrine was exemplified in Scottish instrument-maker James Ferguson's *Astronomy Explained upon Sir Isaac Newton's Principles*, a work adapted across northern India by Hindu and Muslim experts in Arabic, Persian, Bengali, and Urdu versions.[24] Ferguson's Newtonian theology saw everywhere in space astronomically engaged beings who adored their Creator. It helped reinforce the discourses on empire and extraterrestrial life analyzed in much more detail in Edward Guimont's chapter later in this section. The intrinsic piety of all alien souls urged in Ferguson's cosmology was a suggestive notion for Christian Britons in Bengal.[25] Yet their astronomy was not always synonymous with orthodox piety. In 1796 Harington and his Calcutta colleagues debated a new edition of Thomas Paine's incendiary *Age of Reason*, composed in Paris, which used arguments for a plurality of inhabited worlds to demolish biblical authority and the Incarnation.[26]

So to understand orientalists' astrotheologies, it is instructive that few of these debates' British protagonists managed observatories or expertly used astronomical instruments. Rather, the astronomy of advocates such as Colebrooke and Bentley, both of whom professionally compiled figures on Bengali trade, resembled legal or commercial printed tables in ordered accounts.[27] Some British interlocutors then asserted that practices of Sanskritic astral sciences must also be based on ancient celestial codes. There was an equivalence between recovery of astronomical laws and of Hindu law codes that jurists such as Jones and Colebrooke, dependent on legal pandits, sought to formalize. Then it might be possible to understand evolution of Hindu society—and govern its conduct.[28] One of Colebrooke's earliest postings in 1786 was to Tirhut, a center of Sanskrit and astronomical learning that Jones reckoned might host a "university" that "confers degrees in Hindu law." Colebrooke reported that local Brahmins possessed astronomical tables "which have internal evidence of antiquity, and prove their knowledge of many points of astronomy and mathematics, which became involved in the darkness of declining literature, and have been revived by modern discoveries in Europe." He had "hopes of further information, which is, however, extremely difficult to procure, after the long period of disuse and neglect of more modern Brahmins." Colebrooke's report soon reached the Royal Society in London, with promises to investigate "any point relating to this Extraordinary People."[29]

Ancient Sanskritic astral sciences were thus evaluated against principles of recent European celestial mechanics, codified in *Principia Mathematica* and revisions by Parisian astronomers, embodied in ephemerides in nautical almanacs and transit catalogs. According to some Newtonian creeds, an ancient theology spread across Europe and Asia. Having initially supposed Newton would "be adored in Hindustan" were his works known to Benares pandits, Jones eventually claimed "the whole of his theology and part of his philosophy may be found in the Vedas." Overlaid by myth, orientalists argued, those sciences had degenerated and all but vanished. In his Vedic studies, which inaugurated British accounts of what were taken to be the oldest Hindu scriptures, Colebrooke insisted on their authenticity, setting out proper relations with pandits. "The practice of reading the principal Vedas in superstitious modes tends to preserve the genuine text." Through pious recitation and legal commentary, he argued, "forgeries have become impracticable."[30] He held that these codes testified to an original society comparable to an idealized Britain, an ancient Asian land of egalitarian beef-eaters and monotheist astronomers that had declined such that modern Hindu practitioners "seem to misunderstand the numerous texts" on which their religion was founded.[31] They were to be recovered by enlightened sciences of European modernity: India's past was Britain's present. Other British commentators countered that such sciences had never been developed in ancient India, which under priestcraft remained primitive: India's present was Britain's past. Then it would fall to British rulers to redeem celestial morality in territories now governed by grace of providence.

THEIR HIGH ANTIQUITY AND THE ORIGINALITY OF THEIR KNOWLEDGE

Radical French arguments affected versions of what these Europeans understood of Indian astral sciences' past. India was the "debatable ground of imagined antiquity," so anti-Jacobins supposed. Asiatic Society reactions to these French works were never unambiguous, sometimes hostile. Models of devout sciences and interpretative authority could block notions peddled from Paris.

Using documents assembled principally by Jesuit missionaries in southern Asia and Siam, relying on indigenous interpreters and experts, French astronomers admired Brahmin exponents of *jyotiḥśāstra* who with speed and precision used cowrie shells and palm-leaf manuscripts to calculate eclipse times and planetary positions. According to the errant Guillaume Le Gentil (1725–1792), billeted in a ruined fort at Pondicherry on the Bay of Bengal while seeking to observe the transit of Venus, such astronomical calculation "is a secret locked up in the Brahmin caste, a secret that forms part of their religion, of which only the Brahmins are the depositaries; and no-one except these philosophers can be involved in these calculations." Thanks to the mediation of a Tamil Christian convert who'd studied with the Jesuit mission, Le Gentil trained with Nana Motu, an expert *jyotiṣa* who "gave me all the eclipse phases in less than three quarters of an hour," a result Le Gentil promptly checked against his almanac.[32]

French astronomers sought to evaluate Indian solar and lunar tables, reckoning they might have been assembled almost five millennia before. Such chronologies were common in Indian traditions, summarized in the sixteenth-century Persian classic the *Ain-i-Akbari*, composed for the Mughal emperor Akbar by his *vazīr-e azam* and historian Abu'l Fazl (1551–1602), published in an English version in Calcutta in 1783.[33] Chronological inquiries were also familiar to European astronomers such as Newton and his interlocutors, who scoured classical histories to correlate astronomical events with Holy Writ. Unknown to early modern European chronologists, Sanskritic epochs transmitted via Islamic tabulations had been adopted in some Christian astronomical schedules for the Deluge. Astronomical chronology required judgment of reliability of past observations, of astral sciences as marker of a civilization's sophistication.[34] "If we are obliged to allow that the Hindoos were so far advanced in the science of astronomy as to make the observations which they appear to have made about 4890 years ago," wrote wealthy nabob Quintin Craufurd (1743–1819), based in revolutionary Paris in 1790, "we must necessarily suppose many ages previous to that in which a great and numerous nation might gradually proceed to that degree of refinement and knowledge which the Hindoos seem to have enjoyed." Edinburgh mathematics professor John Playfair (1748–1819), admirer of French astronomers and correspondent of Calcutta orientalists, insisted the same year that astronomy could uniquely unlock history since it was independent of culture or creed. The next chapter of this book shows how for protagonists such as Sayyid Ahmad Khan, the great Delhi-born scholar and reformer in the East India Company's service, the sciences offered ways of winning legitimacy for a carefully defined faith. The science of the sometime Kirk minister and pugnacious Enlightenment scholar Playfair was a way of producing reliable history seemingly independent of confessional loyalty: "It is through the medium of astronomy alone that a few rays from those distant objects can be conveyed in safety to the eye of a modern observer free from the false colourings of vanity and superstition."[35]

Commentaries on the antiquity of Sanskritic astral sciences came from Craufurd's and Playfair's source Jean-Sylvain Bailly (1736–1793), protagonist of astronomical analysis in Parisian academies and observatory manager at the Louvre. His vast histories used ancients' astronomical prowess to calibrate cultures. Bailly declared that underlying extant tabulations lay principles established at Benares, judged the religious center of Indian astral sciences: "A method named *souria siddantam* served as a rule. This rule was once understood but today no-one understands it." The term Bailly used for this allegedly primordial and forgotten science was adapted from the *Sūrya Siddhānta* [*Doctrine of the Sun*], one more recent version of which was assembled in an astronomical verse compilation around 400 CE with instructions for calculations of planetary and lunar orbits, much revised until at least 800 CE, its provenance and scope summarized by the Ujjain astronomer Varāhamihira (505–587). Few details of this history were known to eighteenth-century European astronomers, who instead stressed achievement and loss of a much older tradition. Bailly inferred that "this astronomy that exists at Benares, which served as a rule and which is no longer understood, seems therefore to be the original and primitive Astronomy of India, on which these different tables have been built."[36]

Common measures of zodiac division, precession of stellar positions, and establishment of a founding epoch of 3102 BCE (identical to the cosmological start of the Kali Yuga, the age of darkness in which humanity now lived) seemingly testified to a shared Oriental astral culture much older than biblical chronology, with its tales of Fall and Deluge. Bailly was convinced that planetary conjunctions at that distant epoch were established by observation: "To suppose they got these mean motions without ancient observations would be an absurdity," since it would suppose Indian knowledge of Parisian celestial mechanics. Matches with eighteenth-century European calculations appeared impressive. The annual period according to Sanskrit tables through which the Sun returned to a given position viewed from Earth differed only 96 seconds from best French estimates. Bailly urged that while Indian calculations of eclipses were directed by religious needs, nevertheless "they possessed that part of the science much more completely than all the astronomers who followed them and preceded us; the riches of the Indians in this respect, as well as the accuracy of their mean motions, demonstrate simultaneously their high antiquity and the originality of their knowledge." Embodied in Sanskritic tradition and calibrated against enlightened European sciences, astronomy could allegedly separate religious prejudice from celestial truth. As Bailly insisted, "I have pleaded the cause of Asia and the ancient inhabitants of the Earth."[37]

Bailly's arguments were soon approved by French academies. A report coauthored by Pierre-Simon Laplace (1749–1827), doyen of Parisian exact sciences, noted accurate Indian estimates of the Moon's mean motion, though they lacked the law of gravity, and deduced that Indian astral sciences fixed values no later than 4100 BCE, long before even the start of the Kali Yuga. "This agreement testifies with even more verisimilitude to the high antiquity of the astronomical determinations of the Indians, the theory which led us to them having been unknown." Laplace told Bailly that the coincidence of ancient Indian and modern French values for Saturn's mean motion showed the superiority and antiquity of astral sciences.[38] According to journalist Guillaume Dubois de Rochefort (1731–1788), such astral sciences challenged scriptural history. Since "astronomy can rightly be regarded as one of the Sciences which do most honour to the human mind," demonstration of its Oriental antiquity "is to put the Indians above all educated peoples" and would "give to this Nation the right of primogeniture over all the others."[39]

THE MOST STRIKING EXAMPLE OF THE PROGRESS OF SCIENCE

Parisian claims about Indian antiquity were attentively read in Edinburgh, whose literati ardently debated novel French doctrines and were deeply invested in the East India Company's regime.[40] From 1783 to 1784 the Edinburgh MP Dundas helped overhaul company administration under a Board of Control with himself as president. The project coincided with the impeachment of the company's governor, Warren Hastings (1732–1818), a trial of what his prosecutor Edmund Burke described as "geographical morality, by which the duties of men . . . are not to be governed by their relations to the Great Governor of the Universe, or by their relations to men, but by climates, degrees of longitude and latitude." Legitimacy of British rule in Bengal seemed to hinge on relations with an ancient Indian constitution.[41] Dundas was lobbied by Playfair, Robertson, and the law professor Allan Maconochie (1748–1816), Edinburgh academics impressed by the antiquity of Indian astral sciences. They had familial Indian connections: their sons served the company, Robertson's in war against Mysore, and Maconochie's hired with Dundas's patronage as master of the Madras Mint.[42] Robertson headed Edinburgh University and the Scottish Kirk. He encouraged his friend Adam Smith (1723–1790), Glasgow philosophy professor, to compose a history of astronomy that argued "heathen antiquity" mistakenly treated irregular events as the direct results of divine powers, "the invisible hand of Jupiter." With social progress, according to Smith, celestial phenomena would gradually be integrated into rational astronomical systems.[43] Robertson was also a colleague of Edinburgh natural philosopher James Hutton (1726–1797), whose Earth history allegedly challenged scriptural chronology. Beginning in 1785, Robertson redrafted Hutton's statements that "it belongs not to religion to teach that natural order of events, which man in his science may be able to unfold," mitigating their polemic and advising Hutton to consult Smith. Hutton edited Smith's history of astronomy and primitive religion, while Playfair produced a widely read summary of Hutton's theory and its implications. Some critics held that it was Playfair's long-term geohistory that made him foolishly endorse Indian antiquity; others damned what seemed the anti-scripturalism of Laplacian *theophobia gallica*.[44]

The Edinburgh lobby used conjectural histories of evolution through stages of hunter, herdsman, farmer, and merchant to modernity emergent in Scottish civil society.[45] Robertson's kinsman and client Maconochie lectured on evolution from the "savage state" through "the rise of religion and mythology" to pastoral societies such as India, thence to Scottish civility. He read Bailly's account of Asiatic development and told the new Royal Society of Edinburgh about "the religion, political institutions and sciences of the Hindoos," using stadial theories of social evolution "to promote enquiry into the subjects on philosophical principles."[46] The lobby's aim was patronage for Jones's new Asiatic Society in a historical program of consequence for astronomy and imperial governance. In a 1783 memorandum for Dundas, Maconochie summarized Le Gentil's demonstration of antiquity and accuracy of Indian astronomy, stressed relations between their astral sciences and religion, and claimed that "a sort of College of them subsists at Benares where the Bramins, that possess in other parts of India the Astronomical calculus, pretend that a much superior astronomy to theirs is understood and the principles known on which their calculations are founded." Maconochie urged funds for Jones in Calcutta so "this Nation, when Sovereigns of the East," could secure knowledge of "the only institution subsisting among mankind that has preserved by legitimate descent the sciences and opinions of the ancient world."[47] After the opening of Hastings's trial and Bailly's new accounts, in 1788 Maconochie told Dundas that Bailly's work "has been considered with much attention at Edinburgh" and reiterated that Indian astronomy was ancient, autonomous, and more exact than any other nation, its observational basis dated back to at least 3102 BCE, and it was "taught at Benares by learned Bramins" where "very ancient works in astronomy are still extant." Maconochie set out what he understood of the relation between these works and religious traditions. "It is an object worthy of our Monarch, the Sovereign of the Banks of the Ganges, and the distinguished patron of Literature and Science, to give such directions as may be necessary for discovering, collecting and translating whatever is extant of the ancient works of the Hindoos."[48]

The campaign was reiterated in a university memorial to Dundas, with a summary of Le Gentil and Bailly. Claiming it was "infinitely more probable" that Indian astral sciences were autonomous from any other classical tradition, the memorial drew the minister's attention to the *Sūrya Siddhānta* as the oldest version preserved by "the Brahmins of Benares and over all Bengal." It appeared unlikely to the Edinburgh scholars that any "books that explain the principles geometrically and by means of diagrams" survived: there was no Sanskrit *Principia Mathematica*. But it seemed in Bengal, under British rule, there must be tables that would not only convey these principles but also "contain facts which ascertain their own date and authenticate themselves." Astral sciences had a unique capacity to establish chronology and prediction together. These sciences must be purified from religious superstition: "It is of consequence that the one should be distinguished from the other," Dundas was told, and the charge of the British state through company officials was to achieve this end.[49]

The lobby's arguments about antiquity of Indian astral sciences and patronage of Oriental scholarship became public at the height of Hastings's trial and crises of French politics. Before assuming the Edinburgh mathematics chair, Playfair experienced fieldwork with the Astronomer Royal Nevil Maskelyne (1732–1811) and with Reuben Burrow (1747–1792), who later surveyed in Bengal and joined the Asiatic Society.[50] In early 1789 Playfair lectured the Royal Society of Edinburgh in support of Le Gentil and Bailly, echoing Maconochie that Sanskritic astral sciences were original to India, founded on celestial observation conducted before the start of the Kali Yuga at 3102 BCE. "That observations made in India, when all of Europe was barbarous or uninhabited, and investigations into the subtle effects of gravitation made in Europe near five thousand years afterwards, should thus come in mutual support of one another, is perhaps the most striking example of the progress and vicissitudes of science, which the history of mankind has yet exhibited."[51] Yet Playfair reiterated that Indian astronomy "gives no theory, nor even any description of the celestial phenomena." Its pious practitioners allegedly acted by rote, not reflection. "The Brahmin seating himself on the ground and arranging his shells before him repeats the enigmatical verses that are to guide his calculation." Coincidence between Indian measures of precession of the equinoxes, obliquity of the ecliptic, and changes in motions of the Moon, Jupiter, and Saturn and the most recent Paris results by Laplace and Lagrange was impressive because Sanskrit tradition lacked understanding of celestial mechanics, the European theory that defined it.[52]

The most widely read versions of these arguments appeared in Robertson's

Historical Disquisition Concerning the Knowledge which the Ancients Had of India, composed from 1788 and prompted by the Hastings trial. It was published in 1791 in a large print run. Copies were distributed by the East India Company and presented to the governor-general in Calcutta and to Dundas, who responded that "nobody can read your work with attention without perceiving in it traces well-calculated to aid our Indian statesmen of modern times."[53] This conjectural history followed Smith in asserting that primitive societies had been polytheist and superstitious. Robertson held that while Brahmin priests were rational monotheists, they had maintained the populace in ignorance, insinuating similar contrasts between Scotland's moderate churchmen and benighted populace.[54] Scottish civic humanist cosmology was analogized with Playfair's summaries of French accounts of Sanskritic astronomy. Robertson's manuscript drafts on Indian astral sciences exploited Playfair's analysis, indicating Sanskrit tables' "precision not inferior to what could be attained by consulting the most perfect Tables now known in Europe or by attending to all the improvements in Astronomical Science from the age of Kepler to the present time." Le Gentil, Bailly, and Playfair all confirmed, he wrote in draft, that Indian observations "were made above three thousand years before the Christian aera." But in print, "to bring the chronology of the Hindoos to accord somewhat better with that of the Old Testament," he moderated his apparently anti-scriptural claim from "above three thousand years before the Christian aera" to "a very early period."[55] Robertson published the program of his fellow lobbyists, notably that what his draft called "pretious remains" of astral methods were preserved in the *Sūrya Siddhānta* by Brahmins at Benares, "the Athens of India." This sobriquet for Benares assimilated it somehow to Robertson's literary audience of polite classicists. With a complementary view to the Asiatic Society and the Board of Control, the Edinburgh principal argued, "Those entrusted with the administration of the British empire in India should enable some person, capable by his talents and liberality of sentiment of investigating and explaining the more abstruse parts of Indian philosophy."[56]

IN GREAT MEASURE INDEPENDENT OF THE PANDITS

Exchanges between Edinburgh and Calcutta on Indian astral sciences intensified through the principal project launched in the later 1780s on Sanskritic astronomy at the Asiatic Society, involving Jones and the company administrator and surveyor Samuel Davis (1760–1819) at Bhagalpur. They knew Le Gentil's reports but had not yet seen Bailly's latest account. In 1788 Davis completed an initial study of the *Sūrya Siddhānta*, a version of which, together with an invaluable commentary, had been offered to him via officials from "the Athens of India," Benares. Since Davis barely understood Sanskrit and knew little of the history of these astral sciences, the input from the pandit Rādhācaraṇa Śarmā was decisive. Śarmā was a "lively young Brahman" who passed the Sanskrit text to Jones in April 1788 with guidance on its terms.[57] The result was a paper copyedited by Jones, which sought principles of *jyotiḥśāstra*, explained how it used eccentrics for unequal planetary motions, computed eclipses from its methods calibrated against those of the Board of Longitude's *Nautical Almanac*, and, from estimates of changing obliquity of the ecliptic, dated its observations to a remarkable 2050 BCE.[58]

Calcutta orientalists acted as though spokesmen for Indian astral sciences. Davis sent his paper to the Royal Society in London and soon joined its fellowship. The Scottish scholars fitted Davis's work into their favored frame of ancient lost astronomy concealed by religious tradition. In May 1792 Playfair told Robertson that while Davis "does not appear to be a deep scholar in the Mathematical sciences and indeed does not pretend to be so," nevertheless, the Edinburgh lobby should continue to demand subsidy for Jones and Davis to render the *Sūrya Siddhānta*, since "as pointed out both by internal characters and by the traditions of the country that Book is the most antient and also the most profound." Just before his death in June 1793, Robertson added to his history, distinguishing between the "accompaniment of fiction and extravagance" and what is "left behind, a very rational and elaborate system of astronomical calculation." He reckoned that material in *Asiatick Researches* "is extremely favourable to the hypothesis which ascribes a very high Antiquity to the Astronomy of the Brahmins."[59] Meanwhile, Playfair used Scottish stadial theory to argue that trigonometry's invention—the "application of arithmetic to geometry," evident in *Sūrya Siddhānta* extracts from Calcutta—represented a millennial advance in ancient India from land surveying to celestial sciences. So if the *Sūrya Siddhānta* were dated to 2000 BCE, then "the origin of the sciences in Hindostan" was one thousand years earlier and Bailly's chronology vindicated. "The *Caly Yug*, or the year 3102 before our aera, marks a point in the duration of the world, before which the foundations of astronomy were laid in the East, and those observations made, from which the tables of the Brahmins have been composed."[60]

Copies of Bailly's history of Indian astronomy belatedly reached Calcutta in late 1789 and a year later Playfair's summary.[61] Davis and Jones soon sought more evidence for Indian astral sciences' antiquity. Accounts of changing equinoxes and solstices, provided by the pandit Rādhācaraṇa Śarmā from a corrupted passage of Varāhamihira's sixth-century divinatory survey *Bṛhat-Saṃhitā*, according to Jones, must refer to astronomical observations in 1181 BCE, which Davis sought to correct to 1391 BCE. Either date, unlike Bailly's five millennia, could fit scriptural chronology.[62] Alongside subtle alterations in chronology, these orientalists denounced what seemed "gross errors" in

Bailly and, by implication, in Playfair. Jones repeatedly expressed the hope of Davis's "triumph over M. Bailly."[63] "It is from you alone," Jones wrote in early 1791, "not from Bailly or Playfair, that we must expect accurate information on those subjects." The key was relation with the pandits: dialogue with their work gave Calcutta orientalists authority. "By translating the *Sūrya Siddhānta* from the original with the assistance of the Brahmens, you will have a great advantage over Le Gentil and all the other writers on Indian astronomy." Pandits' expertise was essential, yet the orientalists must somehow win autonomy. "I am very anxious that (if you do not already read Sanskrit with ease) you will learn enough of it to be in great measure independent of the Pandits."[64]

In early 1793 Jones told the Asiatic Society that "an accurate knowledge of Sanskrit and a confidential intercourse with learned Brahmins are the only means of separating truth from fable."[65] Soon after, queries from Playfair arrived in Calcutta, requesting translation of the *Sūrya Siddhānta* and catalogs of Sanskrit astral sciences, along with records of instruments and observatories "certainly known to be of Hindu origin" and examination of the heavens in the company of "a Hindu astronomer." This "shows how curious they are in England [*sic*] on every subject relating to Indian astronomy," Jones remarked, grumbling that though he'd asked Rādhācaraṇa Śarmā to point out Canopus (Agastya), he had allegedly never managed to recruit "an intelligent Hindu astronomer" in Bengal "who could name in Sanscrit all the constellations."[66] Jones answered Playfair in early 1794, in his final annual Asiatic Society discourse: "We must suspend our opinion of Indian astronomical knowledge" until the *Sūrya Siddhānta* and its companion texts were translated. It was crucial to discriminate between what was introduced from Islamic traditions in central Asia and Persia, what from Europe, and what was "clearly of Indian origin."[67]

A salient example of the complex work on the sources of ancient Indian astral science and its alleged decay was carried out in Benares, where in 1771 the East India commander Robert Barker (1732–1789) scrutinized the remains of an impressive astronomical observatory on the banks of the Ganges. Barker's account revealed the observatory's remarkable precision, so he inferred its very considerable age followed by inevitable decline. In 1788 the astronomer Reuben Burrow again surveyed it and asserted its antiquity, while William Jones recommended making a model of the structure for the Asiatic Society. In fact, the Benares Observatory had been commissioned by the Jaipur ruler Jai Singh II (1681–1743) between 1728 and 1730, along with several other astronomical establishments across northern India. Though it also recruited Jesuit experts from Portugal and France, Jai Singh's program was at first seemingly unknown to the British scholars, who instead assumed that great precision must be a sign of antiquity subsequently degenerated. As late as 1794, the

East India Company soldier and Bengali world traveler Dean Mahomed (1759–1851) affirmed that the Benares Observatory was centuries old. When in 1796 the Calcutta scholars at last acknowledged it had been more recently built for the raja and sent Playfair a copy of Jai Singh's astronomical tables, they then guessed that it was rather "of European as of Hindu construction." Yet they continued to hold that "the Hindu astronomy from the learned disquisitions of Mr Bailly and Professor Playfair appears to carry internal marks of antiquity which do not stand in need of confirmation by collateral evidence." The observatory in "the Athens of India" thus provided a complementary indication of conflicted assumptions in histories of the evolution of astral sciences in northern India.[68]

Debates between Paris, Edinburgh, and Calcutta scholars showed that pandits' skills as what have been called "colonial antiquarians" were decisive.[69] Calcutta-based diplomat and Islamic scholar Tafazzul Husain Khan (1727–1801), commissioned by the Edinburgh lobby to "make enquiry among the Brahmans of Benares" about "the ancient astronomy of the Hindus," and who interpreted descriptions of Jai Singh's instruments, nevertheless warned of "the present ignorance of the pandits and the necessity of employing a translator." In a review of Bentley, Benares official and orientalist James Prinsep declared, "In a study connected so closely with the comparative age of nations and their priority in the enlightenment of civilization," authority lay with "pundits and astronomers alone."[70] Northern Indian patronage for these experts was undermined with crises of landed wealth and destructive effects of company rule. While the Asiatic Society's fellowship remained barred to indigenous members until the end of the 1820s, the regime drew pandits to Calcutta from Bengali centers of doctrinal practice and from Benares, identified as headquarters of Sanskrit astronomy, where titular British administration was established by 1795.[71] A Sanskrit college was set up there "for the cultivation of the Laws, Literature and (as inseparably connected with the former) the Religion of the Hindoos." With pandits' aid, such an institution might extract "complete treatises" in "Hindoo Religion, Laws, arts or sciences."[72] Colebrooke had experience with those pandits, notably in projects to summarize Hindu legal codes that might displace these experts' advice by judicial texts. Residence in Benares in 1797–1798 with funding for two pandits allowed him to start legal summaries of Hindu religious duties. He found that his pandits offered as their own what amounted to plagiarisms of legal writings and concluded he should do the work himself, "since no Pandit is able (or adapted by his habits of thinking) to compile a digest in the form I require." Nevertheless, it seemed to Colebrooke, "the publick have no doubt more confidence in the Pandits than in me." By May 1803 he urged a Sanskrit department at the College of Fort William, to be managed by another of his Benares pandits, Manirama Tara.[73] The same month, Bentley's charge that canonical Sanskrit texts were forgeries revived questions about relations between British orientalists and those who provided and interpreted those writings.

THEY TURNED THE PAST INTO THE FUTURE

"Mr Bentley was, as his writings evince, a good hater." So Colebrooke summarized polemics the Calcutta calculator issued from 1796. Bentley allegedly knew no Sanskrit. Though he relied on pandits and "Hindu astronomers," no names survive. Rather, he employed the hermeneutics of suspicion to detect a vast astronomical conspiracy to make Hinduism an ancient and thus legitimate religion.[74] Bentley stockpiled Sanskrit manuscripts from bazaars and libraries, including versions and commentaries on the *Sūrya Siddhānta* transcribed in Bengal between 1796 and 1805. He chatted with Calcutta colleagues about them, passing on a two-rupee copy of the *Sūrya Siddhānta*, which Bentley claimed must have been composed only eight centuries ago.[75] He also asserted that Colebrooke's aim was somehow to disrupt the Asiatic Society. Bentley claimed that a key work Colebrooke studied, the *Brāhmasphuṭasiddhānta*—composed by the great mathematician Brahmagupta (590–668), a copy of which lay in the East India Company library—"is a complete forgery probably framed among many other books during the last century by a junta of Brahmins for the purpose of carrying on a regular systematic imposition." Bentley alleged having been approached by an astrologer "who told me his profession was book-making, and that he could forge any book whatsoever to answer any purpose that might be required." When Bentley refused to hire him, this astrologer supposedly then went to work for Colebrooke, who had thus been deluded by a pandit practicing what Bentley saw as the ancient Hindu art of religious deceit. Colebrooke denied the story.[76]

Yet Bentley was certainly not alone in urging a much more recent period for Sanskritic astronomy than those proposed by Bailly and the Edinburgh lobby. Under the First Empire, Georges Cuvier (1769–1832) composed a widely read discourse on Earth history that compared geohistory with astronomy in revealing past mechanisms of the world. As "a new kind of antiquarian," Cuvier challenged Oriental antiquity: he argued Indian astronomers made all their progress in just a few centuries, not millennia. The eminent naturalist claimed that Indians lacked history, their chronologies were myths; and though their theology seemed to preserve memories of Earth's successive revolutions, including the Deluge, Cuvier cited Bentley's arguments that Indian astronomical tables were artifacts calculated retrospectively.[77] Contemporary arguments to make ancient traces fit orthodox history extended across empires, from Central America to India and Egypt. Both Cuvier and Bentley, for example, denied radical Parisian claims that astronomical imagery in Egyptian temples showed they were of vast antiquity.[78]

Cuvier dedicated his masterpiece to Laplace, who gradually abandoned initial approval of Bailly's claims about the antiquity and accuracy of Indian astral tables. The first 1796 edition of Laplace's prodigious *Exposition du Système du Monde* commented that "the remarkable accuracy of the mean motions [the Indians] assigned Sun and Moon necessarily demanded very ancient observations": but from 1798 this sentence disappeared from all later editions. Laplace convinced himself his own celestial mechanics ruled out the planetary conjunction imagined at the start of the Kali Yuga, so he came to see Sanskritic astral sciences as more recent, certainly post-Ptolemaic.[79] At the Paris Observatory, Jean-Baptiste-Joseph Delambre (1749–1822) also criticized Bailly and endorsed his own patron Laplace's new argument about the Kali Yuga: if "the tables in fact indicate this conjunction, the epoch is therefore a fiction, and Indian astronomy is much more modern." He charged that "the mysterious form of their tables and methods would suffice to produce suspicions about their veracity."[80] Bentley declared that Maskelyne and Delambre, preeminent state observatory managers, backed his views. These were weapons against Scottish reviewers to help unmask Brahmin conspiracy. Any attempt "to uphold the antiquity of Hindu books," Bentley thundered, "supports all those horrid abuses and impositions found in them under the pretended sanction of antiquity," from Hindu widow-burning and infanticide to Edinburgh skeptics' subversion of the Bible.[81] (See figure 1.1.)

Bentley's attacks were presented at the Asiatic Society after Jones's death, in a series of papers delivered from 1796. He endorsed the late president's efforts to reconcile Indian with Mosaic chronologies and cut seemingly vast Hindu timescales by converting one thousand "poetic" years into one "astronomical" year.[82] He charged Bailly and Playfair with ignorance of "the principles of the Hindu systems of astronomy." According to Bentley, from current planetary positions a Brahmin astronomer set planets' mean motions to concoct a great conjunction back at the start of the Kali Yuga in 3012 BCE, deceiving ill-informed European readers into thinking the conjunction had been observed and "assign a degree of antiquity to the work it never possessed." The difference of year length between modern and Sanskritic estimates was apparently due to corrections to maintain the deceit.[83]

This was "not the only pious fraud committed by the sons of Brahma." Bentley argued the true period of the Sanskrit tables was when their errors were least. By calibration against Paris ephemerides, he calculated the tables' date as 1068 CE. This matched what his versions of the *Sūrya Siddhānta* seemed to imply: it had been authored by Varāhamihira, whom Bentley believed lived in the eleventh century. "Any Hindu work in which the name Varaha is mentioned

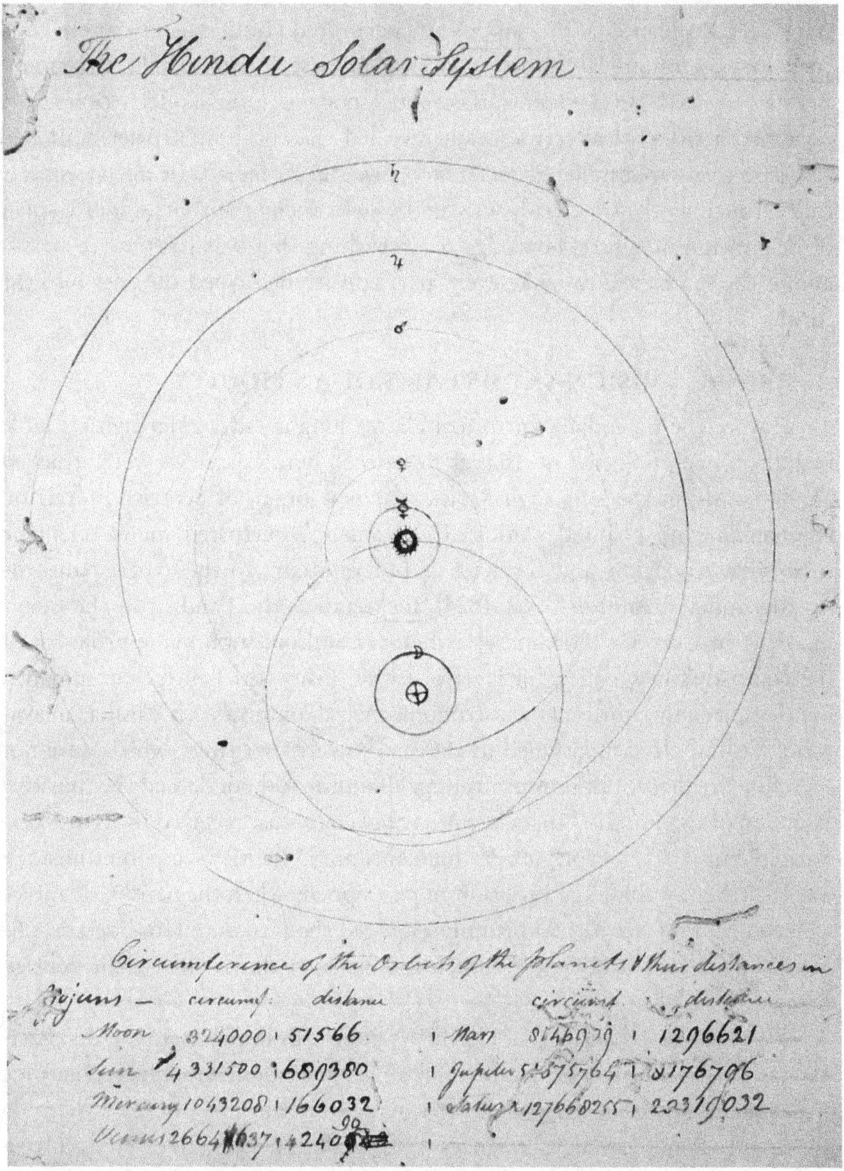

Figure 1.1: John Bentley, *The Hindu Solar System*. Wren Library, Bentley papers, Add. Ms.a.617, by permission of the Master and Fellows of Trinity College Cambridge.

must evidently be modern, and this circumstance alone destroys the pretended antiquity of many of the Puranas and other books, which through the artifices of the Brahminical tribe, have been hitherto deemed the most ancient in existence."[84] At stake were rival forms of hindsight and foresight. Modern astronomical tables let analysts apparently retrodict positions and movements of planets. Prophetic visions supposedly permitted Hindu sages to foretell historic and astronomical events. In a long manuscript on Hindu astronomy, Bentley insisted "an ancient author could not write on modern times." He reckoned astronomical retrodiction revealed that Brahmins' priestcraft had retrospectively manipulated texts and "gave out *as they do at the present day* that all such works were truly written by such ancient authors under a spirit of inspiration or a foreknowledge of everything that was to come to pass—and at the same time to make every part consistent turned the past into the future."[85]

VISIONARY DREAMS OF ANTIQUITY

These were the incendiary historical claims fought out at Harington's table in 1803. Once published in *Asiatick Researches*, Bentley's views were attacked anonymously in the *Edinburgh Review*, the new organ of Scottish literati on commercial and colonial worlds. The *Edinburgh* entrusted much scientific reviewing to Playfair and coverage of Indian affairs to the expert Sanskritist Alexander Hamilton (1762–1824), nicknamed "the Pandit." In the opening issue in October 1802 and elsewhere, Hamilton, soon made professor at the company's new college in Hertfordshire, dismissed Bentley's arguments. Bentley allegedly confused the astronomer Varāhamihira with Varāha, an avatar of Vishnu often mentioned in classical Sanskrit writings, which were not therefore artifacts of Brahmin cunning. Hamilton did not defend the "positive assertion of the Pundits" that the *Sūrya Siddhānta* was "received by divine revelation"; but it was "a work of very high antiquity." Bentley's apparent mistake was that there was a huge variation in past epochs when the Sanskritic tables seemed accurate, yet he had brutally averaged them to date Hindu astral sciences. Either these tables were much less exact or much older than Bentley claimed.[86]

The issue was credibility: of Calcutta pandits, Brahmin astronomers, Parisian savants, and British orientalists. Hamilton maintained that interpolations of recent materials in texts presented by pandits did not discredit them. It was certain Sanskritic astral sciences were ancient. Bentley countered, "I have nothing to do with visionary dreams of antiquity, nor with the ideal systems of the Edinburgh Reviewer, my object is truth."[87] Bentley held that astral chronology Brahmins peddled had been invented around 500 CE; all previous astronomical and scriptural systems had conveniently vanished; there were allegedly even stories told by "learned Hindus" that the fearsome Mahrattas "destroyed all the works of the ancient astronomers they could meet with."[88] According to Bentley, his pandits admitted that in the Kali Yuga, the age of darkness, literary forgeries exaggerated timescales because only the ancient was trusted. The astrologer who reportedly deceived Colebrooke was such a forger. Calcutta orientalists' sense that pandits might be untrustworthy was ripe for exploitation. "We may now plainly perceive the little dependence there is to be placed on what is usually called the universal or general opinion of the Hindus."[89]

Bentley ultimately published in Calcutta an astonishing history of Hindu religion and astral sciences, a work soon posthumously reprinted in London, to show how Indian priests forged an entire tradition. Varāhamihira was now judged not an eleventh- but a sixteenth-century astronomer because of his observation of Canopus as the morning star over Ujjain in what Bentley reckoned must be 1528 CE. That would make him a contemporary of Akbar, apparently duped into believing that a more ancient Varāha helped author Sanskritic astral sciences, a fable he claimed was inserted in the *Ain-i-Akbari*, thence transmitted to British orientalism.[90] Translating poetic into astronomical chronologies, Bentley could date Hindu creation to 2352 BCE, matching the Deluge. With the arrival of Western astronomy and Christianity in India around 51 CE, Bentley guessed the "Hindus found they were far behind in point of antiquity" and supposedly feared they "would be considered a modern people in respect of the rest of the world." So, according to Bentley's detective story, in 538 CE Ujjain Brahmins invented a new system of astral sciences, which they pretended had immense antiquity.[91] This would explain the expulsion of the Buddhists, whose history Bentley discussed with Bengal missionaries, as well as some Indian refugees' flight to Java, where there were apparently preserved chronology's true traces, newly presented in Stamford Raffles's *History of Java*, compiled after the company's invasion of that island.[92] The forgery allegedly applied both to chronology and to creed. From the Bengali reformer Ram Mohan Roy, formerly a company pandit, Bentley obtained a commentary giving planets' alignments at Krishna's birth, allowing an estimate of its date as 600 CE. This was "consequently the time the story of Chrisna was invented": a faked charismatic figure of similar name to Christ allegedly helped Brahmins explain away the Incarnation. Fabrication of a new astronomy and cult in northern India "was doubtless intended as a blow on Christianity."[93]

Bentley sought forensic identification of covert authors. It was telling that he was misled by the anonymity of the *Edinburgh Review*'s writers, key to their assumed position as adjudicators. One Bombay reader supported "the high antiquity of Sanscrit science" and identified Hamilton as Bentley's anonymous antagonist. But because Hamilton had quoted Colebrooke on pandits' advice and cited Colebrooke's claim that modern interpolations did not disqualify ancient Sanskrit texts, Bentley guessed Colebrooke was author of the fiercest attacks. He wasn't.[94] In fact the most sustained riposte to Bentley's claims about Brahmin forgers of Sanskritic astral sciences was published incognito in the *Edinburgh* by Playfair. He knew Bentley's views in *Asiatick Researches* were endorsed by eminences such as Delambre and Cuvier. He cautioned against Laplacian revisions and denied Cuvier's claim that Indians evolved their astronomy in a few centuries. Playfair reiterated Bailly's veracity, proposed that Indian astronomy might include relics of antediluvian science that survived the Flood, and denounced the "tedious and obscure argumentation of Mr Bentley."[95] After Colebrooke returned from Bengal to London, the Sanskritist published a major edition of writings in Indian algebra, the result of work with pandits to authenticate mathematical materials from Sanskritic astronomical tradition. Bentley produced his own manuscript version of one of these mathematical classics, while Playfair discussed Colebrooke's work in the *Edinburgh Review*, where he again referred to Bailly's claims for antiquity of Indian sciences. Though "Eastern Astronomy" was preserved as "a wreck which has survived the memory of its authors," this did not justify claims by Bentley and his allies that astral science was but a retrospective forgery.[96]

Playfair's fullest answer to Bentley appeared in the *Edinburgh Review* in summer 1807. He began by insisting that an astronomer's principles were "everywhere the same, whether he has gone to work on the banks of the Ganges or the shores of the Atlantic, has lived in the antediluvian ages or in the nineteenth century, has been instructed by the philosophy of Newton or amused by the fictions of Varáha." Astronomy's alleged spatiotemporal independence made it a tribunal. "When we know the astronomical system of any nation, we must be enabled to judge with some accuracy its state of refinement." It also meant Bentley, whom Playfair castigated as "amateur," could not counter astronomical professionals such as Bailly.[97] Playfair conceded Bentley's claim that retrospective calculation of astronomical alignments was "probably connected with religion" and gave "the same inactivity and inertness to their science that pervades among them the whole of the moral and intellectual world."[98] But there had been observations of great antiquity. The Moon's position over Benares at the start of the Kali Yuga from the Board of Longitude's best lunar tables differed by only thirty-seven minutes from that found in the Sanskrit tables studied by Le Gentil and Bailly. These could have "no source but actual observation." Playfair argued, "A science must always be older than the books that treat of it." So "the art of astronomical observation is of high antiquity in India, and goes back not less than 3000 years before the Christian era."[99]

PUTTING SCHOLARS ON THEIR GUARD

This long dispute did not cease with Bentley's death. As Hamilton stressed in an essay on Sanskrit poetics in the *Edinburgh Review*, these polemics focused on *histories* of ancient sciences. In this crucial respect, Bentley's astronomical computations differed less dramatically from those of critics such as Playfair or Colebrooke. In 1826, for example, Prinsep's Benares Literary Society set out the details of Bentley's complex chronological scheme. But it did dismiss his stories of forgery and instead recommended "Pundits of integrity from Benares or elsewhere."[100] Issues of rival historical authority were decisive. The Edinburgh-trained journalist James Mill closely followed Bentley, "who has paid more attention than any other European to the books of Hindu astronomy," alongside Laplace and Delambre, in long sections added to his *History of British India*, composed between 1806 and 1817. Mill correctly reckoned the *History* would get him hired by the East India Company, which assigned the book as reading for its trainees. He also exploited his erstwhile friend Playfair's condescension about the stasis and non-theoretical character of Indian sciences, added an appendix on "arguments for the antiquity of the Hindu astronomy," and concluded that the science was in a primitive state, that it had always been so, and that it was in the interests of Brahmin priests to delude their flock and Europeans who consulted them. In his new *Westminster Review*, Mill argued that Bentley's book should "meet with a favourable reception from every man who feels an interest in the early history of that extraordinary people, the Hindus." The anticlerical Mill held that Bentley had "done great service by putting scholars on their guard against the dishonest inventions of the priests."[101]

Which priests? It is tempting to treat each orientalist project on Indian astral sciences assembled between 1785 and 1825 as reflexive. Whether in Calcutta, Tirhut, or Benares, in London, Paris, or Edinburgh, orientalists projected onto an often ill-understood northern India principles they sought to celebrate or damn in their own lands. This was the concern of missionaries and officials who fearfully established the anti-Jacobin College of Fort William, of Asiatic Society fellows who debated religious and political meanings of Sanskritic chronologies, and of abrasive metropolitan periodicals that set standards of literary and philosophical morality. Robertson insinuated similarities between Brahmin rationality and his moderate Kirk; Playfair used Sanskritic astral sciences to free natural knowledge from overweening scriptural discipline; Colebrooke located an ancient Indian culture matching his own ideal of religious and social virtues; Mill sustained utilitarian campaigns against landed aristocrats and the established church by venomous condemnation of what he claimed to see in Hindu sciences.

Yet the temptation should be carefully scrutinized. Astral sciences reorganized measures of position and of time; evolutionary histories of these sciences thus intervened in the politics of the past. This politics centered in complex networks in Calcutta, across northern India, and beyond. Indigenous experts, many drawn from elite groups, were crucial, as has been convincingly suggested, in the forging of novel accounts of religions and sciences, which would not have been constructed without these protagonists.[102] Definitions of what counted as sacred and what as scientific were dramatically overhauled there. Burke might deny that morality should have a geography; Playfair might insist astronomy was the same along the Ganges and the Atlantic; but sciences of their time depended intensely on these defined conjunctures and locations. The aim of this analysis has been to situate more precisely these projects and the religious and scientific histories they evolved.

2

Debating Evolution and Religion in Nineteenth-Century South Asia

Tamara Fernando and Sarah Qidwai

Seven months after the publication of Charles Darwin's ***On the Origin of*** *Species by Means of Natural Selection, or the Preservation of Favoured Races in the Struggle for Life* (1859) in a room "crowded to suffocation" in the village of Oxford, "700 persons or more" sat at what would become a famed meeting of the British Association for the Advancement of Science.[1] Although Darwin himself did not attend the meeting, Thomas Henry Huxley (1825–1895) served as his advocate, reportedly facing off against the bishop of Oxford, Samuel Wilberforce (1805–1873), on the theory of natural selection, evolution, and the relationship of science and religion. The exchange, best known in academic shorthand as the Huxley-Wilberforce Debate, features in most histories on the nineteenth-century reception of evolutionary thought. Despite being subject to critical scrutiny, it is still described by historians as a "touchstone" event, "an iconic moment," or, in Darwin's own words, an episode "of enormous importance."[2] Of course, even apart from this single episode, six editions of the text would be issued within Darwin's lifetime, all part of "the first truly international scientific debate in history."[3] The supposedly fierce clash of Darwin's evolutionary ideas with the beliefs of members of the Anglican clergy has become a paradigmatic case for historians of science thinking through the relationship of science to religion.[4] Building on the wealth of existing literature, scholars continue to offer fresh analyses of the event using new methods and tools in reception histories and digital humanities.[5]

But the history of debate stretches long before the scientific revolution of the seventeenth century (whether we choose to accept that terminology or not) or, indeed, the remarkable discoveries of the nineteenth century in the life sciences. It is to this central concept of debate that we will return throughout this review. But first, some words on historiographic matters that motivate this chapter: a few decades ago, a straightforward narration of the history of evolutionary biology in the nineteenth century would have presented a predominantly British or Western European story, and events like the above would understandably take center stage. In this telling (as with much of the history of science), science was the preserve of Europeans, and it filtered—in stadial fashion—outward to other places and peoples. If South Asia featured at all in these narratives, it was primarily as an extractive site or a theater of experience and knowledge gathering for Europeans involved in these debates. This is particularly the case because prominent actors in the history of evolutionary thought, including botanists such as Joseph Dalton Hooker (1814–1879) and John Scott (1836–1880), the animal trader Edward Blyth (1810–1873), and Ernst Haeckel (1834–1919), certainly traveled in and through South Asia. It is no coincidence that imperial expansion and developments in the history of the life sciences overlapped. In turn, scholars have chosen to focus on how these individuals' experiences in South Asia shaped their intellectual output, or they might attempt to shine light on their local intermediaries and interlocutors.[6] Of course, the history of science is no longer content with narratives based firmly in Europe that make little account for knowledge making and science in other regions of the globe. The question, then, is how best to approach these intellectual traditions and the work that non-European actors were doing in the nineteenth century within this field of thought?

Here we want to pose a provocation as a thought experiment. What if we narrated our histories of religion and science not from the steepled towers of Oxford but rather from pavilions made of beams of tropical wood, open to the warm air of the town of Pānadurē, on the island of Ceylon, eight thousand miles away from the imperial metropole? What if we set our sights, imaginatively and analytically, on a scene dated to August 26, 1873, when instead of men (and some women) dressed in black, two thousand Buddhist monks draped in robes of saffron, and a lay audience of several hundred others, gathered to watch the Sinhalese Wesleyan minister David da Silva debate the Buddhist monk Migettuwatte Guṇānanda on the place of science within Christianity and Buddhism?[7] In short, would the history of the reception of evolutionary thought look different—force scholars to ask divergent questions—from the viewpoint of Pānadurē rather than Oxford?

As is now well established, in the mid- to late nineteenth century and well into the twentieth, debates on human origins and the descent of man played out in different parts of the world, involving new publics and the periodical press but reaching different answers to questions of science and religion—answers embroiled deeply in the politics of contestation in colonial societies progressing shakily toward nationhood.

Seventeen years after the Oxford debate, the above exchange, known as the Pānadurē Vādaya (Pānadurē debate), forms one of five major confrontations, the Pañca Mahā Vāda, between Buddhists and Christian missionaries on the island of Ceylon (contemporary Sri Lanka). On the surface, the two episodes share certain notable similarities, despite their geographic and cultural distances. In Pānadurē as in Oxford, codes of polite or appropriate behavior, rhetorical prowess, and audience responses shaped public perceptions of the debate.[8] Similarly, in colonial Ceylon as in Britain, print and face-to-face encounters were entangled, and influenced and inflamed one another. However, crucially, in Ceylon the contest was *not* between men of science and men of faith (this is, of course, a gross oversimplification of the Oxford debate) but rather between learned men from two distinct religious orders—Buddhism and Christianity—in the context of a British Crown colony.

Local religious traditions had long held their own ways of making sense of time, species, and questions of design, but in the latter half of the nineteenth century they met evolutionary ideas and Darwinism in the context of colonial rule and the globalizing currents of scientific modernity and reform. These ideas, we now know, were absorbed and recast in local veins. In the Pānadurē debate, Christian missionaries, who had spent the last seven decades attempting to discredit and wrest souls away from the preexisting Buddhist orders on the island, directed their attention to the question of Buddhist cosmology, arguing that traditional Buddhist theories were "in antagonism to the most obvious teachings of science."[9] Beginning in the 1860s, Buddhist monks turned this discourse on its head, arguing that in fact it was Christianity that was unscientific in its approach. We find echoes of this same strategy across religions in Asia, including Islam, Hinduism, and different strands of Buddhism. In the early twentieth century, science, as historians Dhruv Raina and Irfan Habib have cataloged in the case of colonial India, served as a battering ram in anticolonial discourse as well as in contests of colonial legitimacy.[10]

As outlined in the introduction to this volume, imperialism casts a long shadow over the narratives of the emergence of nineteenth-century science. In a South Asian context, important developments included the decline of the Mughal Empire and the gradual consolidation of land in India by the East India Company (established in 1601), which ended with a shift to oversight and direct governance by the British Crown. The end result was the independence of India and creation of Pakistan in 1947. In Ceylon, meanwhile, the latter half of the nineteenth century saw the strengthening of the export economy, particularly of coffee, with a plantation economy run under a single government for the first time in the island's history.[11] Within its hybrid colonial society, taking inspiration from the Pañca Mahā Vāda, debate came to occupy a newly central place in contests for legitimacy in colonial (and later postcolonial) South Asia.[12] This chapter will argue that *vivadaya*, or debate, as concept, from the Buddhist example (or *ijtihad* in the Islamic context), is a productive analytic to import into the history of science through which to read the nineteenth-century reception of evolutionary thought in South Asia.[13]

In foregrounding debate (Sinhala: *vivādaya*, Arabic/Urdu: *tafsir*/*ijtihad*), we are indebted to scholars of religion, both Buddhism and Islam, who have generally not been in direct conversation with historians of science. An overview of the history of science literature can help clarify what is to be gained by foregrounding indigenous ideas of debate. The historiography on the reception of evolutionary thought is changing rapidly. In the last decade, the reception history of Darwinism and the global history of evolutionary thought has moved beyond simple models of diffusion toward accounts of "appropriation." Instead of tracing the movement of specific texts, like *Origin*, and charting linear timelines of when an idea reached a certain host site, in a South Asian context, important new scholarship increasingly attends to the full flowering of the creative, inventive, and combative ways that non-European actors used (and abused) this knowledge. This approach within studies of South Asia shares much with the work of scholars writing on the Middle East and often involves stretching timelines prior to the nineteenth century to shed light on longer local lineages of thought.[14] Although both "science" and "religion" have particular linguistic and etymological histories in the English language and particular meanings within the English-speaking academy, turning to vernacular terms on their own merit has intellectual payoff for the history of science in South Asia. It is in this regard that we highlight "vivādaya," "tafsir," and "ijtihad," glossed here in English as "debate."

In thinking through the history of evolution in South Asia, it is helpful to establish some key parameters. First, unlike the predominantly Christian society of mid-Victorian Britain, South Asia was distinct in that it included within its bounds the plural religious worlds of Parsis, Muslims, Sikhs, Christians, Jews, Hindus, and Buddhists, each with their own distinct but also overlapping cultural, literary, and intellectual traditions (the three are obviously not so clearly parsed apart). This means that, compared to the prominent place of natural theology in mid-Victorian Britain or the relatively settled status of theological doctrines of divine providence, salvation, and the afterlife, South

Asians historically subscribed to a plethora of religious beliefs and doctrines, depending on which religion or religious subgroup they belonged to.

It is not enough to broaden the history of science canon to include scholarship by historians of Buddhism or Islam; we must also ask how these (often separate) works speak to one another—even obliquely, or in the background. In South Asia, heterogeneous religious worlds brushed up against one another and often competed with each other in the late colonial or high imperial period, with "science" playing a key role in these same contests. Any account of the responses to evolutionary thought that do not recognize this diversity—the fact that actors *were always in dialogue* (even if only implicitly) with other religions—is one that risks masking this heterogeneity. Thus, in framing South Asian religious responses to evolutionary theory, the core question is one not just of the relationship between religion and science as was presumed in Europe but of myriad religion(s) under empire to science.

We also want to forward some notes of caution about our own terms of scholarly inquiry. The first is the notion of a "Darwinian Revolution," which, while problematized, continues to dominate historical narratives of evolutionary thought in the nineteenth century.[15] What is edged out of the frame when we focus exclusively on Darwinism? But there are promising avenues of inquiry. Mackenzie Brown's *Hindu Perspectives on Evolution*, for instance, presents an account of Hindu intellectuals in the nineteenth century responding to, or navigating around, Darwinism, coupled with older indigenous ideas derived from the Vedas and other texts.[16] Similarly, Brown's 2020 edited collection, *Asian Religious Responses to Darwinism: Evolutionary Theories in Middle Eastern, South Asian, and East Asian Cultural Contexts*, also explores Darwinism in Asia from a variety of geographical and religious viewpoints.[17] Another option is presented by one of the most exciting monographs to emerge in South Asian studies on this topic recently, when Luzia Savary demonstrates "how European racial concepts gradually made their entry into Urdu and Hindi Language," tracing vernacular concepts as a way to read the broader public reception and use of these notions of racial difference and human origins.[18] Brown and Savary demonstrate the rich narratives that can be unpacked when the topic of evolution is told from the perspective of historical actors from South Asia.

The second term is "South Asia" as an analytic and geographical placeholder.[19] This warrants some explanation, especially because area studies approaches, seen largely as geopolitical outcomes of the Cold War and American strategic interests, have received much critical scrutiny in the last two decades, not least from scholars of religion.[20] Despite this criticism, historians of science have been slower to catch up with the dismantling of easy regional or geographic blocks.[21] Unlike other terms used by authors in this volume, "South Asia" was not an actors' category in the same way as was "the Ottoman

Empire," "Spain," or "Japan." What are the implications, then, of holding Aligarh and Colombo together under this rubric? One way to approach this question is to investigate more robustly the categories that historical actors used. Anagarika Dharmapāla (who we will meet below) for instance, in one letter, described himself to King George V of England as "born in Ceylon, of Sinhala parents, working for the welfare of all Buddhists in Asia, especially Ceylon, and trying to resuscitate Buddhism in India."[22] He spoke often of "civilization," from which he excluded several persons, particularly Muslims, and his vision was of a pan-Asian community of Buddhists.[23]

What then are the merits of using the term "South Asia" to approach the history of science in this region? It is certainly still useful to hold Ceylon and India's trajectories together: they had a shared experience of British colonialism, functioned as extractive sites for the collection of scientific data and specimens for European naturalists, and saw similar agitations for reform in the late nineteenth and early twentieth centuries.[24] Their respective elites often negotiated similar subject-positions, being English-educated but racially inferior, fighting for their place in a colonial order. Proceeding in this vein, we could compare Buddhism to Hinduism, Dharmapāla to Vivekananda, and colonial policy or elite local actors and printed publications across Bengal, Colombo, or Delhi.

But one should not extend the comparison too far. Although India is Sri Lanka's closest proximate region, particularly from a Buddhist perspective, Ceylon's responses to science and evolutionary thought are closer to those found in Japan, Siam, or Burma, as the chapters on Japan in this volume show. In fact, Buddhist studies scholars have pointed out that nineteenth-century monks or figures like Dharmapāla find closer comparisons not to Indian reformers but instead to figures like the Burmese philosopher Shwe Zan Aung, who drew parallels among Buddhism, Darwinism, and science.[25] A focus on debate, by comparison, allows us to view communities in motion and exchange of ideas across time and space. This means that engagements with the history of science in South Asia might make allowances not just for comparisons between India and Ceylon but also between Lahore and Cairo or Galle and Rangoon.

This chapter uses two intellectuals to anchor an inquiry into the narration of the history of biology and evolutionary thought in the region: the Muslim reformer and polymath Sir Sayyid Ahmad Khan (1817–1898) and the Buddhist reformer and "world renouncer" Anagārika Dharmapāla. Reading the evolutionary writings of Sayyid Ahmad Khan and Anagārika Dharmapāla also allows us to choose non-Hindi, non-Bengal-based figures who claimed to speak for Buddhist and Islamic traditions. Despite this, the approach we have taken here once again reproduces the much-criticized focus on literate men;

and a more somber warning is that it also poses the danger of reinscribing nationalistic or chauvinistic visions onto history.

In accounts of Buddhism, for instance, which depict it alternatively as Westernized, demythologized, rationalized, Protestantized, or psychologized, there is little mention of other religious actors who shared space in colonial Ceylon. Although these accounts then successfully move us beyond Europe in the history of science, they require us to constantly interrogate their implications more locally: In this case, how can we narrate Buddhism in Lanka without reifying Buddhist Lanka? Or Islam in northwestern India without reifying the destiny of Pakistan and single-religion nation-states?

Regardless, a focus on Dharmapāla and Sayyid Ahmad does allow us to make our primary intervention in terms of foregrounding the intellectual history of debate as a key conceptual category in understanding conflict or contestation over ideas of human origins.[26] Debate moves us beyond the teleology of linear diffusionist narratives and allows different temporal, spatial, and cultural traditions to emerge. Religious studies scholars have long recognized that debate as a concept has a much older indigenous legacy, whether in Buddhist polemics or exegetical debates in the Muslim tradition, and this is amply clear in Dharmapāla's and Sayyid Ahmad's writings. A focus on evolutionary thought means that this must necessarily be a nineteenth-century inquiry, hence the prominent presence of "modernist" or "reformer" figures such as the above two men; it is no accident that they formulated their ideas during a period of high imperialism and contests over scientific racism and civilizational superiority. Moreover, while there is much still to be learned from more straightforward reception histories, especially the exciting new work on translation and the vernacular in the history of science in the subcontinent, debate also allows other influences in, whereas translation can appear unidirectional.[27]

EXEGESIS, INTERPRETATION, AND NEW SCIENTIFIC PUBLICS IN ALIGARH

In April 1854, several hundred Muslims and Christians gathered at the Church Missionary Society of Agra to attend a public debate between Karl Gottlieb Pfander (1803–1865), a missionary associated with the Basel Mission, and the Muslim scholar Rahmatullah Kairanawi (1818–1891). The topic of this public debate addressed the authenticity of Christian scripture.[28] "The Great Debate" took place over two days and ended with both parties declaring victory. Whereas Pfander defended the accuracy of both the Old and the New Testament, Kairanawi insisted on their falsehood. Bubbling under the surface of this debate was the charged environment of unrest with the East

India Company's gradual expansion and the concern of missionary activity in the subcontinent.

The concept of a legal debate, or *ijtihad*, is an integral part of Islamic legal thought. Thus, viewing the relationship of evolution and the Islamic context as a debate can demonstrate methods certain individuals used to make sense of the theory and its relationship to Islamic law. The textual sources in question would lead to new avenues of investigation. While the debate between Pfander and Kairanawi was described as contentious, other sources demonstrate a different relationship between religious exegesis, Islam, Christianity, and imperialism. For example, between 1862 and 1865, Sayyid Ahmad Khan published *The Muhammadan Commentary on the Holy Bible* (*Tabyin-ul-kalam*) and presented a new avenue of discourse. This publication is unique in its historical setting. It was an interpretation of the Bible from the perspective of the Qur'an. In this text, Sayyid Ahmad even discussed topics such as astronomy, geology, and the origin of life. Thus, the theme of debate is drawn out with references not only between missionaries and Muslims but among the Indian Muslim intellectual community as well.

Returning to the topic of evolution in this context, there is no monolithic Islamic or unified "Muslim view" on evolution, but we can still draw some conclusions by investigating key historical actors.[29] From as early as the eighth century, when Arab traders sailed to the coasts of Sri Lanka, Muslim communities settled, integrated into, and influenced South Asian lands and society.[30] Historiographically, Sayyid Ahmad's views of human evolution were first told not through the Indian context, nor through his publications, but as a response to other nineteenth-century Muslim reformers. In *Reading Darwin in Arabic*, Marwa Elshakry introduces a debate between Jamal al-din al-Afghani, a pan-Islamic polymath, and Sayyid Ahmad Khan. Elshakry explains how, while in Calcutta, al-Afghani published *The Truth about the Naturalists and Their School* (*Haqiqat-i madhhab-i nichari va bayan hal nichariyan*).[31] While the target of this publication was Sayyid Ahmad, Elshakry, writing about other geographic contexts, introduces Sayyid Ahmad's ideas through a one-sided debate by those responding to his positions for political clout but not directly engaging with him.

Sayyid Ahmad wrote several articles on human evolution and alluded to the theme in several of his publications.[32] For example, in 1865, a discussion of the origin of life is tucked into *The Muhammadan Commentary on the Holy Bible*. Sayyid Ahmad wrote: "It is this, that all created things have but one source of existence and owe their being to the same substance or essence. The stone, the tree, the animals all share equally with ourselves in this respect. The only difference between us and them is in the diffusion among us of the substance which we all have in common in a proportionately greater or less degree."[33] In this text Sayyid Ahmad did not elaborate further, but it is a not so subtle nod referencing the debates at the start of this section—not debates between scientists and defenders of religion per se, but between a Muslim and a Christian missionary. Both scholars are cited by Sayyid Ahmad in *The Commentary*. While he was dealing with the Qur'an and the Bible, unlike Pfander and Kairanawi, Sayyid Ahmad considered the truth of both the Qur'an and the Bible, and in doing that he did strike a more conciliatory tone.

Sayyid Ahmad's clearest discussion about human evolution is found in the article he published in 1896 titled "Adna Halat se Aala Halat par Insaan ki Taraqqi" (The stages of human development from an inferior to superior state) in the Urdu periodical *Tehzib-ul Aklaaq* (A refinement of manners).[34] In the article, Sayyid Ahmad raises his first point, that the universe is composed of *hajr* (rocks or minerals), *shajr* (trees/plants), and *haivan* (animals), already presenting an interesting account for the origin of life on Earth. He went on to explain his views on how humans were part of the animal kingdom, how they evolved over time, and how this process was not antagonistic toward Islamic practices.[35] Sayyid Ahmad does mention Darwin in this text, but it was not the focus. Instead, we see an account of human growth and development that was compatible with Sayyid Ahmad's religious worldview.

Another perspective from which to examine Sayyid Ahmad's views, as Savary and Pernau have, is through the concept of "civilizations," which will also become important to Dharmapāla, as in the following section. One of the earlier uses of the term is found in a record of Sayyid Ahmad's speeches (*Khutbat*). In his speeches and publications, Sayyid Ahmad does not provide direct references to his sources, but we can tie him to both Islam in India and the wider Persianate context. Though his views should not be taken as representative of all Muslims in the subcontinent, we can use him as a starting point. The fact that his family was a part of the Naqshbandi Sufi order means that certain of his views were extensions of or responses to some theories posed by the influential Shah Wali Ullah (1703–1762), who wrote extensively on a concept called *irtifaqat*—the idea that human civilizations go through different stages toward development.[36] Wali Ullah argued that humans inched toward development by attaining knowledge. According to him, there were four steps in this process of human development. First, man was a nomad. Then, by gaining the ability to speak, he formed civilized towns. Third, he eventually formed city-states. Fourth, he attained universal presence, which was oneness with God.[37] Here, by asking about the significance of Darwin in the nineteenth century, we are omitting a fascinating account of the origin of life that was developed within the Islamic faith.

While it is important to account for Sayyid Ahmad's views on topics such as the origin of humans and civilizations, the other issue at hand is the reception of Darwinism or lack thereof. By framing the investigation as a form of debate, we can examine one such case where the *Origin* is not deemed important enough to translate for pedagogical purposes. For this, we can turn to the Scientific Society, established by Sayyid Ahmad in 1864. Since its inception, members of the society discussed and debated what English texts should be translated from English to Urdu. A letter dated January 28, 1864, from Charles Alfred Elliott (1835–1911), a civil servant in British India and the future lieutenant governor of Bengal, suggests numerous texts for the society to translate, many of which made their way onto the final list. Elliott wrote that one of the works that should be included was "[Charles] Darwin's great and suggestive work which could not fail to interest every reader."[38] Ultimately, after debate and discussion, the Scientific Society did not translate any of Darwin's publications. However, this letter demonstrates that the society's Indian members were certainly aware of them and discussed their utility. Not all discussions have to be hostile or about science and religion. Some are pedagogical. The above exchange helps us shift our focus from the intellectual merits or stakes of the reception of evolutionary thought toward the seemingly more mundane yet equally important: the growth of institutions, the training of students, and the public sphere more broadly.

Another concern is creating a linear account of these debates rather than one-off examples in isolated contexts. Further work is required to enhance this narrative. We need to consider not just contemporaneous writers but also diverse perspectives in the context we are examining. For instance, after Sayyid Ahmad, we can look a generation later, at Muhammad "Allama" Iqbal (1877–1938), an Urdu poet and an individual often deemed one of the architects of the Muslim nation-state now known as Pakistan.[39] In *The Reconstruction of Religious Thought in Islam*, a compilation of lectures published in 1930, evolutionary ideas were littered across Iqbal's various speeches.[40] Two references in particular stand out. In his second lecture, titled "The Philosophical Test of the Revelations of Religious Experience," Iqbal discussed and presented critiques of the three arguments for the existence of God: the cosmological, the teleological, and the ontological.[41] In the essay he wrote, "The discoveries of Newton in the sphere of matter and those of Darwin in the sphere of Natural History reveal a mechanism. All problems, it was believed, were really the problems of physics." Here, he evoked Darwin's name to claim that the battle for a mechanism was raging in the field of biology, but it did not affect the religious experience. The next reference to Darwin is found in lecture 7 and is a critique of Nietzsche. This lecture, "Is Religion Possible?," focused on that very question. Iqbal wrote, "Yet Nietzsche was a failure; and his failure was mainly due to his intellectual progenitors such as Schopenhauer, Darwin, and Lange whose influence completely blinded him to the real significance of his vision."[42] Here, he evoked Darwin's name as a form of critique. Darwin is pictured as someone who led Nietzsche astray, which implies that Darwin did not offer useful insights. Thus, whereas Sayyid Ahmad presents a view in which Darwin's ideas were not important to him, Iqbal uses evolutionary ideas to support his own view of religious thought. Moreover, we can expand this context further and examine Iqbal's relationship with Henri Bergson (1859–1941). In this "debate," new historical actors are introduced to the story, but their debates and discussions are not about evolution alone; they are philosophical to an extent.[43]

In short, here we see that evolutionary ideas influenced the political and religious landscape of British India and the 1947 partition that led to the creation of Pakistan. These ideas were not necessarily related to Darwin but were a mix of Islamic views alongside what was introduced by Europeans. This is an area for future research. By focusing on the reception or appropriation of evolutionary ideas in the long nineteenth century solely as a story about Darwin and Darwin's ideas and followers, we might omit alternative and diverse historical accounts that drew inspiration from alternate traditions such as *irtifaqat*. Also, we miss out on fascinating historical debates among figures such as Iqbal and Bergson that add a fresh angle to the historical narrative.

VIVĀDAYA, COLONIALISM, AND INTER-RELIGIOUS DEBATE IN LANKA

In a Lankan context, Anushka Kahandagama has recently argued that Sayyid Ahmad Khan finds his equivalent in Anagārika Dharmapāla.[44] This assumed kinship is based on the roughly overlapping time periods in which each of these men was active, but also on a set of broader political and cultural forces that shaped the trajectories of the two men. Indeed, in Sri Lankan scholarship, no figure has been more central to histories of late nineteenth-century "reformed" Buddhism than Don David Hewawitharana, better known by the moniker he adopted for himself, Anagarika Dharmapāla (1864–1933).[45] Alternatively recast by contemporary scholars and historians as a proto-nationalist/militant Buddhist or a spiritual seeker/world renouncer, Dharmapāla continues to receive scholarly attention both within and outside Sri Lankan studies, ranging from biographies to histories of Sinhala nationalism.[46] Although he is largely unknown to historians of science, for scholars of Buddhism, Dharmapāla marks a key turning point in the development of narratives about Buddhism's supposed consilience with scientific practice, and his archive includes numerous references to evolution.[47] For the purposes of this review then, we can use Dharmapāla as a means of thinking through the science-religion encounter in Ceylon.

The son of a wealthy Sinhalese bourgeois family, educated at Christian missionary schools in Ceylon, Dharmapāla became "modern" or "reformed" Buddhism's champion par excellence, a fact that holds much of the key for how he used evolutionary thought and to what ends. Dharmapāla engaged (if somewhat lightly and superficially) with some scientific literature, published in Sinhala and English, and spoke at venues outside of Ceylon. Continuing a tradition begun in the mid-nineteenth century, from the 1890s to the 1920s he used the periodical press in both Sinhala and English to disseminate his views of Buddhism, echoing with the use of printed materials in other chapters in this volume. In 1906 Dharmapāla set up the *Sinhala Bauddhaya* newspaper, to which he contributed a weekly column, "Karunu Denagatuyuttu" (Sinhala for "things one ought to know") to instruct lay Buddhist readers on how to live a pious life. He also established the English-language *Journal of the Maha Bodhi Society* and *United Buddhist World* in 1911, both of which had a readership beyond Ceylon.

Dharmapāla's development of a public persona in a world of global, universalizing spiritual movements deserves attention equal to his role as a figure in the contested world of religious identities in Ceylon.[48] Let us begin with the global, which distinguishes him from Sayyid Ahmad Khan: at the 1893 Parliament of the World's Religions in Chicago, for instance, where other individuals from this volume also make an appearance, Dharmapāla was provided "a platform on which the leaders of Christianity could meet the leaders of the other major religious traditions of the world."[49] In America, Dharmapāla impressed upon his listeners the idea that Buddhism was a "religion of Science." He linked Buddhism's quest for truth to the scientific method: "The strongest emphasis has been put by Buddha on the supreme importance of having an unprejudiced mind before we start on the road of investigation of truth," he explained.[50]

Thus, in Dharmapāla's view, the work is not of *distinguishing* science and religion but rather of subsuming science *into* newly globalized "universal" religions. That is, it is the same process Marwa Elshakry cogently described as the historic reception of Darwin outside of Europe being "as much a force for religious insurgence and revivification as for religious skepticism."[51] Dharmapāla is hardly unique. He joins a host of figures ranging from the Grand Mufti of Egypt, Muhammad Abduh, to those in Korea or Japan.[52] The cultural authority of Buddhism validates science, not the other way around.

Of particular importance was the Buddhist idea of *hethu pala wadaya* (cause and effect), which had a long tradition in Buddhist writing around moral activity and pious actions in a karmic universe; this concept was now recast as foreshadowing the scientific method.[53] In this he echoed the posturing and rhetorical work of another participant, Sōen Shaku (1859–1919), a representative of the Japanese delegation. Shaku likewise insisted that "Buddha's teachings are in exact agreement with the doctrines of modern science," which, for our purposes, is more interesting from the perspective of finding global patterns and in line with problematizing South Asia within the inquiry into the history of the reception of evolutionary thought.[54]

It is exceptionally difficult to parse Dharmapāla as a thinker in the history of science from Dharmapāla as polemicist arguing against European ideas of civilizational superiority over the Orient. Like many elite, educated subjects, he bristled at British attitudes toward their colonial "inferiors."[55] Although he was committed to the path of the brahmacharya, that is, a world renouncer on a quest for salvation, and he had established a branch of the Maha Bodhi Society in Madras in 1900, his approach to Buddhism was largely gleaned from English sources.[56] His Pali language skills were self-taught, and doctrinal points mattered less to him than his insistence on establishing Buddhism as a superior world religion.

We might say the same for his use of "science" or Darwin, which served instrumentally to advance his politics rather than serving as an intellectual inquiry within a scientific or pedagogic or institutional context. In taking him as the case study for Buddhism and science in Ceylon, it is important to be very clear that he was not a scientific figure: not employed by a university or scientifically trained like Fujikawa Yū in Tokyo, or even like Sayyid Ahmad Khan, with his scientific publications and journal articles. This means that his engagement with science operates outside of the context of scientific societies, universities, or public institutions. Mackenzie Brown's observation that the transfer of knowledge of Darwinism and Western evolutionary theory to Asia was "quite random, piecemeal, and unsystematic" rings true.[57]

In Steven Kemper's reading, Dharmapāla "acquired Western assumptions about Buddhism as a world religion."[58] Many Buddhist reformers and European adherents sought to model the Buddha on the ideal Victorian gentleman.[59] Victorians frequently characterized Buddhism as atheistic, nihilistic, pessimistic, and idolatrous,[60] and in the late nineteenth century the Buddha was presented as a kind of "freethinking empiricist." Older Buddhist literature such as the Pali *Kālāma Sutta* were marshaled as evidence for Buddhism's support for free inquiry.[61] In this new Buddhism, for instance, the super- and sub-human beings of pre-scientific Buddhist cosmology, the *yakṣas* and *gandharvas* mentioned in sutras, are notably absent. McMahan explains that "what some scholars have called Buddhism modernism—has been, therefore, a cocreation of Asians, Europeans, and Americans." In other words, Asian reformers like Dharmapāla and Daisetz Suzuki refuse to be placed solely in their respective "area studies" boxes.[62] McMahan describes these positions as a "facet of a more global network of movements that are not the exclusive product of one geographic or cultural setting."[63]

Asian articulations of Buddhist modernism began as resistance to Christian missionaries, co-eval with the colonial project of improvement and "uplift" in places like Ceylon. Alongside his role in the propagation of Buddhism as a world religion, Dharmapāla has also been described as an emblem of a "harder-edged Sinhala Buddhist nationalism, which sat cheek-by-jowl with a culture of cosmopolitan connection" conspicuously present in his own life.[64] He called the British "demons," "vampires," and "sea wolves" but also was fiercely critical of Muslims and Christians at a time when the period of self-rule saw the hardening of identities and new delineations of who belonged and who could be represented in Lankan politics.[65] These reformulations of Buddhism in line with science, then, were, in McMahan's words, a kind of "mimetic forms the colonized developed in order to subvert the agenda of colonizers." The resuscitation of the *Kālāma Sutta*, for instance, was "an anachronistic attempt to establish a kind of Buddhist 'Nullius in verba,'" the credo of the Royal Society in London.[66]

There is certainly scope for exploring evolutionary ideas beyond the remit of someone like Dharmapāla. These programs of research would focus on pirivena education, or on the nascent universities of Ceylon, or on other public spheres. These kinds of investigations might reveal interesting overlaps in terms of thinking about race and caste through evolutionary language.[67] Consider how in public discourse around Sinhalese castes in local-language newspapers through the 1870s and 1880s contributors related "the Devolepment [*sic*] Theory of Dr. Darwin" to "the sad havoc, our 'Carawe' friends have made on the innocent fishes of the Ocean for successive ages."[68] Here Sinhalese caste structures, that of *carawe* (fisher) and *govigama* (farmer), and their sociopolitical struggles drew on Darwinian notions of the survival of the fittest. This mirrors the process that Savary has described for colonial India, where Darwinian ideas inflected the "racial [or casteist] thinking in the colonized," although secondary literature on similar processes taking place in Lanka is virtually nonexistent.[69]

In Ceylon a different view also emerges if we read outside the corpus of prominent, universalizing Buddhists. Newspapers published by the American Mission in Jaffna reveal a range of Spencerian, Lamarckian, and Darwinian stances on evolution, corroborating what we know from other contexts, in that the reception and reappropriation of evolutionary thought was varied and diverse. The Jaffna missionary paper, the *Catholic Guardian*, for instance, carried articles well into the twentieth century claiming that "it is a grave scientific error to believe that the evolution of species from one another has been proved, or that it is generally admitted."[70] In 1902 the bilingual Tamil/English Jaffna newspaper *Utayatarakai* reprinted Herbert Spencer's "The Views of Modern Science," compiling a set of scientific writers who argued that science and Christianity do not contradict one another.[71] More evidence is found when in April 1912 the Point Pedro Literary Club in the town of Jaffna hosted a meeting where J.V. Chelliah (a professor employed at Jaffna College, the foremost educational institution in the north) presented a paper titled "The Ascent of Man." Chelliah took issue with Darwin's concept of the survival of the fittest, turning instead to the theories of Scottish evangelist Henry Drummond (1851–1897).[72] Chelliah's use of Drummond's theistic model of evolution prompts us to dig deeper into the print educational and intellectual currents that adopted Christian evolutionism to the context of a Jaffna literary club.[73]

CONCLUSION

In this chapter, we have shown that the role of imperial groups or other intellectual strands from the West affected the degree to which local actors assimilated, or responded to, evolutionary ideals. Sayyid Ahmad's world precedes that of Dharmapāla by several decades, not yet a space of global religious faiths on equal par that would be debated in parliaments. A second major difference pertains to Lankan Buddhism's openness to outside influence: to the Americans and Europeans who adopted, popularized, and de-mythologized Buddhism, such as the Theosophists Henry Steel Olcott and Madame Blavatsky.

Compared to Dharmapāla, Sayyid Ahmad was more firmly embedded in the context and politics of Aligarh. Sayyid Ahmad left India only once, for a trip to England in 1869. After 1870, he stayed in India. Compare this to Dharmapāla, who, as Kemper points out, spent more of his life outside Ceylon than within it. By comparison, Sayyid Ahmad's scholarly output was primarily in local languages like Urdu, not in English, and his inputs in terms of Islamicate or other texts came primarily from material in Arabic, Urdu, and Persian, not English. Dharmapāla, then, we might say represents the globalized world of the twentieth century, while Sayyid Ahmad represents a moment before it was certain that imperial, global hegemony would dominate.

All in all, using these two actors as our case studies, we have crossed temporal categories in narrating the history of evolutionary thought in South Asia by using debate as a means of grappling with various pressures and influences—both the new, emerging from colonialism and other politics, and the old, in terms of *longue durée* religious traditions of thinking through design, evolution, and the question of species. Other scholars' works, especially from religious studies and intellectual history, have much to offer the history of science if it is to become truly global. Historians' work is cut out for them: broader social histories that use material in Malay, Tamil, Arabic, Persian, Urdu, and Sinhalese language (to name but a few!) are still to be written, continuing the work of scholars like Savary. Newspapers, however, exist in all of the above, not to mention school curricula and other printed material.

It is important to clarify that this chapter has not been about pinning down a singular Buddhist or Islamic approach to Darwin and evolution. The Pānadurē debate or even textual debate between al-Afghani and Sayyid Ahmad demonstrate how using debate as an analytical tool can lead to richer accounts in the global history of evolution and religion. There was no singular Buddhism in the nineteenth century, just as there isn't one now: Gelukpa differed from Zen, Thai, Tibetan, Taiwanese, Japanese, or Sri Lankan traditions. The same can be said for various Islamic societies. Much depended on the social and political conditions of regions. This chapter is, then, a model for reading modes of rhetorical sharpening and debate in the context of the pressures of colonial society. South Asia should not be synonymous with India or Hinduism, nor with a site where Europeans extracted spiritual and scientific inspiration. Debate helps us move past both of these.

3

Evolution in Colonial Australia

Institutions, Religion, and Moral Formation

Joel Barnes

Historians of Darwinism in Australia have often described the hostile reception that Darwinian ideas originally received from the mainstream of Australia's scientific and educational establishment, with their widespread take-up occurring only when a new generation of scientists and scholars made its mark from the 1880s. Although there were earlier exceptions, acceptance of Darwinism by prominent Australian scientists lagged a decade or more behind Britain.[1] This delay was not a problem of the so-called "tyranny of distance" but has generally been connected with questions of local institutional development.[2] The emergence of universities and other scientific institutions produced an establishment socialized in pre-Darwinian natural history that initially resisted the transformative implications of evolutionary theory.[3] The most important period of institutional formation was the 1850s, which saw the founding of the Universities of Sydney (1850) and Melbourne (1853) and of related scientific institutions such as the National Museum of Victoria (1854) and the Royal Society of Victoria (1859, with predecessor organizations from 1854). This moment marked the end of what Colin Finney has labeled the "heroic age" of natural history in Australia—heroic because its achievements were mostly developed with limited institutional support, by individuals and by a succession of small, voluntarist, often highly fractious and short-lived scientific and philosophical societies.[4] When the arguments of Charles Darwin's *On the Origin of Species* (1859) arrived in Australia in March 1860, they were announced into a different context, one of emergent institutional maturity in which a small but growing scientific establishment set the terms of engagement with evolutionary ideas.[5]

This chapter revisits the institutional settings of early anti-Darwinism in Australia's "colonial" period—before national federation in 1901—in light of a rethinking of the structures of empire and, within that, of closer attention to the character of the relevant institutions, namely the new universities. While institutions are indeed critical to understanding the initial resistance of Australian scientists to evolutionary thought, prevailing explanations for that resistance warrant reconsideration. I argue that the forms of thought found in and given legitimacy by the authority of the new scientific institutions have to be understood by reference to the purposes of those institutions themselves. The universities were brought about and shaped by religious, moral, and civic concerns, in contexts of sectarian conflict and contests between religious and secular models of education. They were not research enterprises at the forefront of the exploration of new knowledge but conservative institutions of colonial governance that developed in the compromises and interactions between competing sectional interests. As Richard Selleck, historian of Melbourne University, puts it, the universities were to be a "force for stability, not an irritant in the body of society."[6]

Much of Australian science in the several decades from midcentury was thus conducted not, as in Britain, by a class of leisured gentleman naturalists but by the appointees of institutions whose responsibilities in the first instance were for the intellectual and moral cultivation of an elite of the male colonial youth. By present-day standards, this student body was young. At the University of Sydney, ten of twenty-four matriculants in 1852, and thirteen of nineteen in 1857, were between fifteen and seventeen years of age. At Melbourne, 83 percent of those entering the university between 1855 and 1864 were aged between fifteen and nineteen.[7] Universities' relations with their often young students were typically understood to be in loco parentis, providing surrogate guardianship outside the family home, an understanding that persisted until the middle decades of the twentieth century.[8] Early professors, who were appointed by London-based selection committees on the basis of locally devised instructions, were considered in terms of not only their disciplinary skills but also their moral and religious characters. Those who wrote testimonials on behalf of applicants addressed both the expertise and the moral standing of prospective appointees, and committees rejected candidates who were given to such vices as unpredictable tempers or playing cards with students.[9] Examining the universities' foundational logics—the competing religious and civic interests involved in their formation and the processes by which early appointments were made—shows the importance not only of disciplinary expertise but also of ecumenical moral values in the constitution of Australia's midcentury scientific establishment.

AUSTRALIAN ANTI-DARWINISM AND ITS INSTITUTIONAL CONTEXTS

The growth of institutions, especially the universities, in the 1850s deepened and extended the basis for science in Australia, and it was from the new professoriat, especially at Melbourne, that the most dramatic public controversies over evolution issued in the 1860s. At Melbourne University, key anti-Darwinians included Frederick McCoy, professor of natural science and the long-serving director of the National Museum, and George Britton Halford, professor of anatomy, physiology, and pathology. At Sydney University, John Smith, professor of chemistry and experimental philosophy (i.e., physics) held similar views. Other prominent scientists who rejected the transmutation of species included, in Sydney, the naturalists William Sharp Macleay and the Reverend William Branwhite Clarke, and in Melbourne, Ferdinand Mueller (later von Mueller), director of the Royal Botanic Gardens.[10] Mueller's Sydney counterpart, Charles Moore, studiously avoided the issue, sticking to matters of botanical description and classification rather than theory.

The significance of institutions is also evident in the inverse case of what Barry Butcher has called a "heterodox" minority of colonial Australian scientists who were persuaded by Darwinian ideas in the decades before its more general take-up in the 1880s.[11] Those who did so—including in the 1870s the anthropologist Alfred Howitt and the botanist Robert David Fitzgerald, who worked on Australian orchids from the mid-1860s—operated largely outside formal institutional structures. The institutional thesis is confirmed by the case of Gerard Krefft in the 1870s, whose downfall as curator of the Australian Museum in Sydney was caused in part by conflicts with the museum's trustees over his acceptance of evolutionary theory. The museum was the creation of an earlier generation of the Sydney scientific elite associated with Macleay, who died in 1865, and the trustees included surviving members of his family. To this pattern of heterodoxy might be added Richard Davies Hanson, chief justice of South Australia, who in the mid-1860s argued that theology must adapt itself to scientific, including Darwinian, insights or risk descending into irrelevance. Hanson was of the Adelaide establishment, but his scholarly authority depended on his status as a jurist rather than as a scientist. As Finney notes, views of the Darwinian hypothesis were "often relative to the distance of an individual's vocation from the dispute."[12]

The institutional argument has generally been framed in terms of the control of Australian science by British anti-Darwinians, since leading figures of key institutions were typically appointed to their positions on the strength of the recommendations of prominent British proponents of natural theology, and oriented their work within that tradition. Butcher, in many respects the most sensitive historian of Darwinism in Australia, has situated the growth of Australian scientific institutions as the "entrenchment" of a "British structure of science" that was "creationist" in outlook, at a time when scientific naturalism in Britain was starting to challenge older traditions of natural theology.[13] In particular, the grip on Australian science of Richard Owen, Britain's leading comparative anatomist and paleontologist, who worked extensively on Australian specimens for decades and who after 1859 became a prominent anti-Darwinian, has been seen as particularly firm and long-lasting. Ann Moyal has argued that Owen's influence was "monolithic."[14]

These arguments are problematic both empirically and conceptually. Empirically speaking, there is an issue of chronology. With the exception of Halford, who was appointed in 1862, all the aforementioned institutional critics of Darwinism migrated from Britain (or, in the case of Mueller, from Germany) and were appointed to their positions before the publication of the *Origin*—that is, not within a context shaped by the politics of heightened debates over evolution. The issue is one therefore not of the external control of appointments but of why scientists in Australia did not prove as readily persuadable on the question of evolution as the metropolitan British scientific establishment, which was likewise largely committed to an anti-transmutationist natural theology before Darwin's influence was felt. More significant, however, is the broader conceptual issue: much more recent writing on the structure of imperial relations has recast understandings of empire in ways that make core-periphery models of one-way control appear inadequate.

The alternative explanatory framework developed here is built up by combining insights from several literatures. First, in his study of Darwinism and place, David Livingstone has argued that the "encounter with Darwin needs to be understood in terms of domestic circumstances" and that "specific venues condition what can and cannot be said about new knowledge claims."[15] If Livingstone's work emphasizes the significance of locality, the "new imperial history" has re-envisioned empire as a web or network of interconnected systems and meanings and processes.[16] As Tony Ballantyne has written, the metaphor of the web "underscores that the empire was a *structure*, a complex system of overlapping and interwoven institutions, organizations, ideologies, and discourses." The web "captures the integrative nature of . . . cultural traffic," the connecting together of "disparate points in space into a complex mesh of networks."[17] The most thorough application of this understanding of empire to the institutions at the heart of this chapter—universities—is Tamson Pietsch's

Empire of Scholars, which seeks to understand academic networks across the empire in terms of "relationships between specific sites, imperial rule and the production of knowledge and culture."[18] Combining Livingstone's arguments about locality with the outlook of the new imperial history suggests the importance of the local *within* empire, the local as something distinctive that nonetheless participates in wider imperial structures, existing as a node in complexes of broader relations.

The web model of empire aligns well with concepts of "moral economy," as defined by Lorraine Daston and recently rehabilitated and adapted by Rob Boddice.[19] Daston's "moral economy" refers not to the production and exchange of material resources but to "a web of affect-saturated values that stand and function in well-defined relationship to one another." These values form "an organized system that displays certain regularities" and are tied to "activities . . . which anchor and entrench [the moral economy] but do not determine it."[20] A moral economy is thus something more than culture or ideology: it is a structure of values grounded in practices and constituted by action.[21] Its model of power is "of the microscopic, internalized Foucauldian sort, rather than of the political (or martial), externalized kind."[22] This distributed or weblike understanding of power mirrors that of the new imperial history. Although Daston's model originally provided an internalist language for the practices of science rather than for its institutional or cultural contexts, Boddice's recent adaptation of the concept gives it a looser formulation applicable to society at large—including, feasibly, at the level of empire—while retaining the dual focus on values and practices.[23]

Imagined as such, a moral economy of empire is visible in the processes by which its structures and institutions were created and perpetuated, within the web of connections linking localities across space. Universities as nodes of the network of empire were designed to meet local needs, based on models elsewhere within the empire. Initial appointments were made in the imperial metropole according to locally devised instructions. Appointment processes are practices that reveal the texture of the moral economy, in terms that were explicitly normative given appointees' roles as pastoral supervisors of youth. Consistent with Daston's point about systemic regularity, this economy of values, as will be seen, was ecumenical and struck a negotiated balance between secular and religious imperatives; it placed a high value on masculine conceptions of "character" and "manners"; it entailed loyalty to empire; and, finally, it was racially and nationally exclusive, cultivating a whiteness in the colonies defined by Britishness. As Warwick Anderson has argued, whiteness in the Australian colonies until the 1880s was imagined in national terms, in relation to a British lineage; after that it became a broader racial type.[24] This shift aligned with the wider take-up of evolutionary reasoning in Australia from the 1880s, but in complex ways: on the one hand, the later perspective might appear as a more biologized vision of the human organism; on the other, the imagining of whiteness as a "type" could insulate heredity from environmental influence, limiting earlier concerns about racial degeneration in new climatic conditions. But in the 1850s what mattered was not whiteness as a broad type but Britishness and imperial loyalty.

THE UNIVERSITY OF SYDNEY

The Universities of Sydney and Melbourne developed in the context of debates between secular and religious models of education, in mixed denominational environments in which Protestant and Catholic sectional interests both competed and collaborated. In Sydney, disputes between secular and religious education were fiercely fought out and on several occasions nearly derailed the founding of the university. Indeed, the structure and design of the University of Sydney, as its major institutional history notes, can only be understood against the backdrop of ongoing religious dissension and controversy over the first half of the nineteenth century.[25] As the penal colony founded in 1788 grew to become the colonial outpost of Sydney and the nucleus of New South Wales, the authorities both in Britain and in Sydney itself assumed that the Church of England was the established church in the colony. The early development of schooling was accordingly placed under the direction of Anglican chaplains. This assumption was soon resisted by other denominations, and in 1836 the Church Act established subsidies to support the Anglican, Presbyterian, and Catholic churches on an equal basis, nullifying the idea of the Church of England as the established church and reconfiguring church-state relations in the colony. This development was to feed directly into debates over the place of religious education in the founding of the university.

In 1851, 53 percent of the New South Wales settler population was Anglican, and 30 percent was Catholic. Most of the rest were of other Protestant denominations, with Presbyterians the largest group at 10 percent.[26] In this environment, the university's founders struggled to balance several competing imperatives. On the one hand, there was general agreement that private efforts would not suffice and that the new institution would require state endowments.[27] On the other, in a mixed religious community with no established church, in an age in which education was predominantly denominational—and indeed in which the education of a substantial proportion of university students served as training for holy orders—public funding for education was unavoidably contentious. The solution was some form of federal system, in which separate denominational colleges were linked by an overarching university.[28] To employ a distinction used by Rowan Williams, the colonial state pursued a secularism that was *procedural* rather than *programmatic*: a negotiated balance between the positive liberties of multiple confessional communities rather than policies of negative liberty in which religious belief was a matter of private interest.[29]

Positive state commitments to multiple religious communities proved difficult to negotiate. Between the initial university structure first proposed by the university's motivating spirit, William Charles Wentworth, in 1849 and a final settlement in university-college relations reached in 1858, several versions of quasi-federalism were tried out. Wentworth had the federated models of Oxbridge, the Queen's Colleges in Ireland, and the University of London to draw upon, but none suited Sydney's needs exactly. The University of London, the awarding body for the Anglican King's College and the secular University College, was the closest fit in that it likewise bridged disparate institutions, but it did so by serving a purely administrative certifying function, with teaching and corporate life devolved to the colleges themselves. Both in Wentworth's original vision and in the model that eventuated, Sydney University was to have fuller teaching rather than just accrediting responsibilities. Following a number of years of contest and tension between secular and religious interests, and between different religious communities, the institution that eventuated comprised a cluster of geographically proximate denominational colleges organized around a secular central university. To establish the colleges, the colonial government made grants of land and for building costs to each of the four main churches—Anglican, Catholic, Presbyterian, and Methodist. Student attendance at professorial lectures was made compulsory, in effect centralizing all teaching, while a short-lived religious observance requirement was abolished by parliament in 1858 on the grounds that it functioned as a de facto religious test, which was prohibited by the university's act of incorporation. The 1858 settlement represented the victory of the central secular institution over the colleges and ended their ambitions to act as teaching institutions.[30] The reduction in status to non-teaching residential facilities was resented by some in the colleges well into the twentieth century.[31]

This, then, was the religious and institutional context in which the university's three founding professors were appointed in 1852. John Woolley was made professor of classics, Morris Pell professor of mathematics, and John Smith professor of chemistry and experimental philosophy. To make the appointments, the university senate commissioned a London-based selection committee comprising Sir John Herschel, the astronomer and mathematician who would later refer to natural selection as "the law of higgledy-piggledy"; George Biddell Airy, the Astronomer Royal; Henry Malden, professor of Greek at University College, London; and Henry Denison, a relative of future governor Sir William Denison who had resided for some years in New South Wales and was familiar with members of the senate from his time there.[32] By way of advertisement, the committee printed a pamphlet containing a number of documents related to the university's establishment, governance, and regulations, as well as extracts from a long letter from three members of the senate containing instructions to the selection committee.[33]

The senate's instructions to the selection committee emphasized the importance not just of expertise and academic abilities but also of moral and religious character. The preamble to the enabling legislation, which was reproduced in the pamphlet, opened by stating that the university was established "for the better advancement of religion and morality," and this phrase was likewise used in the first sentence of the senate's letter of instruction.[34] The senate advised that the committee should consider "qualities of mind, character and manner," as "we require for our purposes gentlemen scholars, men of ability and vigor of character."[35] Against what Wentworth had originally intended, being in holy orders was not considered a disqualification, but in general layman were thought preferable.[36]

In the case of Woolley, a liberal Anglican clergyman, the committee opted to set the senate's preference aside. Inevitably, a professor in holy orders provoked some consternation in an environment of religious tensions, but the only other candidate for the classical professorship with equally high academic accomplishments was rejected on account of a confession that he did not observe the Sabbath. Woolley came with recommendations from high quarters, principally from the Reverend Adam Sedgwick, professor of geology at Cambridge and subsequently a prominent anti-Darwinian. Pell was likewise Anglican, American-born and a graduate of Cambridge.[37] Smith was a committed Presbyterian who late in life developed interests in Theosophy and spiritualism, though seemingly without any loss of faith in traditional Christianity.[38]

The moral and religious characteristics that the senate sought featured prominently in testimonials written in support of candidates, as can be seen in the case of Smith. George Dickie, professor of natural history at Queen's College, Belfast, commented that Smith's "exemplary deportment, in a religious and moral point of view, renders him well fitted to be an instructor of youth."[39] Dickie, like Smith Scottish-born and Scottish-educated, was subsequently the coauthor with the Reverend James McCosh of the natural theology treatise *Typical Forms and Special Ends in Creation* (1856).[40] William McGillivray, professor of natural history at Marischal College, Aberdeen, commented approvingly on Smith's "moral and religious character," attesting that Smith "professes what I esteem scriptural principles of religion, and has not, to my knowledge, in practice, deviated from a high standard of Christian profession." The Reverend John Croumbie Brown of Aberdeen similarly wrote of Smith that "his manners and bearing are those of the gentleman—his sentiments and conduct those of the christian [*sic*]."[41] Such concerns with morality and religion in the selection processes were even more prominent in the case of the founding appointments at Melbourne.

THE UNIVERSITY OF MELBOURNE

Whereas Sydney was more than sixty years old when its university was founded, just eighteen years had elapsed between the first European settlement in Melbourne in 1835 and the establishment of the University of Melbourne in 1853. It was a move that many at the time considered premature but one underpinned by the discovery of gold in central Victoria, which transformed the colony into a booming destination for young men on the make and led the European population of the district to triple between 1851 and 1854.[42] Victoria's religious demographics were similar to those of New South Wales, though with a slightly smaller proportion of Catholics. Around half of the Victorian population was Anglican, a quarter of other Protestant denominations, and a quarter Catholic.[43]

Although there was less contention than at Sydney University between denominations and between secular and religious educational ideals, the onset of the gold rushes raised the specter of "materialism," making the teaching of science and its relations to faith matters of special concern. The pursuit of material riches was seen by many in the churches as a threat to Christian values, being a "state of excitement . . . most unfavourable to spiritual prosperity," and as an all-absorbing interest that led the "great mass of people . . . to have forgotten the existence of Christianity." That the reverse was true—the proportion of the population that were regular churchgoers doubled from an eighth to a quarter between 1850 and 1861—is not the significant point; however empirically unjustified, anxieties about supposed "materialism" were themselves shaping factors in the politics and social relations of Victorian colonial society, including in the organization of the new university and the role of science within it.[44] Melbourne's first professors also arrived just as thirteen men were charged with treason for staging an armed rebellion against the colonial government at the Eureka stockade, on the Ballarat goldfields in central Victoria.[45] The defendants won wide public sympathy, and all were acquitted, most in trials before Justice Redmond Barry, also chancellor of Melbourne University. The acquittals signaled the developing egalitarian sentiments of the colony and threatened the social order that privileged colonial elites.

The university-college division of labor worked out at Sydney was replicated at Melbourne, with the variation that whereas the Sydney colleges were specifically legislated for by parliament, at Melbourne the university's act of incorporation empowered its governing council to affiliate other colleges or educational establishments as it saw fit. The Melbourne legislative model was then copied at the Universities of Adelaide (founded 1874), Tasmania (1890, in Hobart), Queensland (1909, in Brisbane), and Western Australia (1911, in Perth).[46] In the planning and establishment of Melbourne University, there was relative equanimity and cooperation among the parties involved. Learning from mistakes at Sydney, the founders established a twenty-man council—the equivalent of Sydney's senate—comprising four clergymen (Anglican, Catholic, Presbyterian, and Congregationalist) and sixteen laymen (twelve Anglican, two Catholic, one Presbyterian, and one Quaker), including the Anglican Barry as chancellor. The local press praised the "absence of sectarianism" in the choice of clergy represented, and though some of these sectional representatives were given to fighting sectarian battles in other contexts, Selleck comments that "church leaders seem to have reached a tacit agreement that efforts to establish a university would founder if religion became an issue."[47]

The relative calm between different religious interests in the making of Melbourne's university was both enabled by and reflected a series of differences from the situation in Sydney. For one, there was to be no provision like that at Sydney mandating compulsory worship.[48] Most importantly, in Melbourne there was no public funding for residential colleges, though the government did set aside land adjacent to the campus with the development of colleges in view. It was not initially envisaged that these would necessarily be church establishments, but in 1860 the Catholic Church applied for a portion of the allocated college land, and by October 1861 a scheme of division had been agreed whereby the same four denominations as at Sydney each received a little over ten acres.[49] The absence of state funding, with the colleges relying instead upon the private resources of their respective confessional communities, meant that the Melbourne colleges were not established for another decade or more. The Anglican Trinity College, founded in 1870, was the first, followed by Ormond College (Presbyterian, 1879), Queen's College (Methodist, 1887), and finally Newman College (Catholic, 1916) (see figure 3.1). The university-college division of labor was protected in the incorporating act of the university, which established that there be no religious tests for entry to the university but also put a restriction on the capacity of the university to make "statutes" regarding affiliated colleges; it could make any such statutes so long as these did not "affect the religious observances or regulations enforced" in the colleges.[50]

As at Sydney, to make founding appointments the Melbourne council employed a London-based selection committee, comprising Herschel, Airy, and Malden, all of whom had been on the Sydney committee, plus Robert Lowe, a former resident of New South Wales familiar with colonial conditions, and Victorian chief justice Sir William à Beckett, who happened to be

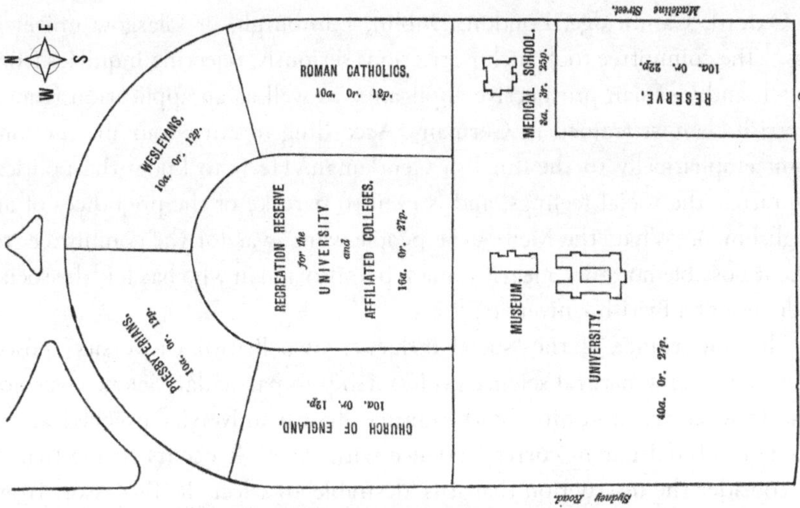

Figure 3.1: Map of the University of Melbourne campus and affiliated college grounds, showing college land allocated to the four main Christian denominations of Victoria. From the University of Melbourne *Calendar*, 1877–1878, 33. Reproduced with the permission of the University of Melbourne Special Collections, Baillieu Library.

visiting Britain at the time.[51] The committee was to appoint four professors in classics, mathematics, natural science, and a composite post encompassing "Modern History, Modern Literature, Political Economy, and Logic." Again as it had for Sydney, the committee printed a pamphlet advertising the positions, this one containing the enabling act of the Victorian parliament, the minutes of a meeting of council, and two letters from Barry to committee members setting out requirements for appointees and the social and institutional contexts in which they would work.[52]

In his instructions to the selection committee, Barry described the university as having been "established on a basis of comprehensive liberality, with a due regard to the conscientious scruples of the different bodies of which the community is composed." He made clear that—unlike at Sydney—appointees should not be in holy orders but should be of "approved worth and moral standing; and of such stability of character as to command respect." "Literary attainments" were desirable, but these had to be matched by "amenity of disposition" and a "total abstraction from political or sectarian interference." A further requirement was for "such habits and manners as to stamp on their future pupils the character of the loyal, well-bred English gentleman." The prevailing structure of values demanded loyalty to empire and a commitment to a gentlemanly "Englishness" (in fact, a Britishness). On the nationality restrictions, Barry noted that it was "desirable" that candidates be graduates of Oxford, Cambridge, London, Dublin, Edinburgh, or Glasgow universities.[53] The committee took such restrictions seriously, rejecting inquiries from Dutch and German prospective applicants, as well as an application from a Scottish chemist trained in Germany. According to Airy, "our instructions point emphatically to the 'English Gentleman.' He is to know the politics, the tastes, the social feelings, and is even to partake of the prejudices of an Englishman." What "the Melbourne people want" was for the committee "to take if possible not only a learned man, but also a man who has felt the social influence of a British University."[54]

The importance of the "social influence of a British University" posed problems for the natural science professorship in particular. Science was not exactly what the mid-nineteenth-century British university excelled at, as Malden pointed out in correspondence with Airy: "It occurs to me that if we consider the instruction that it is 'desirable' to select the Professors from Graduates of certain universities to be an absolute prescription, we shall be in great danger with regard to the Professor of Natural Science of not having the best men to choose from." The kind of scientific knowledge required, Malden noted, was as likely to be found among non-graduates as among graduates. In reply Airy suggested that it would be better to act without too restrictive an interpretation of the instructions, though keeping the value of a university background in play: "We must have the science; but it will be ten times as valuable if it comes with University education. . . . So the thing will require good judgment."[55] In the end the committee half-adhered to the council's instructions, appointing as professor of natural science Frederick McCoy, who had studied medicine at the University of Dublin but never graduated. On the strength of subsequent geological and paleontological work, McCoy was in 1849 appointed the first professor of geology and mineralogy at Queen's College, Belfast, experience that the committee considered outweighed the technicality of being a non-graduate.[56]

What is evident in these discussions about the significance of experience in a British university is the importance of socialization in its moral norms, in addition to disciplinary expertise. Patterns of scientific development in the United Kingdom, frequently taking place outside universities' walls, posed with particular sharpness the question of the relative importance of each. The problem was further exacerbated by the gold rush context, which raised in addition the threat of "materialism." Questions of morality were especially pertinent in the case of the teaching of science, as a counterweight to a purely material acquisitiveness to which technical knowledge was otherwise likely to be put. In instructions to the committee, Barry reflected upon the significance of the "remarkable discovery" of gold in central Victoria: "In a country where material wealth is so engrossing an object of pursuit, and where the realities of tangible demonstration are so much heeded," it was the sciences, rather than other fields of study, that were likely to be most attractive to students. Moreover, as the advancement of the colony depended upon the exploitation of gold and other resources, there was a "great necessity for instruction in Geology, Mineralogy, Metallurgy, Chemistry, Natural History, and Physiology" to build up the stock of local expertise. At the same time, the moral threat represented by the unalloyed pursuit of wealth meant that much depended on "the mode in which young men are swayed and stimulated by their instructors." For the youth of the country, "the first great effort to reclaim the intellect, create the taste, form the manners and confirm the loyalty, depends completely on the temper and spirit with which its University opens its doors and conducts its operations." Science teaching must not only be technical but had also to produce gentlemanly tastes and temperaments and confirm imperial loyalties.[57]

Barry expounded further on these themes in a high-flown address at the inauguration of the university in April 1855. The state, he proclaimed, recognized as a "great moral principle" the need for "learning, knowledge, and science" to displace ignorance and prevent vice and crime. Without such learning, he claimed, alluding to the rush for goldfields wealth, "worldly riches sit, like undeserved honors, gracelessly and unprofitably on the possessor." The required thing was not merely disciplinary expertise but "an education learned and liberal," and with this in mind the university council had attempted to design a system of education that would "instil a sense of self-reliance and a high moral perception of the obligations of a spontaneously imposed responsibility."[58]

At the opening ceremony were three of the four founding professors appointed by the London committee: McCoy for natural science, William Wilson as professor of mathematics, and Edward Hearn for the composite post, soon after reduced to just history and political economy. All three came from previous professorships at the Queen's Colleges, McCoy and Wilson from Belfast, and Hearn from Galway. Henry Rowe, the appointee in classics, died shortly after his arrival in Melbourne and was replaced in 1856 by Martin Irving.[59] Herschel commended the initial four appointments to Barry as "one and all . . . men of the highest & most unblemished moral character & reputation."[60] Wilson and Hearn were both Anglican and involved in the subsequent establishment of Trinity College. Wilson was instrumental to the college's development, viewing the secular university with no formalized space for religious and moral instruction as "incomplete." Hearn was chancellor of the Anglican Diocese of Melbourne from 1877 until his death in 1888.[61] McCoy, who like Woolley at Sydney was supported for the natural science professorship by Sedgwick, was raised Catholic in Ireland but shortly before his appointment to Melbourne converted to Anglicanism over difficulties accepting the doctrine of transubstantiation. When he died in 1899, a controversy erupted over a claim that he had reconverted to Catholicism on his deathbed. Though McCoy's grandson denied the allegation, the priest who allegedly took his confession reaffirmed that McCoy had done so.[62] Irving came from a complex religious tradition, his father, Edward Irving, having split from Presbyterianism in the 1830s to found the Catholic Apostolic or Irvingite Church, of which the younger Irving remained a member. Irving had nonetheless signed the Thirty-Nine Articles of the Church of England at Oxford, maintaining that doing so was not incompatible with his membership of the Catholic Apostolic Church.[63] The other figure crucial to subsequent debates over evolution was George Britton Halford, who joined the university in late 1862 as professor of anatomy, physiology, and pathology. Halford came from a distinguished research career at the Liverpool Royal Infirmary and Lunatic Asylum and at the Grosvenor Place School of Medicine.[64] Delivering the Halford Oration at the Australian Institute of Anatomy in 1940, A.L. Kenny described him as "of religious mind, he was humble before his God. . . . intolerant of religious discussions he would say to his family: 'Just have faith, my children.'"[65]

The moral economy visible in the formation of the first generation of the Australian professoriat emphasized values of character, gentlemanly manners, imperial loyalty, racial Britishness, and a balancing of secular knowledge with an ecumenical Christianity. Although none of the appointees, except conceivably Halford, could have been selected with anti-Darwinism specifically in view, the overall conservative character of the new Australian scientific establishment was a product of the institutions they represented and of the values guiding the selection process. These values were at once local—reflecting complexes of secular, religious, and sectarian influences at play in the colonies—and imperial, as the new institutions were created through and as part of wider networks of empire. Seen in these lights, the largely negative response of the Australian scientific establishment to the intellectual upheaval of Darwinian evolutionary theory in the 1860s is only to be expected. Daston notes that a moral economy comprises "an organized system that displays certain regularities, regularities that are explicable but not always predictable in their details."[66] This was the case in the pattern of early Australian academic responses to Darwinism—a pattern of regularities, with an exception that proves the rule.

EARLY RESPONSES TO DARWINISM

The most prominent public controversies over evolution in the 1860s came from McCoy (figure 3.2a) and Halford (figure 3.2b) in Melbourne. The stories of these disputes have been well told by a number of historians.[67] In 1863,

Figures 3.2a–d (*clockwise from top left*): Frederick McCoy (1817–1899), professor of natural science, Melbourne. Reproduced with the permission of the State Library of Victoria (image 1708227); George Britton Halford (1824–1910), professor of anatomy, physiology, and pathology, Melbourne. Reproduced with the permission of the University of Melbourne Archives (image 2017.0071.00595); John Smith (1821–1885), professor of chemistry and experimental philosophy, Sydney. Reproduced with the permission of the University of Sydney Archives (image G3_224_1906); William Hearn (1826–1888), professor of modern history, modern literature, political economy, and logic, Melbourne. Reproduced with the permission of the University of Melbourne Archives (image 2017.0071.00621).

Halford delivered several lectures attacking Thomas Henry Huxley's Darwinian views, in one of which he stated that Huxley's works "might have been written by the devil." Halford's lectures prompted a protracted dispute with William Thomson, a local medical doctor and a former student of Huxley's. Thomson sent Huxley one of Halford's lectures and published his reply in the *Argus* newspaper. In principle, professors were to abstain from political and religious controversies, but the university council was content to give Halford free rein so long as he entered debates in defense of orthodoxy. Even still, the intemperate character of his attack on Huxley constituted a considerable embarrassment for the fledgling university. Halford had the backing of McCoy, who tried to prohibit students from reading Darwin's works and refused to have any of his books in the university library. McCoy spent several years trying to acquire stuffed gorillas for the National Museum, believing that seeing them in the flesh would persuade the public of "how infinitely remote the creature is from humanity." In 1865, McCoy finally succeeded in obtaining gorilla specimens, which became a significant public attraction. At the end of the decade the issue was reignited by Huxley's assertion that "protoplasm" was the formal basis of all life in his essay "The Physical Basis of Life" (1869). In response, McCoy delivered several lectures on *The Order and Plan of Creation* (1869–1870), arguing against Huxley and the possibility of transmutation.[68]

At Sydney, controversy was more muted, largely because the university remained for some years committed to an Oxbridge model of classical-mathematical liberal education, with limited teaching in natural history until the 1870s, a fuller expansion of which followed in the early 1880s.[69] The response of Smith (figure 3.2c), the only professorial representative of natural science in the early university, is nonetheless of note. Although Smith's chair did not encompass natural history, in his previous post at Aberdeen he had been a lecturer in agriculture, and testimonials in support of his application for the post indicate a proficiency in botany, geology, mineralogy, and some branches of zoology.[70] While he was not a vigorous anti-evolutionary partisan, at a scientific society meeting in 1863 Smith discussed the French archaeologist Jacques Boucher de Perthes's findings of ancient human remains in the Somme Valley. Conceding the vagueness of the language of the Old Testament from which the origin of humanity had traditionally been dated, Smith argued that there was no difficulty in adding "a few thousand years to the generally recognised six thousand." The Somme findings were, however, evidently much older, and to explain them Smith speculated about the possibility of an earlier pre-Adamite race of humans that had become extinct long before the biblical chronology.[71] There was no prospect of Smith countenancing a more ancient, evolutionary lineage for contemporary humans.

Hearn (figure 3.2d), a moderate conservative and free trader, constitutes the notable exception to the rule of professorial rejection of Darwinism in 1860s Australia. Several years before emigrating to Melbourne, he completed the unpublished "Essay on Natural Religion," arguing in favor of a benevolent designer and against evolutionary doctrines in the forms given them by Jean-Baptiste Lamarck and Robert Chambers, the then still anonymous author of *Vestiges of the Natural History of Creation* (12 editions, 1844–1884). However, Hearn absorbed and came to accept the arguments of the *Origin* quickly, and his *Plutology* (1863), a textbook of political economy, shows the application of evolutionary ideas to social and economic problems. Though in the earlier study Hearn rejected evolution, under the influence of his former teacher Richard Whately at Dublin, he had already then conceptualized society in organic terms, and this emphasis provided the basis for his subsequent acceptance of Darwin's reworking of evolutionary theories.[72] Darwin's use of economic metaphors in the *Origin*, as well as the influence of Herbert Spencer, provided the "common context" for Hearn's adaptation of biological ideas into social and economic realms.[73] At the same time, there were limits to that commonality. Hearn, like Hanson, was an expert in constitutional law and political economy; both accepted Darwinism from vantage points of significant institutional positions but from outside the realm of the biological sciences.

In the pattern of moral and religious values set by this early generation of Melbourne academics, the place of Hearn's political economy is instructive. Hearn was the most comfortably "establishment" of the founding professors, and his post-*Origin* acceptance of evolutionary theory apparently did not entail any loss of faith or alienation from the Anglican Church. Nor did it seem to prompt much pushback from clergymen or lively public controversy of the sort pursued by Halford.[74] This may be because *Plutology* hardly sold, but there were also broader ideological congruences at play.[75] Whereas Huxley's Darwinism, to which McCoy and Halford objected, represented a radical breach in the biological sciences that had far-reaching implications for questions both of faith and of scientific method, Hearn's use of evolutionary theory underpinned a restatement of orthodox political economy. That notionally competing readings of the plausibility of evolution could be put to such ideologically comparable ends shows, as Butcher argues of the Halford-Huxley dispute, that what were really at play were moral and philosophical issues, rather than ones resolvable in scientific terms.[76]

THE COMING OF EVOLUTIONARY SCIENCE

The eventual take-up of evolutionary science in Australian universities represented a restructuring of the moral economy prevailing at midcentury. By the 1880s evolution no longer represented the moral schism it had in the 1860s, as established religious interests came to terms with and found ways to accommodate its challenges. The muting of the moral threat was a precondition of the expansion of evolutionary science in the universities; such expansion in turn further reshaped the economy of values.

At the level of university teaching, these changes are detectable from the mid-1870s, initially at Adelaide, Australia's third university, where the evolutionist Ralph Tate was appointed in 1874 as the first Elder Professor of Natural Science. What is notable is not just the appointee chosen but also the role of Huxley in the appointment process. In the 1850s Huxley had been a potential applicant to Sydney University. He had stayed in Sydney intermittently throughout his four years on HMS *Rattlesnake* (1846–1850), and his fiancée Henrietta Anne Heathorn was still living there. He would gladly have applied to Sydney had it included among its founding professorships one in natural history, in which case he could have expected the support of the influential Macleay.[77] In the 1860s, as debates over evolution grew, he became the target of Halford's and McCoy's attacks from Melbourne. His relationship with Australian universities changed again starting in the 1870s, as first Adelaide and then Melbourne called on him to sit on selection committees for scientific professorships, all but ensuring evolutionary-minded candidates would be selected. Having taken advice from Melbourne on the general process for the selection of candidates, the Adelaide council followed the by then established process of employing British selection committees.[78] Huxley was among the membership of the committee for the professorships in natural science and in mathematics.[79]

It is likely relevant that the University of Adelaide's chancellor at the time was Chief Justice Hanson, who had revealed his sympathy for Darwinian ideas in the mid-1860s. But this was no clean break: the University of Adelaide, like Sydney and Melbourne before it, was a product of diverse and sometimes competing interests. The university's vice-chancellor was the Reverend Augustus Short, Anglican high churchman and Lord Bishop of Adelaide, who viewed science in a natural theological mode and held that geology could be harmonized with Genesis. Hanson and Short had been frequently opposed disputants in Adelaide Philosophical Society debates on science and religion throughout the 1860s.[80] When Hanson died shortly before the formal opening of the new university, the task of delivering its inaugural address fell to Short, who gave a natural theological account of the evidences of science. In the published text of Short's speech, he appended a footnote that cast doubt on "the evolution theory of Mr. Darwin."[81]

In late 1886 Huxley was called upon by Melbourne University to help choose McCoy's successor. The appointment of Walter Baldwin Spencer as professor of biology was a pivotal moment in Australian science.[82] An advocate of evolution trained in up-to-date experimental methods, Spencer updated McCoy's outmoded curriculum, established modern laboratories, and along with several lecturer associates made Melbourne the main center of evolutionary science in Australia. Appointments at other Victorian institutions, including P.H. MacGillivray at the Bendigo School of Mines and Daniel McAlpine at Ormond College, strengthened the basis for evolutionary science. These figures were all active in the formation of the Australasian Association for the Advancement of Science, which after a decade of planning finally came into being in 1888.[83] The *Melbourne University Review* in 1887 heralded these developments—the "outburst of scientific zeal, within and without the Universities, throughout the Australian Colonies"—as a "scientific *renaissance.*"[84]

Similar moves were made at Sydney in the 1880s. The university expanded and reorganized its science teaching from 1882, appointing William John Stephens as professor of natural history and shortly thereafter establishing a full faculty of science.[85] As the faculty was being established, Edward Meyrick, the late master at Sydney Grammar School, lamented in the *Sydney University Review* the slow take-up of Darwinian thinking in Australia and, attending especially to problems of education, argued that "anyone who professes to teach biology, rejecting the theory of evolution, will be compelled to pass over as inexplicable what is elsewhere considered as the very essence of the subject."[86] This argument, from a representative of one of the university's key feeder schools, represented a felt need within educational circles for serious engagement with evolutionary theory. That need was met in the university not so much by Stephens as by William Haswell, demonstrator in comparative anatomy and physiology. The major breakthrough at Sydney of evolutionary biology is usually dated to Haswell's subsequent appointment in 1890 as the first Challis Professor of Biology, though he was already teaching evolutionary science in other roles throughout the 1880s.

The full development of evolutionary biology in Australian universities did not always mean the acceptance of Darwinian theory. Even in the 1920s and 1930s, Frederic Wood Jones, professor of anatomy at Adelaide and then at Melbourne, was an avowed Lamarckian and a critic both of Darwin and of the new theories of genetics.[87] Nor was the coming of evolution an innocent development. Just as biological science in the last decades of the century enabled a recasting of logics of whiteness, so too did it reorient ideas about Indigenous peoples in complex ways to which the historiography has not always done justice. The waters have been muddied by vague and overly capacious uses of "social Darwinism" by late twentieth-century historians of frontier conflict such as Henry Reynolds. Such uses have attracted appropriately sharp critique from Mark Francis, who challenges the "Darwinian" character of many ideas and policies labeled as such.[88] On the other hand, Butcher, drawing on the arguments of Robert Young and James Moore that Darwinism was always social, highlights Darwin's own use of Australian—including Aboriginal—evidence in his theory building. "Social Darwinism" in Australia was not, in Butcher's account, a crude socialization of imported biological models but part of a feedback loop in which Darwinian concepts provided new ways of seeing and describing Australian phenomena that had themselves shaped those concepts.[89]

Steering between these accounts, what is certain is that the scientific ascendency of the 1880s—which included evolutionary if not always Darwinian orientations—underpinned the late nineteenth-century growth of fields such as physical anthropology. Figures such as Baldwin Spencer and, in Adelaide, the anatomists and anthropologists Archibald Watson and Edward "Ted" Stirling were pioneers in the making of academic knowledge about Indigenous peoples and supporters of assimilationist policies that separated mixed-race children from Aboriginal mothers, creating what today is known as the Stolen Generations.[90] From 1911 Baldwin Spencer was himself involved in administering such policies in the Northern Territory—though Francis rejects the view that his doing so was in any way "Darwinian."[91] These ideas and practices then opened out into later forms of eugenic thought, which overtook much of the University of Melbourne, at least, in the first half of the twentieth century.[92] Though debatably "Darwinian," such ideas were unthinkable without a basis in evolutionary biology.

CONCLUSION

The growth of evolutionary biology, and science generally, in the Australian universities from the 1880s opened up a new vista quite different from that of the 1850s. The moral economy that had structured the appointments of figures such as McCoy, Halford, and Smith—with its concerns for Britishness, character, and gentlemanly manners—had much receded in the face of more strictly scientific considerations. The world of emergent eugenics is quite different, both in tone and in purpose, from the midcentury natural theological style of scientific practice, though as the Adelaide case shows, the progression from one to the other was not always linear. The earlier framework was constructed in empire but reflected local needs and peculiarities—and local participation in wider imperial structures rather than external control. Although this moral economy can be characterized as "creationist," it may more usefully be seen as having developed not so much out of a scientific opposition to naturalism as such but through the interaction of institutional values and practices that reflected pastoral concerns for the moral and religious formation of an elite of the colonial youth.

4

As Above, So Below

The Role of Astronomical Evolution, Imperialism, and Religion in the Long Nineteenth Century

Edward Guimont

The idea that intelligent life might inhabit other planets in the universe predates modern scientific thought, including the development of core aspects of biology and astronomy. Among its earlier advocates, a number of ancient Greek philosophers considered it; the Qur'an has passages generally interpreted as endorsing the notion; and the Dominican monk Giordano Bruno (1548–1600) was burned at the stake by the Roman Inquisition for proposing what was then referred to as the "plurality of worlds" concept. The first two chapters in this section illustrate the long history of the various religions in South Asia debating how notions of astronomy, deep time, evolution, and religion—with their cosmic entities—were related. The European notion of pluralism was seemingly buoyed by the nebular hypothesis of solar system creation proposed in the eighteenth century by Immanuel Kant (1724–1804) and refined in the nineteenth century by Pierre-Simon Laplace (1749–1827), which argued that the solar system formed as planets condensed out of the primeval solar nebula as it cooled from the outside of the solar system toward the center. Therefore, as Mars is further from the sun than Earth, it would also be older than Earth, and its older celestial inhabitants would correspondingly be more philosophically advanced than terrestrial humans. From the early 1800s, proposals circulated—commonly attributed to Austrian astronomer Joseph

Johann von Littrow (1741–1840)—to make contact with these advanced Martians by digging giant Pythagorean shapes into the Sahara Desert of North Africa, filling them with oil, and lighting them at night so that Martian astronomers could see that their inner neighbor was home to cousins who, inferior as they might be, were still advanced enough to have mathematical knowledge and large-scale engineering.[1] Perhaps the conservatism of Australian universities, as discussed in the previous chapter, explains why the Great Sandy Desert was never proposed as a site for this early form of SETI (search for extraterrestrial intelligence).

Celestial beings older and purer than Earthly humans—this evolutionary view of extraterrestrial life has clear religious overtones. Indeed, there is an overlap between religious millenarianism and the hope for first contact with alien life, with Walter Stephens identifying twentieth-century efforts to communicate with extraterrestrials on a continuum of efforts to communicate with angels and devils dating back to the *Dialogus miraculorum* of the German Cistercian monk Caesarius of Heisterbach (ca. 1180–ca. 1240).[2] Similarly, von Littrow's proposal for communication with said heavenly beings requiring the European occupation of Africa to engage in major public-works projects—this is the use of extraterrestrial life and astronomical requirements to justify imperialism, decades before the 1884–1885 Berlin Conference and subsequent Scramble for Africa, the period from the 1880s to World War I when approximately 90 percent of the continent was divided up between European nations. African locales will be returned to throughout this chapter as the basis for both comparison to and observation of planetary locales, as well as inhabitants. In particular, as the religious foundation for cosmic pluralism declined at the same time Africa came to symbolize imperialism in Western minds, the translation to what might be called imperial pluralism was perhaps a logical one.

Those concepts worked in tandem, as they all emerged from a similar intellectual wellspring. From the second half of the nineteenth century, the writings of Charles Darwin (1809–1882) and others ushered in new public debate about the evolutionary origins of life on Earth, including humans. The concurrent Afrocentric New Imperialism both built off and stimulated said debate, providing biologically racist justifications for European domination of supposedly less-evolved non-Europeans, a debate whose Egyptian context is discussed in chapter 7. In the midst of this, beginning in 1877, public debate on extraterrestrial life exploded due to the claimed observations of canals on Mars by Italian astronomer Giovanni Schiaparelli (1835–1910) and his American counterpart Percival Lowell (1855–1916), leading to discussions on the relative evolutionary states of human and Martian.

These three trends of public intellectual debate, while seemingly disparate, were not isolated from each other, nor did they remain static. As stated above, pluralism was debated within Christianity for many centuries, with astronomers reflecting that Christian focus. But by the end of the nineteenth century, astronomy, re-envisioned using evolutionary theory, provided a scientific justification for largely secular imperial and socialist ideas. It was during this period, when the first ostensible evidence for extraterrestrial life was observed, that pluralism evolved beyond its religious roots.

Rather, notions of evolution, astronomy and intelligent Martians, and the racial ideologies of empire informed and built off each other. This was reflected in astronomers' belief that Mars had formed before the Earth had and was therefore a model for the future of human civilization, both biologically and politically; in the debate on whether life in the cosmos proved or precluded evolution on Earth, and the religious implications therein; and in how empires provide metropoles with not only ideal locations for observatories but also exotic cultures on which to model the populations of the worlds observed through those telescopes. These reflect the fact that the idea of evolution is not just applicable to biology; the nebular theory was also an evolutionary model that was seen as applicable to political and social development. In challenging the status quo, by providing seemingly universal laws and maps of progress, the nebular theory's value was made obvious to radicals before Darwin made the connection between the nebular hypothesis and biological evolution.[3] After Darwin, figures like Richard Proctor (1837–1888) integrated evolutionary theory and astronomy even more closely, not only through discussing the evolutionary courses of planets, stars, and solar systems but by connecting the notion with the religious model of pluralism as well. Neither Proctor nor Darwin were socialists; however, the socialist use for evolutionary astronomy could not be done until the link between evolution and astronomy was made, something Darwin himself did not do.

This chapter will focus on how that combination of secularization of religious pluralism and its use for radical political projects emerged, via a transnational genre of work featured by astronomers such as Schiaparelli, Lowell, and Garrett P. Serviss (1851–1929); socialists such as Friedrich Engels (1820–1895), Louis-Auguste Blanqui (1805–1881), and Aleksandr Bogdanov (1873–1928); and in particular, such authors of popular fiction as Jules Verne (1828–1905) and H.G. Wells (1866–1946), who played vital roles in promoting the idea, even among scientists, of what life in the universe—and their political beliefs—might look like. To a degree, the categorization of these figures is imprecise, given that, for example, Serviss was an astronomer as well as a fiction author while Wells and Bogdanov were both socialists and fiction authors. However, this is an indication of the extent to which these various interests overlapped and were expressed in many different mediums.

DESTINATION MOON

It is perhaps useful to begin by recounting an anecdote from the era of lunar mission planning. In the sixties, in preparation for the launch of the first three-man capsule around the Moon, the architects of the American space program were concerned with the potential for communication with our nearest neighbor. In the recorded annal of their conversation, the head of the program described a historical idea by which

> a German geometrician proposed to send a scientific expedition to the steppes of Siberia. There, on those vast plains, they were to describe enormous geometric figures, drawn in characters of reflecting luminosity, among which was the proposition regarding the "square of the hypothenuse," commonly called the "*Ass's Bridge*" by the French. "Every intelligent being," said the geometrician, "must understand the scientific meaning of that figure. The Selenites, do they exist, will respond by a similar figure; and, a communication being thus once established, it will be easy to form an alphabet which shall enable us to converse with the inhabitants of the moon." So spoke the German geometrician; but his project was never put into practice, and up to the present day there is no bond in existence between the Earth and her satellite.[4]

Of course, the decade in discussion was not the 1960s but the 1860s, and the above transcription was not from a NASA publication but from Jules Verne's 1865 novel *From the Earth to the Moon*, the first serious writing—albeit in fictional form—of how such a titular expedition could be conducted. The character quoted—Barbicane, president of the Baltimore Gun Club and future commander of the capsule—is explicitly stated as being religious, frequently speaking "God preserve us!" and other religious invectives.[5] More interesting, however, is the passage of the novel about how the Gun Club raises the funds needed for its expedition. There not being enough capital in the United States, they go international, including soliciting funds from the central bank of the Ottoman Empire. As it turns out, the Ottomans are quite willing to fund a voyage to the Moon: "Turkey behaved generously; but she had a personal interest in the matter. The moon, in fact, regulates the cycle of her years and her fast of Ramadan."[6] To Verne, it seemed clear that a journey to the Moon would have vast implications for Muslim authorities. Further, the internationalist aspect of the American lunar expedition is in line with Verne's early, positive view of the United States as a contrast to the European empires of his day, before his attitude soured later in the nineteenth century, leading to such cynical satires of American capitalism as his 1895 *Propeller Island*.[7] It should also be noted that of all the European countries' contributions to the Baltimore Gun Club, the only indication of their wealth deriving from empires is a reference to the contribution from "Holland and her colonies."[8]

From the Earth to the Moon was not the only novel in Verne's lifetime to make a connection made between "Selenites" and the Ottomans. His great British counterpart, the socialist H.G. Wells, wrote *The First Men in the Moon* (1901), about an expedition to the Moon by two Englishmen, the scientist Cavor and the businessman Bedford—demonstrating the unification of Victorian science, capitalism, and imperialism. On landing, Bedford recounts how, in drunken celebration, "in some way that I have now forgotten, my mind was led back to projects of colonisation. 'We must annex this moon,' I said. 'There must be no shilly-shally. This is part of the White Man's Burden.' . . . I embarked upon an argument to show the infinite benefits our arrival would confer on the moon. I involved myself in a rather difficult proof that the arrival of Columbus was, on the whole, beneficial to America."[9] This Kipling-esque call to literally ensure the sun never sets on the British empire—coming from the businessman, no less—is complicated by their discovery that the moon is populated by a race of insectoid drones called Selenites. While individually mindless, the Selenites are lorded over by the Grand Lunar, named in clear reference to the "Grand Turk," used to describe the Ottoman Sultan.[10] The term "Grand Turk" had been used in England since the sixteenth century as a "powerful, arrogant, bombastic and boastful stock figure," one originally introduced as a replacement for a similar stock figure of "Mahomet," thereby transforming the leader of the Selenites into the literary descendant of English depictions of both Muhammed and the despotic Ottoman ruler.[11] While Bedford returns to Earth near the end of the novel, Cavor is left behind on the Moon and radios back his attempts to negotiate with the Grand Lunar.

In describing how the Selenites bring food from the lunar surface to their subterranean abodes, Cavor states, "I have as yet scarcely learnt as much of these things as a Zulu in London would learn about the British corn supplies in the same time."[12] At the end of the novel, Cavor's transmissions end, apparently cut off by the Grand Lunar as part of preparations for a preemptive strike at Earth after learning from Cavor of the human (or perhaps specifically British) propensity for conquest.[13] The reference to Zulus—an expansionist kingdom in southern Africa that the British empire went to war with in 1879—and their attack on the remote British outpost that Cavor symbolizes is particularly notable. It not only transposes the Battle of Rorke's Drift to another planet but also takes on new context given that Wells wrote the novel during the Second Boer War (1899–1902), during which the British secured control of southern Africa. While Cavor's transmissions are eventually disrupted, they are picked up by a Dutch radio operator who was inspired to listen to the sky by the actual claims of scientist Nikola Tesla to have picked up signals from Mars.[14]

Mars was of course the focus of Wells's prior novel, the 1898 *War of the*

Worlds. Islam is also present there; during the beginning of the titular conflict, the Martians destroy the Shah Jahan Mosque, the first dedicated center of Muslim worship in the United Kingdom, built in 1889.[15] The novel memorably begins with the narrator comparing the Martian invasion of Britain to the British genocide of the Tasmanians but ends with Martians killed by bacteria, "the humblest things that God, in his wisdom, has put upon this earth."[16] Satirical or not, it is the divine presence that saves Earth from the supremely evolved technological colonialists. *The War of the Worlds* was compiled into novel form in 1898, the same year as the Battle of Omdurman resulted in the end of the Mahdist revolt, an Islamic uprising in Sudan that had annihilated Charles Gordon's Anglo-Egyptian army at Khartoum in 1885 and ruled Sudan as an independent theocracy since. Just as Wells's Martians had heat rays and war tripods that the British could not match, Omdurman was the debut of the machine gun in colonial wars. Forty-seven British were killed, compared with over ten thousand Sudanese.[17] In this way, not only is the idea of the plurality of worlds turned into a metaphor for imperialism, but it was immediately connected to pro-imperial narratives for the triumph of Christianity over Islam.

Nor are *The War of the Worlds* and *The First Men in the Moon* the first works of Wells to center on colonialism; imperialism is also prominent in his prior novels, *The Time Machine* (1895) and *The Island of Doctor Moreau* (1896), and dominates his post–World War I nonfiction.[18] A particularly relevant example is his 1894 story "In the Avu Observatory," where a British astronomer in an observatory in Dutch-controlled Borneo is attacked by a relict pterodactyl-like creature. Here, colonialism, observation of the cosmos, and evolution—or one who has escaped from it—are all present.[19] Indeed, from 1884 to 1887, Wells studied biology under Thomas Henry Huxley at the Normal School of Science. Huxley instructed Wells in Darwinian evolution, and Wells based his Martians on his interpretation of what the natural endpoint of Darwinian selection would be for humanity.[20]

SPACE: THE FINAL FRONTIER

These works of fiction are important not just due to their immense popularity but also because they demonstrate that tropes of colonization, religious conflict, capitalist exploitation, and space exploration were clearly present in European popular culture across the last third of the nineteenth century. While it may have taken European authors to hone the convergence of these topics, it is perhaps no surprise that the logical conclusion of their merging—the notion of a permanent, large-scale human settlement of space—would emerge from an American author, given that American Manifest Destiny was the epitome of the combination of religious and racial convictions from which space exploration tropes derived.[21] The first literary depiction of a space station—and for over eighty years, the longest piece of fiction about a satellite of any sort—was the 1869–1870 serialized story "The Brick Moon" by Edward Everett Hale (1822–1909).[22] Hale was a Congregationalist minister who would become chaplain of the US Senate from 1903 until his death in 1909. The Brick Moon he describes is a 200-foot-diameter sphere that is accidentally launched into orbit with several dozen families, who create a miniature world on the Moon's surface. The unwilling space colonists are kept sane by church attendance, the narrator describing how "we all came to the Chapel, Sunday, kept awake there, and taught our Sunday classes special lessons on Perseverance."[23] In Hale's view of science, the Brick Moon's spherical structure allows for the full range of environments to be experienced on its surface (which an atmosphere clings to, perhaps due to an inaccurate understanding of gravity by Hale). This environmental range means that evolution proceeds rapidly on the surface. The inhabitants signal down to the surface to "write to Darwin that he is all right. We began with lichens and have come as far as palms and hemlocks."[24]

Nor is this rapid evolution limited to plant life; the narrator later records, "They had plenty of vegetables to eat, with prospect of new tropical varieties according to Dr. Darwin. Rob Shea was sure that they carried up hens; he said he knew Mrs. Whitman had several Middlesexes and Mrs. Leonard two or three Black Spanish fowls, which had been given her by some friends in Foxcroft. Even if they had not yet had time enough for these to develop into Alderneys and venison, they would not be without animal food."[25] Although Hale wrote the first space settlement story, perhaps the first individual to seriously explore the philosophical underpinning of expansion into space as a logical outgrowth of both human exploration and evolution was the Russian schoolteacher Konstantin Tsiolkovsky (1857–1935), who in 1911 wrote his famous aphorism that "Earth is the cradle of humanity, but one cannot live in a cradle forever," part of his wider view that only space travel could prevent the eventual extinction of humanity from a disaster on Earth.[26] Tsiolkovsky lived for most of his life in the Russian Empire, when he wrote most of his important works, inspired by Verne's novels. However, it was in the final years of Tsiolkovsky's life that his work became heavily popularized by the Soviet state, which saw it as a way to integrate traditional pre-revolutionary religious cosmism and a particularly Bolshevik interpretation of Marxist belief in dialectical materialism supporting extraterrestrial life.[27] This leads to a new avenue of analysis: the degree to which, from the late nineteenth to early twentieth century, the belief in life elsewhere in the universe, especially Mars, was linked to belief in a socialist future here on Earth, a counterpart to the rise of "imperial pluralism" and in its own way both supplanting and adapting the religious pluralism through adapting astronomical evolution as a sign of laws of nature guiding humanity. In other words, how did terrestrial reds look to the red planet—or rather, the pop-cultural view of it as home to a dying civilization of canal builders—as inspiration for revolution?

That there might be a link between astronomy and socialism was advocated by no less of a notable socialist than Albert Einstein, who wrote in 1949, "It might appear that there are no essential methodological differences between astronomy and economics: scientists in both fields attempt to discover laws of general acceptability for a circumscribed group of phenomena in order to make the interconnection of these phenomena as clearly understandable as possible."[28] The special theory of relativity proposed in 1905 by Einstein was in a way a refutation of industrial capitalism and imperialism, which introduced and required the concept of rigid, universal timekeeping into a standardized workforce that prior to factory labor had acknowledged time as relative and its measurement as atomized—but also that all space was equal, with there being no privileged absolute space.[29]

As historian of science Peter Galison has noted, "turn-of-the-century time synchronization never inhabited a place isolated from industrial policy, scientific lobbying, or political advocacy"—but ironically, the scientific and educational apparatus, both physical and institutional, that allowed Einstein to combat industrialized and imperial time were products of those frameworks.[30] This could be seen as an argument that radical notions can never escape the conservative institutions that give birth to them. But to look at it from a Marxist point of view, this is simply dialectics in action: the thesis giving rise to the antithesis, which is destined to replace it. Nor was this conceptualization new with Einstein. In both his statement and his theorization, Einstein was only one in a long line of political radicals before and after him whose observations of the relations binding humans and the relations binding stars informed each other.

Nineteenth-century radicals particularly made use of astronomical metaphors, just as conservatives used them to oppose radical politics. An example is the Irish astronomer Robert Stawell Ball (1840–1913), who in the 1860s worked with Lord Rosse (1800–1867), former president of the Royal Society, to attempt to denounce the nebular hypothesis due to its implications for providing support to the notion of universal law and progress. Despite this, Ball still believed in extraterrestrial life and would go on to lecture about the potential for life on Venus.[31] At the same time, the fact that the nebular hypothesis had resulted from the work of "heroic" astronomers, including William Herschel (1738–1822), discoverer of planet Uranus, imbued a level of patriotic pride into the hypothesis for Victorian Britons.[32] However, by the mid-nineteenth century, the nebular theory had firmly become the realm of radicals. The Scottish astronomy popularizer John Pringle Nichol (1804–1859) and the radical MP for Bath John Arthur Roebuck (1802–1879) saw the nebular hypothesis as a scientific example of the laws of natural reform that political progress should be based on—a building block in the road leading from religious to secular pluralism.[33] So successful was Nichol in this popularization of the nebular hypothesis that one of his converts was American author Edgar Allan Poe (1809–1849), who cited both it and Nichol in his 1841 "Murders in the Rue Morgue"; Poe's work would go on to be one of Verne's inspirations for *From the Earth to the Moon*.[34]

Nichol would adapt his theory of nebular hypothesis radicalism to support the 1848 revolutions in Europe, from which socialism in its modern form would emerge. Leading socialists—Louis-Auguste Blanqui, Karl Marx (1818–1883), and particularly Engels—would identify the development of socialist ideology as being intrinsically linked with their interest in the cutting-edge astronomy of their day. In this, the socialist interest in astronomy was another inheritance from the philosophy of Georg Wilhelm Friedrich Hegel (1770–1831), whose opinions on the cosmos in *The Phenomenology of Spirit* (1807) have generally (though not universally) been interpreted as viewing the solar system as a system of order preceding intelligent biological organization of matter, with such advanced development present only on Earth.[35] Nineteenth-century socialists, however, were more willing to incorporate the concept of the development of both life and intelligence into their comparisons of how the heavens related to Earth.

Blanqui was a notable early example, a pre-Marxist socialist active in France across the nineteenth century, denounced by Marx at the time, and by Marxists after, as having no true ideology beyond revolution for revolution's sake. In 1872, while imprisoned by the French Third Republic for his actions during the 1870 collapse of the Second Empire, Blanqui wrote *Eternity Through the Stars*, his last major work. Ostensibly a pure discussion of astronomy, it reflects a development of revolutionary theory stemming from both his lifelong ups and downs and the cosmology of his day. In the book, Blanqui discusses the nebular hypothesis of solar system creation, which replaced the earlier notion that the solar system was eternal and unchanging. Instead, the nebular hypothesis proposed by Kant argued for a solar system that had formed from a cloud of gas and dust collapsing in on itself, forming the sun and the planets. Blanqui argued that stars had to be constantly collapsing in on themselves as they ran out of heat and energy, the collapse giving them renewed energy to rebound and be reborn and to create new types of life at higher-organized rates—something that is not only dialectical but reflects later Soviet interpretations of Marxist dialectical materialism.[36]

Although sometimes viewed as a sort of alternative history where the Paris

Commune was successful, *Eternity Through the Stars* uses the science of astronomy to argue that multiple worlds must exist, each carrying out our own history but with different outcomes—including ones where the forces of revolution, and not reaction, triumphed.[37] It is also Blanqui's musings on extraterrestrial life that perhaps form the first significant thread of speculation on the matter separate from the religious framework, which was inseparable from the concept of the plurality of worlds. This secular Blanquist thread of cosmic pluralism would exist distinct from, if sometimes entwined with, the religious thread that continued to be promoted by the likes of David Brewster (1781–1868), William Whewell (1794–1866), and Proctor until the end of the nineteenth century, who saw the observable complexity of the planets as seeming evidence that they had been divinely designed for life just as the Earth had been.[38]

In the case of Marx, he and Engels discussed astronomy as far back as the 1846 *German Ideology*, which noted that the organization of labor in an international context was required for contemporary astronomers to carry out their science and highlighted the leading role Newton had in causing celestial mechanics to become the most popular science of eighteenth-century Europe.[39] But among the radical pair, it was Engels who wrote the most on astronomy, particularly in the *Anti-Dühring* (1878) and *Dialectics of Nature* (1883). Engels echoes Blanqui's interest in the Kantian nebular hypothesis, although for him it comes as a result of, rather than leading into, an acceptance of a dialectical worldview. This underlies a key aspect that would come into play in the twentieth-century perceptions of orthodox Marxism: that the organization of matter into forms of higher complexity, ultimately leading to life, proved Marxist dialectical materialism and demonstrated that Marxism reflected the laws of the natural world, rather than merely serving as a theory.[40] In this, the later Marxist perceptions of extraterrestrial life more closely reflect the religious plurality of Proctor.

LINES IN THE SAND

Religious or secular, the debate over extraterrestrial life seemingly took a major step from the theoretical toward the practical in 1877, when the Italian astronomer Giovanni Schiaparelli announced that he had observed *canali* on the surface of Mars, appearing to connect the dark patches on the planet, which at the time were believed to be vast areas of vegetation. The Italian word *canali* translates into English as "channels," denoting a natural formation; however, Schiaparelli's claims were instead mistranslated as "canals," denoting artificial construction. This led to the idea not only that Martians were responsible for building enormous canals but that the project was the result of the supposedly advanced age of Mars. If Mars was an older planet

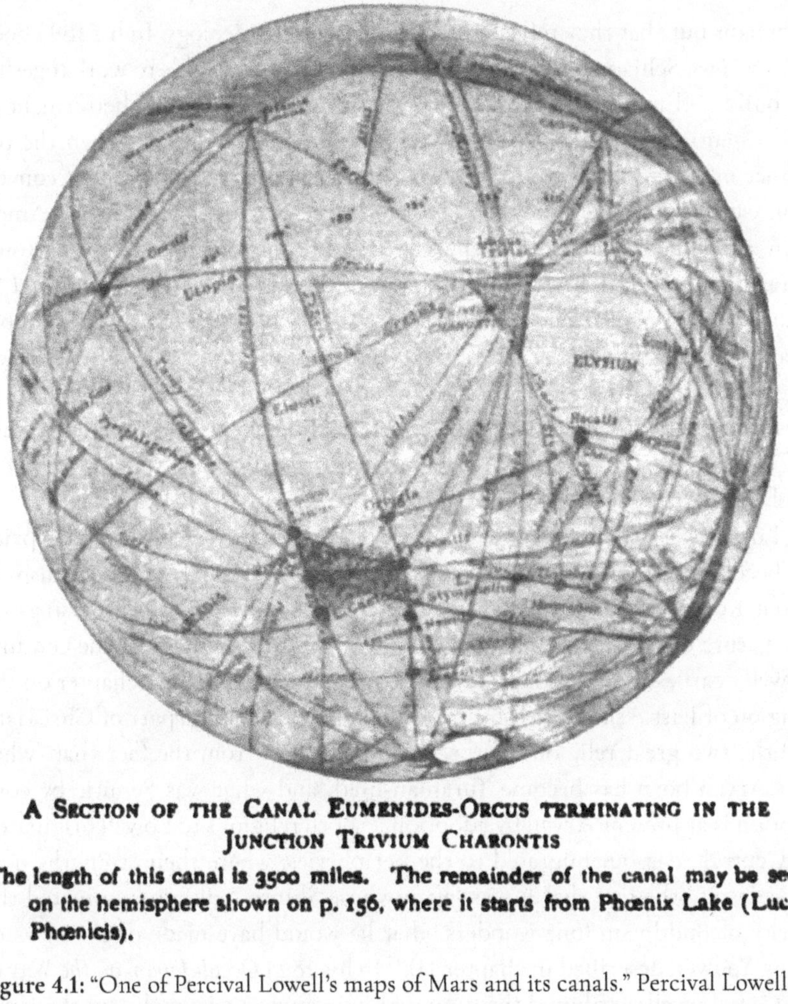

A SECTION OF THE CANAL EUMENIDES-ORCUS TERMINATING IN THE JUNCTION TRIVIUM CHARONTIS

The length of this canal is 3500 miles. The remainder of the canal may be seen on the hemisphere shown on p. 156, where it starts from Phœnix Lake (Lucus Phœnicis).

Figure 4.1: "One of Percival Lowell's maps of Mars and its canals." Percival Lowell, *Mars as the Abode of Life* (New York: Macmillan, 1910 [1908]), 150. Public domain.

than Earth, then not only was its civilization older and more advanced than humans, but the planet itself closer to death, its water slowly vanishing. The canals were therefore interpreted as the last efforts of an elderly race to prevent the extinction of themselves and their entire world by gathering water into the canals. The characterization of the canals also marks an important development of the Anthropocene era on Earth: the Martian canals were the first time scientists seriously speculated that industrial projects could permanently alter a planet's global environment, just as by 1939 climate models constructed around observations of Venus helped provide the first data-based evidence of terrestrial global warming.[41]

Although Schiaparelli originally believed that the canals were natural features, he came around to the view not only that they were built by intelligent

Martians but that they reflected a specific political ideology. In his 1893 book *Life on Mars*, Schiaparelli stated that if the Martians were able to work together to build and maintain a planetary irrigation network for the betterment of all—essentially, the canals as a planetary public water utility—then the red planet might, naturally, be a "paradise for socialism."[42] Schiaparelli's conversion came on the heels of the advocacy of the canal concept by the American astronomer Percival Lowell, who in addition to public speaking tours spread his beliefs in a trilogy of nonfiction books: *Mars* (1895), *Mars and Its Canals* (1906), and *Mars as the Abode of Life* (1908). It was Lowell who, more than any other individual, was responsible for establishing the notion of intelligent canal-building Martians in the mind of the global public (see figure 4.1). Lowell's view of the Martian civilization building canals in an effort to stave off their coming extinction particularly influenced Wells's version of highly evolved Martians.[43]

Lowell was a member of the wealthy Lowell family of Boston, and prior to becoming interested in mathematics, he lived in and wrote extensively about Korea and Japan. As demonstrated in chapter 9, Western musings on the nature of Eastern religion was in vogue around the turn of the century. Lowell's earliest book, 1888's *The Soul of the Far East*, contains a chapter on the religion of East Asia, establishing Buddhism as the counterpart of Christianity: the two great religions of the world, mirrored from the fact that "what was Aryan-born has become Turanian-bred, and what was Semitic by conception is at present Aryan by adoption."[44] Both religions to Lowell originated in Central Asia but migrated to the peripheries, where their strengths now lay—despite the fact that it was Japan whose Shinto "cult" contaminated the purity of Buddhism (one wonders what he would have made of the ideas of Kang Youwei, described in chapter 11).[45] In his 1894 *Occult Japan, or The Way of the Gods*, Lowell attributed the supposed "uncommon personality of the Japanese mind" to the hierarchies of the Shinto religion based around dedication to the emperor.[46]

Occult Japan was written in the period of Lowell's interest in Martian proselytizing, and it is easy to see how Lowell's view of the Japanese culture as ossified reflected his views of an advanced, but declining, Martian civilization. Nor was it only Shintoism that remained reflected in his astronomical views; take for example his description of astronomers from *Mars and Its Canals*:

> Astronomy now demands bodily abstraction of its devotee. Its deities are gods that veil themselves amid man-crowded marts and impose withdrawal and seclusion for the prosecution of their cult as much as any worshiped for other reason in more primeval times. To see into the beyond requires purity; in the medium now as formerly in the man. . . . Thus it comes about that today besides its temples erected in cities, monasteries in the wilds are being

> dedicated to astronomy as in the past to faith; monasteries made to commune with its spirit, as temples are to communicate the letter of its law.[47]

To Lowell, it is clear that observatories are not merely the inheritors of ancient monasteries and temples in the sense that those religious communities had been centers of astronomy themselves. Instead, as centers where the deeper mysteries of life's origins and its connection to the heavens could be discerned, observatories were the direct inheritors of the spiritual mission of ancient temples.

After returning to the United States, Lowell relocated to Flagstaff, Arizona, where in 1894 he commissioned the building of Lowell Observatory specifically to facilitate his efforts to map the Martian canals. Lowell Observatory's location in the recently closed (1890) colonial frontier of the American West—Arizona would not become a state until 1912, the last of the continental forty-eight states to achieve that distinction—is the first of several connections between American imperialism and astronomy. The widespread acceptance of massive Martian canals was itself partly due to their supposed observations occurring in an era of massive canal construction—specifically the Suez Canal (1859–1869) and Panama Canal (1881–1914), both of which were explicitly colonial endeavors, and the former of which was the catalyst for Egypt being taken under British "protection" for ninety years.[48]

The trope of the "dying Martian" has clear parallels with the trope of the "dying Indian," the settler idea that Native American cultures were doomed to vanish due to their inability to adapt to a changing world. In his popular John Carter of Mars series (1912–1948) of novels, American novelist Edgar Rice Burroughs (1875–1950)—whose other major creation, Tarzan, is similarly steeped in colonialism—connects the American frontier with Mars. In the opening story, the titular John Carter, a former Confederate cavalry officer, relocates to Arizona following the American Civil War. Attacked by Apaches, Carter takes refuge in a cave imbued with a form of magic that teleports him to a Mars heavily based on the claims of Lowell. There, Carter forms an alliance with the "Red Martian" race, which has become dominant since the distant extinction of the "White Martians" from their complacent hedonism.[49] In the first sequel, *The Gods of Mars*, Carter helps save the Red Martian civilization from the false religion of the phony goddess Issus, who promises an afterlife in a paradisial river valley, clear allusions to the Egyptian Isis and the Nile.[50] Lowell himself also saw this connection between Martian astronomy and the frontier, discussing in *Mars and Its Canals* how

> the resemblance, distant but distinctive, of the climatic conditions necessary on earth for the best scanning of Mars with those which prove to be actually existent on that other world has a bearing on the subject worth considerable

> attention. It helps directly to an understanding and interpretation of the Martian state of things. Though partial only, the features and traits of our arid zones are sufficiently like what prevails on Mars to make them in some sort exponent of physical conditions and action there. Much that is hard of appreciation in a low, humid land shows itself an everyday possibility in a high and dry one. The terrible necessity of water to all forms of life, animal or vegetal, so that in the simple thought of the aborigines rain is the only god worth great propitiation upon the due observance of which everything depends, brings to one a deeper realization of what is really vital and what but accessory at best.[51]

In other words, to Lowell's mind, the fact that he had relocated to the Arizona desert to study Mars had a direct impact on his ability to understand the planet—markedly unlike the astronomers back home in the "low, humid land" of the East, who notably were unable to perceive the same details of Mars that Lowell was. His comment on the similar appreciation of water for "aborigines" and Martians adds another frontier layer onto the equation, the same one that Burroughs would later build from.

Indeed, Lowell's maps of Mars and its canals occurred in the context of a trend of American mapmaking that emerged around the fin de siècle, partially inspired by the popularity of Verne's novels, as well as by the rapidly changing nature of cities, the political movements that emerged within them, and the new methods of transportation linking them.[52] These new methods of transportation also allowed for the knitting together of metropoles and colonies, along with transportation of soldiers and settlers alike. As can be seen, the idea that alien beings lived (or had lived) on Mars and had left behind enormous structures that could be observed from Earth was widely accepted in both the public and scientific communities by the late nineteenth century. Nor did the idea receive any major opposition from religious figures; especially with Schiaparelli's embrace of socialism, and Lowell's later usage by Bogdanov, the Martian canal issue is perhaps the point at the which the secular pluralism of Blanqui finally overtook the religious pluralism that had been dominant earlier in the nineteenth century.

REVOLUTIONS OF THE HEAVENLY SPHERES

If religion was no longer the foundational basis for speculating on the nature of those extraterrestrial civilizations, and theories on observing and communicating with them, the late nineteenth century had found a new framework: imperialism. Many of the observatories used to examine planets for signs of life were built by European powers in their colonies, while the hypothetical alien civilizations were based on models of civilizational hierarchy developed to categorize the colonized societies surrounding those observatories—and the fear that Earth itself might become a colonized settlement of its own. One of the most striking literary illustrations of this stems from the friendship of Wells and Joseph Conrad (1857–1924), with Conrad's novel *Heart of Darkness* (1899) coming out a year after *The War of the Worlds.*[53] Conrad inverts the imperial criticism of his friend's Mars, with a sailor telling the narrator Marlow his thoughts on beholding the African jungle:

> Somehow it didn't bring any image with it—no more than if I had been told an angel or a fiend was in there. I believed it in the same way one of you might believe there are inhabitants in the planet Mars. I knew once a Scotch sailmaker who was certain, dead sure, there were people in Mars. If you asked him for some idea how they looked and behaved, he would get shy and mutter something about "walking on all-fours." If you as much as smiled, he would—though a man of sixty—offer to fight you.[54]

In this passage, it is easy to turn the Congo River into a proxy for a Martian canal and the sailmaker into the credulous Lowell. Nor was Conrad the only author influenced by Wells who adapted the Martian civilizational hierarchy. In his 1938–1945 Space Trilogy, the British fantasist and Christian theologian C.S. Lewis (1898–1963) established a setting whereby Earth was a "fallen" world due to the original sin of Adam and Eve, while Mars was a world whose elder races had transcended sin and Venus a younger world whose own primeval couple had not yet faced temptation.[55] In this way, Lewis grafted a theological dimension onto the temporal hierarchy of the nebular model.

More abstractly, the nebular hypothesis on which the relative superiority of Martian to human was based found a terrestrial, if inverted, equivalent in what became known as the "Aryan invasion," first proposed by the German philologist Friedrich Max Müller (1823–1900) starting in 1841 and refined by the British anthropologist Thomas Griffith Taylor (1880–1963) in 1921. In Taylor's "lava-flow analogy," race and language were linked, with the most advanced examples of both to be found near what he believed to be the source for humanity, the Caucasus Mountains of Central Asia. Each new "wave" of humanity migrated outward from the existing settled locations, meaning the cultures farthest from the Caucasus—which he identified as "Negro" and "Papuan"—would be both biologically younger and culturally less advanced than the preceding one, with the most advanced being the "Aryans" of the source of humanity (see figure 4.2).[56] For Taylor, identifying parallels in what he saw as universal aspects of religious development, such as the "sun cult" and "serpent cult," were key aspects of identifying the racial stages of cultural development.[57] In this model, the concentric circles of racial evolution invert the concentric circles of planetary development of the nebular hypothesis,

THE LAVA-FLOW ANALOGY

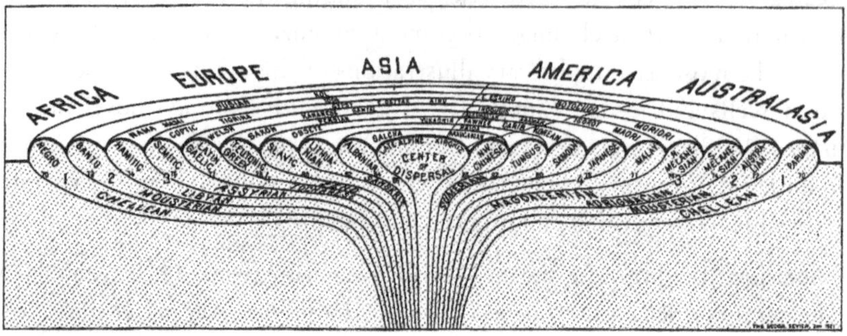

Fig. 9—The lava-flow analogy applied to studies of evolution and linguistics. The center of dispersal in Central Asia is pictured as a sort of "fissure eruption" sending forth streams of lava in all directions. Each new eruption arises from the center and, while covering some portion of the previous flow, pushes most of the previous lavas out to the periphery. Other analogies are represented in Figure 8. Fossil languages, buried under existing tongues, are seen in section. On the surface are shown existing languages typical of the several migration zones.

Figure 4.2: "Griffith Taylor's lava-flow analogy chart for racial and cultural development." Thomas Griffith Taylor, "The Evolution and Distribution of Race, Culture, and Language," *Geographical Review* 11, no. 1 (January 1921): 104. Public domain.

with the innermost Aryans taking the role of the Martians, while the then-colonized peoples of the Earth are the humans.

The Aryan invasion lava-flow analogy was part of a trend of nineteenth-century colonial thought, whereby incredible works of engineering discovered by European colonialists in Africa and the Americas were interpreted as being the products of long-forgotten white explorers, as non-Europeans obviously were incapable of building such accomplishments as Great Zimbabwe or Cahokia. This "foreign builder hypothesis" of colonial revisionist history parallels the idea of panspermia, that life originated elsewhere in the universe and spread to Earth via spores. While often connected to Swedish physicist Svante Arrhenius (1859–1927), who discussed it in his 1908 book *Worlds in the Making: The Evolution of the Universe*, the idea of panspermia predated him, its original DNA stemming from the religious notion of plurality of worlds. The idea was promoted by Lord Kelvin (1824–1907) in his 1871 presidential address to the British Association for the Advancement of Science as an alternative to Darwinian evolution. Fueled by both scientific and theistic opposition, Kelvin argued that life was eternal in the universe and merely traveled between worlds.[58]

Religion, astronomy, and ideas around politics and evolution continue to be intimately linked—particularly in the West, after the start of the Cold War and the space race gave new justification to resurrecting a religious basis for the (American) conquest of space. In the first three-quarters of the twentieth century, there were 250 major scholarly publications on the nature of the Star of Bethlehem, for instance.[59] Following World War II, the German rocket scientist Wernher von Braun (1912–1977) converted to evangelical Christianity after his recruitment by the US Army's Operation Paperclip, a transformation paired with his secular conversion from Nazism to American democracy and embrace of the Manifest Destiny heritage of his new West Texas home.[60] At the dawn of the space race, C.S. Lewis wrote an article for the *Christian Herald* about the impact of space exploration on Christianity, ultimately arguing that neither the discovery nor the absence of alien life would impact the religion's truth.[61] In the early decades of the twenty-first century, both conspiracy theorists and evangelical Christian officials within the US Air Force increasingly believe that UFO sightings are actually demonic visitations in disguise, bent on interbreeding with humans to alter the future of the species.[62] In response to the publicity over UFO studies pushed by lobbyist groups in the United States, a number of Catholic scientists have voiced their support for the belief in intelligent extraterrestrial life; for his part, Pope Francis compared the discovery of extraterrestrial life to the European discovery of the Americas.[63] In all of these religious concerns about encounters with extraterrestrial life, however, the debate about pluralism is entirely absent and, in Francis's case, completely replaced with the imperialist association.

CONCLUSION

Despite Lowell's influential connection of Martian canal builders to frontier imperialists, it was Schiaparelli's socialist interpretation of them that would ultimately make the leap to praxis via no less than the 1917 Bolshevik Revolution in Russia. Wells was widely admired in Russia both before and after the revolution (as was Verne) and journeyed to Moscow to interview Vladimir Lenin (1870–1924) in 1920 and Joseph Stalin (1878–1953) in 1934.[64] Among the Russian admirers of Wells was the early Bolshevik leader Aleksandr Bogdanov. Although Bogdanov was expelled from the Bolsheviks by Lenin in 1909 after mounting an almost-successful leadership challenge against him, Bogdanov remained a fellow traveler and important translator of Marx's work in the early Soviet Union. Prior to the revolution, he wrote two science fiction novels, *Red Star* (1908) and *Engineer Menni* (1913), depicting the creation of a socialist society on Mars that emerged as a result of the failures of the planet's atomized capitalist society to build the canals needed to preserve the world's environment. Centuries after the title character, Engineer Menni, leads the red planet's red revolution, a veteran of the 1905 Russian Revolution is taken to Mars by an undercover emissary observing the early socialist struggle on Earth in order to help educate him in the proper path forward for terrestrial communism—a secular form of Caesarius of Heisterbach's attempt to communicate with angels, or an interplanetary version of the transnational network of revolutionaries that developed in nineteenth-century Europe.[65] In 1908, just after the publication of *Red Star,* Lenin wrote to his mother:

> Today I read an amusing newspaper article on the inhabitants of Mars in connection with a new English book by Lowell, *Mars and Its Canals.* Lowell is an astronomer who has worked for a long time in a special observatory which, I believe, is the best in the world (in America). It is a scientific work. It argues that Mars is inhabitable, that the canals are a miracle of engineering, that people on that planet must be two and two-thirds the size of our people here, and that they, furthermore, have trunks and are covered with feathers or animal skins and have four or six legs. Hmm . . . [Bogdanov] cheated us by describing the Martian beauties only in part.[66]

While Lenin may not have thought much about Bogdanov's novel, he at least was aware of it—as well as at least idly credulous of Lowell's theories of canal-building Martians and the engineering abilities their civilization could achieve.

But by the twenty-first century, it was American socialists who were now envisioning Mars as a pre-capitalist refuge under threat from the likes of billionaire space enthusiasts like Elon Musk and his dreams of terraforming the planet into Earth—the inversion of the Martian decline envisioned by Lowell.[67] In the best tradition of Marxist dialectic, the end of history has been arrived at: Mars now going from an inspiration for the radicals of nineteenth-century Europe to one for the socialists within the victorious capitalist empire. The Red Star of Bogdanov, collapsed in 1991, has been reborn in response to the excesses of the forces that defeated it—the stellar evolutionary cycle of Blanqui in action. In an era in which globalization has to an extent made frontiers global, it only makes sense that the resistance to empires must evolve to survive beyond the globe.

PART II

Authority and Minority

5

Evolution in Times of Political Transformations

Science, Religion, and the Media in Restoration Spain (1874–1898)

Jaume Navarro

Spring of 1880. Madrid. Two long years after the call was opened, the Academy of Moral and Political Sciences finally resolved the contest on the best original book proving the harmony between religion and science. This was a response to John William Draper's 1874 *History of the Conflict between Science and Religion*, which had been translated into Spanish in 1876. The result disappointed many contestants as well as the donor, the wealthy marquis of Guadiaro: the jury could not pick a winner and instead chose four manuscripts, each receiving an *ex aequo* second prize. The episode triggered a fair amount of mockery in both the liberal and the ultramontane press, and the indignant marquis withdrew his donation and diverted the money to a society of Catholic schools. The irony of the whole story is that neither the contestants nor the marquis nor the Academy seemed to realize that a very subtle but poignant criticism to Draper's book was the seventy-two-pages-long prologue to the Spanish edition, written by the philosopher and politician Nicolás Salmerón (1839–1908) from his exile in Paris.[1]

A longtime champion of Krausism, the most influential reformist movement in mid-nineteenth-century Spain, Salmerón was now turning to positivism and, together with other intellectuals, was promoting the Institución

Libre de Enseñanza, an organization to transform the Spanish educational and cultural systems along positivist and liberal lines.[2] It is in this context that he agreed to write a preface to Draper's book, even though he thought the book was full of factual mistakes and errors in interpreting the data. For instance, he disagreed with the excessive weight the book placed on Islamic science, he regretted the absence of Spanish medieval science, and he also thought Draper failed to appropriately portray the distinction between Catholic and Reformed Christianity in their different historical relations to freedom of thought and scientific inquiry. And yet, it was worth advertising Draper's oeuvre since, in his words, "contributing to the promotion of this book is tantamount to working for the redemption of Humanity. I wish this book is widely disseminated in our nation, since it may help shake up the lethargy of our religious and scientific conscience!"[3]

Salmerón was in exile after having actively participated in the politics of the so-called *Sexenio*, the six-year interval between 1868 and 1874, which saw the ousting of the Bourbon monarchy, a number of provisional governments, a monarchic parliamentary democracy under Amadeo of Savoy, and a less-than-two-years-long First Republic that came to an end with the restoration of a Bourbon king in the person of the young Alfonso XII. As we shall later see, historians of Spanish science have often stressed the close connection between the social and political unrest during and immediately after the *Sexenio* and the arrival and reception of Darwinist ideas in Spain. It is, thus, against this backdrop of political turmoil and the restoration of the monarchy that the conflict thesis was appropriated and transformed in Spain. The literary contest promoted by the marquis de Guadiaro and poorly managed by the Academy of Moral and Political Sciences, as well as the impact it had in newspapers and weekly and monthly periodicals, is testimony to the importance of the generalist editorial market as it became the locus for disputes about the relationship between science and religion.

The Bourbonic restoration brought with it a long period of relative political stability, if by that we mean the continuity of the regime. The situation was, however, like in most European countries, one of strong tensions between those nostalgic for the ancien régime, reformists, liberals, Freemasons, Catholic authorities, and, later on, early socialists and anarchists. From the 1880s and well into the new century, the monarchy achieved stability by successively appointing governments led alternatively by the liberal and the conservative parties. This "alternation," as it is known, meant that fundamental policies and reforms were always under scrutiny, challenged by the opposition, and, more often than not, aborted halfway through. The role of the Catholic Church in the fabric of the state and of society, including education and freedom of the press, was one such contested issue.

When talking about science and religion in Spain, one should not forget that many intellectual, cultural, and academic debates were tainted by stances on clericalism and anticlericalism. Religion was almost exclusively shorthand for Catholicism, and this was, more often than not, identified with clerical structures. Yet the Catholic front was far from uniform, and, as discussed later, there were many different stances on evolution in particular and on the science and religion relations in general. The same applies, of course, for the anticlerical front. In some Catholic milieus, for instance, the words "Freemason," "positivist," "materialist," or "Protestant" were synonymous with atheism or infidelity, and they also implied "of foreign origin." Moreover, as in many other languages, the word "science" (*ciencia*) meant "knowledge" and seldom referred primarily to the natural sciences. As a consequence, many of the supposed conflicts between science and religion were the result of academic and political disputes, and their protagonists were mostly philosophers, historians, and politicians, including clergymen—only at times naturalists and hardly ever engineers.[4]

While historians of Darwinism in Spain have often paid attention to books and academic and intellectual works, my goal in this chapter is to focus on the ways the periodical press reflected on the tensions around the introduction of Darwinism and evolutionary theories at large. As James Secord, Bernard Lightman, and many others have shown, the role of popular science in Victorian Britain was essential in the creation and development of conflicts such as the one between science and religion.[5] The decrease in the price of paper and new printing techniques opened up new markets and possibilities. This is partly true also in Spain, albeit with an added complexity: the laws on freedom of the press and freedom of speech were often changing, although not always enforced. Thus, some of the new newspapers, journals, and pamphlets across the political and ideological spectrums benefited from potential persecution and closure: controversies were also a sales strategy.

I aim to explore if and how the periodical press became an important site for the popular debate on evolution in general and Darwin in particular. In tune with the subject of this section of the book, "Authority and Minority," we shall find that the highly politicized climate during the *Sexenio* and the early years of the Restoration led a number of editors and publicists to make political use of Darwin and evolution as a means of fanning the flames of controversy to increase sales, thus preventing, rather than promoting, a calm scientific debate or a serene popularization of the theory. Contrary to what was happening in Britain, evolution and Darwinism were far from being major topics in the public debate, if only because the tradition of science popularization had hardly taken off by the end of the nineteenth century. That is why the Darwin encountered in many periodicals was, more often than not, simply a political weapon in the hands of the many ideologies in the public debate and not the representative of a well-defined scientific theory.

In order to better understand the uses of evolution and Darwinism in the periodical press, we need first to describe the well-known *Polémica* of Spanish science; namely, the heated argument on whether one could talk about Spanish science and the role the Catholic Church had had in it. The following sections will delve into the role of publications in the introduction of Darwinism only insofar as they explicitly dealt with the interface of science and religion, as understood by the actors involved. I shall discuss the role played by periodicals, not by daily newspapers, and focus on three areas: scholars, the popular press at large, and cartoons and other resources in satirical publications.

LA POLÉMICA OF SPANISH SCIENCE

Salmerón's prologue to Draper's Spanish translation should be placed in the context of two disputes that arose at the time of the monarchic restoration of 1874: what is known as the *Polémica* of Spanish science and the so-called second university conflict. The first was a dispute about the intellectual, industrial, artistic, and scientific prowess of the country that from time to time surfaced in the periodical press during the nineteenth century. Since Masson de Morvilliers wrote in the 1780s his article on Spain for Diderot's *Encyclopédie Méthodique*, in which he described it as a "lazy nation," under the influence of monks and nuns and the Inquisition, with nothing to contribute to the world,[6] Spanish intellectuals and politicians often split between those who used the pessimistic diagnosis to promote political reforms and those who regarded Masson's thesis as an unacceptable attack on the "Spanish national genius." The second dispute was the expulsion from their chairs of a number of university professors who had supported the First Republic, accused of teaching what the new establishment regarded as anti-Spanish doctrines—mostly, anti-monarchic and anti-Catholic ideas.[7]

The most interesting and elaborate episode of the *Polémica* took place in the pages of two high-culture periodicals of the time, *Revista Contemporánea* and *Revista Europea*, and brought liberals, republicans, positivists, and Freemasons face-to-face with monarchists, traditionalists, and Catholic intellectuals. It started with the publication of an article on the legal, historical, and philosophical foundations of the modern nation-state by the Krausist professor of law Gumersindo de Azcárate (1840–1917), one of the scholars expelled by the new regime in 1875.[8] Part of his thesis was that both the natural sciences and philosophy had to be constitutive of the state, the former providing it with efficiency, the latter with meaning. A third essential element, religion, should be shaped by a new liberal Christianity, with little or no institutional role, in harmony with liberal politics and scientific rationality. A "doctrinarian monarchy" along the lines of the ancien régime was, thus, incompatible with "self-government."[9]

Azcárate's political ideas were not new. He had advocated similar views during the *Sexenio*. But the publication of this particular article in 1876, at the time a new Spanish constitution for the country was approved, triggered an unexpected public dispute concerning the roles of science and of religion in the configuration of a modern state.[10] The promotion, in the same article, of the Spanish translation of Draper's book and Salmerón's prologue only contributed to turn Azcárate's political diatribe into the target for monarchists and conservative Catholics. The response came from the very young Marcelino Menéndez Pelayo (1856–1912), who—incited by his mentor, the ultramontanist professor of literature Gumersindo Laverde (1835–1890)—produced a series of public letters arguing that the Catholic Church and the Spanish Inquisition *did not* and *could not have* prevented the development of knowledge, especially in the natural sciences and technology.

Both Azcárate and Menéndez Pelayo spoke of a Spanish essence that was intrinsically prone to philosophy, science, and technology. The former held that that genius had not been fully materialized due to the impediments to freedom of thought attributed to the Catholic Church and the Inquisition, while the latter maintained that the Catholic religion was quintessential to Spanish nature and that, if any branch of science had not yet developed, it was reform, not rupture, that was needed. To refute Azcárate's negative picture of the history of Spanish science, Menéndez Pelayo started an encyclopedic survey of past Spanish naturalists, inventors, chemists, medical doctors, and men of science to prove that throughout the centuries there had indeed been much science in Spain. Incidentally, Menéndez Pelayo's monumental work turned out to be a gold mine for historians of science to this day.

Azcárate did not pursue the dispute with Menéndez Pelayo further. Instead, Manuel de la Revilla (1846–1881) first, and José del Perojo (1850–1908) later, took Azcárate's place in the ongoing dispute with Menéndez Pelayo. Revilla was a writer, literary critic, and philosopher, and the *Polémica* triggered his successful activity as a public writer and a journalist. In 1876, he won the chair in philosophy at the University of Madrid, but the appointment did not materialize until 1879 due to his support of Krausism during the *Sexenio*. The tone of the *Polémica*, at times excessively ad hominem, can be easily grasped by the nickname Menéndez Pelayo gave Revilla: a "resurrected Masson," in reference to Masson de Morvilliers.[11] At the time of the dispute, Revilla was shifting, like many of his comrades, from his early Krausist ideas to neo-Kantian and positivist tenets.

Spanish reformist and liberal movements in the mid-nineteenth century coalesced around the Madrid professor of law Julián Sanz del Río (1814–1869) and his particular interpretation of the thought of the German philosopher Karl Christian Friedrich Krause (1781–1832). Krausism was a blend of idealist philosophy with an emphasis on the rational destiny of humanity, on altruism, and on a panentheistic conception of God and of Nature,[12] among other elements. The political activism of its followers prioritized educational reforms, stressing naturalism, secularism, and rationalism. It was not an atheistic movement, though certainly anticlerical, deeply rejecting any interference of confessional Catholicism in the configuration of the state, especially in education.[13] At the time of the *Polémica*, however, and after the failure of the political reforms during the *Sexenio*, a younger generation was seeing Krausism as excessively idealistic, and they turned to neo-Kantian and positivist ideas.

José del Perojo was, perhaps, the most instrumental actor in this transformation.[14] Born in Cuba to a very wealthy landowning family, he relocated to Spain at an early age after the premature death of his father. He studied independently in a number of Spanish universities and moved to Heidelberg, to the neo-Kantian school led by Hermann von Helmholtz (1821–1894), Friedrich Albert Lange (1828–1875), Otto Liebmann (1840–1912), and Kuno Fischer (1824–1907). Perojo, who had political ambitions in Spain, saw in the neo-Kantian approach a midway position between what he regarded as the philosophical flatness of positivism and the sterility of idealism, Krausism included. Thus, on his return to Spain at the end of the *Sexenio*, he became a prolific writer and, more importantly, a publisher. *Revista Contemporánea*, one of the two main journals of the *Polémica*, was his own creation, as were the publishing houses where the Spanish translations of Draper's *A History of the Conflict* and Darwin's *Origin of Species*, to name a few, appeared.

In 1875, Perojo published a number of essays in Spanish journals on German culture and literature, neo-Kantian philosophy, and Herbert Spencer's and Darwin's biological, anthropological, and sociological ideas, compiling them in his first book, *Essays on the Intellectual Movement in Germany*.[15] At the same time, after unsuccessfully trying to purchase the recently created *Revista Europea*, founded by the also Cuban-born and now former priest Tristan de Jesús Medina Sánchez (1831–1886), he decided to create a brand-new publication, *Revista Contemporánea*. With a huge economic investment from his personal fortune, Perojo managed to turn the new journal into the main instrument to promote public debate and to spread foreign modern ideas in all fields: politics, literature, natural sciences, education, and history, among others. His target was not only to attack the very active ultramontanist and conservative Catholic press but also to challenge what he considered old-fashioned ideas of the previous generation of Krausists and early liberals. The collaboration with Revilla, whom he had convinced to evolve from Krausism to neo-Kantian principles, and the *Polémica* were instrumental in the immediate success of

Revista Contemporánea. The commercial competition between the two journals *Revista Contemporánea* and *Revista Europea*, both with similar ideological editorial lines, certainly encouraged the *Polémica*.

Only a year later, in 1877, Perojo started Editorial Perojo, a publishing house devoted to printing Spanish translations of works by foreign authors, including Draper's *Conflict* and Darwin's *Origin*, as well as Walter Bagehot's social Darwinist *Physics and Politics*, Herbert Spencer's *First Principles*, Claude Bernard's *Experimental Science*, and Ernst Haeckel's *The History of Creation*. He also created a short-lived popular science weekly journal, *La Naturaleza: Revista de Ciencias y de su aplicación a las Artes y a la industria* [Nature: Journal of sciences and their application to the arts and industry]. His pet project, however, was to have a Spanish edition of Kant's complete works, of which he managed to produce only a first volume before going bankrupt. Indeed, in spite of his wealth, his publishing zeal exceeded his economic possibilities, and in just two years, all his publishing ventures disappeared, except for *Revista Contemporánea*. The journal survived but was purchased by liberal supporters of the new regime, thus modifying its core philosophy.

To finish with the *Polémica*, note that Menéndez Pelayo compiled the epistolary exchange into a book, *Spanish Science*, a now classic work of continued interest to historians at large and historians of science in particular,[16] partly because the *Polémica* was the excuse to seriously study for the first time the history of science in Spain. The nationalistic and ideological undertones certainly shaped this historical work, and yet, the bibliographic and archival work done by Menéndez Pelayo is still relevant, especially for modern historiographies of science. Interestingly, if liberals and positivists attacked him for his conservatism, for his claims that there had indeed been Spanish science and philosophy, and for his rejection of the supposed catastrophic effects of the Catholic Church and the Inquisition, Menéndez Pelayo's theses were also criticized by ultramontanists. For them, Spanish culture was intrinsically Catholic and Spanish Catholic culture *was* and *had to be* exclusively Thomist. And Menéndez Pelayo did not sympathize with neo-Thomism. Thus, he was referred to as "mestizo," the derogatory nickname with which ultramontanists dismissed any Catholic who did not fully condemn liberalism.

Gumersindo de Azcárate, Manuel de la Revilla, José de Perojo, Nicolás Salmerón, Marcelino Menéndez Pelayo, Gumersindo Laverde, and all other actors in the dispute on the role of science in Spanish society—past, present, and future—were historians, writers, journalists, philosophers, and politicians. Not a single one was a scientist.[17] It comes, thus, as no surprise that the promotion of new naturalistic ideas, such as evolution and Darwinism, as well as the conflict thesis à la Draper, played, for them, the role of ideological tools at this time of political turmoil. My interest in this section has been not only to present the *Polémica* as the stage in which evolution and Darwinism were presented to the public but to do so with particular attention to the editorial marketplace. The commercial competition between two similar journals, *Revista Contemporánea* and *Revista Europea*, or the editorial ventures of people like José del Perojo, played an important role in the spread of foreign new ideas, scientific or otherwise. In other words, it is significant that the promoters of the conflict thesis for political reasons were the same agents spreading evolution and Darwinism to the public. In the next section, I shall give an overview on the early reception of Darwin in Spain.

DARWIN IN SPAIN

As mentioned, Darwin's 1859 *Origin of Species* was first published in Spanish in 1877 with the support of Perojo. It was a year after a publisher in Barcelona had issued the Spanish translation of *The Descent of Man*.[18] This does not mean, of course, that Darwin's ideas were unknown in Spain until that date. But, as historians of science have shown, the arrival of Darwinism had been mediated by French translations and German interpretations.[19] Evolution had been part of the academic and public debates with the creation, in 1852, of a chair of geology and paleontology at the University of Madrid. Its first incumbent, Juan Vilanova (1821–1893), had spent a few years in Paris, and his evolutionary ideas were influenced by Georges Cuvier, Jean-Baptiste Lamarck, and Étienne Geoffrey de Saint-Hilaire, and he argued in favor of uniformitarianism in geology but against transformism in biological species. A fervent Catholic, Vilanova also explicitly stressed in all his books and manuals the harmony between geology and the book of Genesis, though not in a literal form. With them, he influenced a whole generation of Spanish geologists and paleontologists.[20]

If Vilanova was trying to reconcile evolution to the tenets of his Catholic faith, many early promoters of Darwin saw in transformism more of a political and religious weapon than a scientific theory. Thus, for example, the translator of *The Descent of Man*, the young Catalan poet Joaquín María Bartrina (1850–1880), had written in 1869 a pamphlet with the uncompromising title *War to God!* in support of a similar opuscule by the doctor, republican politician, and "angelical atheist"[21] Francisco Sunyer Capdevila (1826–1898).[22] *War to God!* started dismissing "simple men who believe in God out of convenience, used to live in the darkness of ignorance for fear of the dazzling light of truth"; "hypocrites who believe in God so that fanatics may believe in you" even though "you well know that God does not exist"; and "filthy priests who believe in God only because that is your job." This pamphlet was addressed to "those who, convinced that the law of Humanity is progress, do not want to regress so as to believe in God."[23] Bartrina mentioned a dozen books he had used to support his argument, mostly on archaeology and evolution, including Darwin, Lyell, the Belgian d'Omalius d'Halloy, the Italian Giovanni Omboni, and the French Paul-Hippolyte de Jouvencel. This evolutionary, positivist, and anti-religious pamphlet appeared in the same collection as the translations of Ernest Renan's *Life of Jesus*, *The Apostles*, and *Saint Paul* and Victor Hugo's *Speech Against Clerical Education*, to name a few.

The first academic and systematic exposition of Darwin was the chapter written by a disciple of Vilanova, Francisco María Tubino (1833–1888), in the 1872 edited volume *Creation: Natural History by a Society of Naturalists.*[24] Broadly speaking, Tubino stressed Darwin's evolutionism without paying much attention to natural selection. Moreover, a few years later, in *Revista Contemporánea*, he praised the honesty of the English naturalist for rejecting the "hypothetical metaphysics" of Haeckel's materialistic ideas on the origin of life.[25] The first university professor not only to explain but also to promote some Darwinist ideas was the Krausist, and later positivist, Antonio Machado Núñez (1815–1897), professor and rector at the University of Seville during the *Sexenio*. It was in those years when, having created the *Monthly Journal of Philosophy, Literature and Sciences*, he spread many of the ideas of Darwin, Haeckel, and Spencer.[26] While acknowledging the limitations and questions that Darwin's approach left open to further investigations, Machado strongly criticized those who dismissed Darwin only for theological reasons: "The supernatural intervention in inexplicable phenomena . . . should be erased from the language of naturalists. . . . It is a terrible habit to discuss without reasoning, rejecting truths proved by experience with arguments of faith, arguments only acceptable to those who irrationally accept such faith."[27]

With the restoration of the monarchy, Machado resigned from his chair in protest of the expulsion of a number of professors from their positions and joined the team of Krausists to create the Institución Libre de Enseñanza (ILE). Not surprisingly, Krausism was sympathetic with the general idea of evolution, since it seemed to fit into the harmonic continuity in Nature, and between Nature and God, implicit in panentheism. Perhaps the best exposition of evolution by the Krausist school was the 1874 book by the physicist and naturalist Enrique Serrano Fatigati (1845–1918), *La Evolución en la Naturaleza*, in which transformist ideas were the consequence of the "evolution in form." The principle of conservation of matter and of energy, together with the continuity within an individual, within a species, and among species, would show the continuity within Nature and the continuity between Nature and the divine.[28]

While the *Polémica* was in its climax, and Krausists and post-Krausists were creating the ILE using the conflict thesis as one of their most powerful weapons, Vilanova continued the tradition of giving public lectures and promoting the new science of paleontology. In Spain and abroad, paleontology attracted supporters and opponents of evolution and Darwinism in search of evidence that might settle the dispute. In his 1872 book *Origin, Nature and Antiquity of Man*, as well as in all his lectures, Vilanova argued for harmony between the findings of paleontology and the biblical notion of creation. The only evolution he allowed for was intraspecies transformations. His public lectures at the Athenaeum of Madrid were published in the mostly pro-Krausist and pro-evolutionist journal *Revista Europea*, the same where Menéndez Pelayo had published his responses in the *Polémica*. Promoter of geology, paleontology, and prehistory, Vilanova navigated between what he called the extremes of "materialists and atheists," who "thought prehistory was a powerful weapon to destroy the foundations of Catholic dogma," and the "faint of heart and even hypocrites," afraid only of "the very subject the new science intends to study." And to reinforce his point of view, he stressed that the discoveries of Jacques Boucher de Perthes, which had undoubtedly proven the antiquity of mankind, had "nothing to do with Darwinism," nor with the "purely human and in no way dogmatic chronology of Father Petavius," in reference to a chronology devised by this seventeenth-century Jesuit. On the latter, he stressed that "the Catholic Church does not admit that nor any other of the many chronologies as dogmatic, leaving the question completely free."[29]

His successor in the chair of geology, José Solano Eulate (1841–1912), also addressed the possible intersections between the Catholic dogma and the new sciences. His inaugural address of 1880 had the explicit title "Relationship between Geology and Revelation." In it, he moved a step forward and accepted that the transmutation of species was a valid hypothesis, although one he did not share, and blamed religious people's fear of the theory on the materialistic and monistic views of people like Haeckel and Karl Vogt. On this he was echoing the work of the Jesuit Miguel Mir (1841–1912), whose *Harmony between Science and Faith*, one of the four finalists in the anti-Draper contest, clearly stated that one could not see "on which words of Sacred Scripture, on which doctrine and authority could a condemnation of the Darwinian theory be grounded."[30] Here, Mir cited the Catholic biologist and critic of Darwin St. George J. Mivart (1827–1900), whom he had met in England during his exile in the years of the *Sexenio*.

Solano Eulate was, however, right in his diagnosis of the source of some believers' fears. Evolution and Darwinism in Spain had been tainted by German materialist ideas. We have already seen that the *Origin of Species* was published in Spanish only in 1877, a few months after *The Descent of Man*. But, by then, books by Ludwig Büchner (like *Kraft und Stoff*), *Aus Natur und Wissenschaft* by Jacob Moleschott, and *Vorlesungen über den Menschen, seine Stellung in der Schöpfung und in der Geschichte der Erde* by Karl Vogt had already seen Spanish translations. Haeckel's *Natürliche Schöpfungsgeschichte* would be issued in weekly installments within *Revista Europea* throughout 1878. Their explicit anti-religious, certainly anti-Catholic, stance, as well as the political agenda of those promoting their translation, would, as Vilanova and others had argued, put many Catholic people on the defense. To give an example, the translation of Haeckel's 1866 *Generelle Morphologie der Organismen* was promoted in 1887 by Salvador Sanpere y Miguel (1840–1915), a historian and republican politician, only because he thought Haeckel was scientifically proving Krausism.[31] Thus, Krause, Comte, Spencer, Darwin, and Haeckel became, for many on both extremes of the political spectrum, synonymous with monism, republicanism, reform, materialism, atheism, or anti-conservatism.

We have seen how Darwin's ideas were, from the very beginning, mediated by German philosophies and political ideologies. Promoters and translators of Darwin were political actors in the public debates, and this tainted the reception of Darwinism in Spain. Interestingly, however, Darwin was appropriated by both materialists, following Haeckel, Ludwig Büchner, and others, and late-Krausist idealists, who saw in evolution a way to support their panentheistic philosophy and their quest for reforms. As we shall see in the next section, many Catholic periodicals contributed to these materialistic and idealistic appropriations of Darwin by turning him into one of the icons of irreligiosity. In between, the school of geology and paleontology created by Vilanova, influenced mostly by French scholars, debated and progressively incorporated evolutionary tenets in their science but seldom discussed, let alone embraced, natural selection.

CATHOLIC PERIODICALS

What was Darwinism for the average Catholic priest with no training in the natural sciences? *El Consultor de los Párrocos* was a weekly bulletin for priests and written by priests to spread news about their trade. It dealt mainly with clerical matters, but it also contained a few "doctrinal" essays to keep the parish priest informed on contemporary cultural and philosophical matters. The editorial line lay in a middle ground between liberal Catholics and ultramontanists. An article from 1873 epitomizes the idea a parish priest may have had about Darwinism in the early 1870s and well into the 1880s. In order to criticize the new theory, the anonymous reviewer listed four principles to which Darwinism boiled down: "1. That Adam was not the first man; 2. That before, much before Adam existed, other men existed; 3. That man was not formed by God, as Sacred Scripture says; 4. And, last, that man comes from a natural selection or a continuous transformation from simpler beings that have increased in complexity."[32] With this introduction, Darwinism was, obviously, anathema. Of course, things were not always this simplistic in the Catholic press.

One of the three monthly November 1891 issues of the journal *La Controversia: Revista Religiosa, Científica y Política* included a review article about the Catholic efforts to address contemporary controversies in the Spanish periodical press during the second half of the nineteenth century. It praised the large number of books and articles produced to debunk what were perceived as attacks on the Catholic faith in seven areas; namely, (1) Renan's *Life of Jesus*; (2) Darwin's evolution; (3) "the pontifex of contemporary materialism, the dirty Büchner," the "debauched Lamennais," or the "provocations of Dumas"; (4) the "spiritualist superstitions" of Kardec or the "fantastic visions" of Flammarion; (5) the "capricious assumptions and false facts" in Draper's book, on which "impious men had placed all their hopes"; (6) the compatibility between science and the Bible; and (7) the "anti-Christian and anti-Spanish" incursions of Protestantism.[33] While one should not read too much into this particular classification of the controversies affecting Spanish Catholics, it is symptomatic of their concerns. Darwin's evolution was perceived by many Catholics as an attack on tradition, reason, the Catholic faith, and national identity.

La Controversia, previously called *La Lectura Católica*, was a Catholic journal created and directed by the clergyman José Salamero Martínez (1835–1895), known at the time as "the father of the poor" for his activities in areas such as education and the conditions of the working class. Both the journal and his contributions in national newspapers and periodicals were, from his point of view, part of his task to protect the needy from the poverty of error. The "healthy teachings" of the church, he wrote, are "always saving," but especially "in the present circumstances, of so many and so painful rebellions against authority."[34] The motto of the journal, *Spiritus catholicus spiritus Veritatis, Ratio catholica ratio omnium*, shows a core idea of many Spanish clergymen of his generation: that the only possible truth was the one in consonance with the teachings of the Catholic Church, not because of authoritarianism but because of the necessary unity of truth. From that point of view, Catholics *had to* oppose liberalism "in all its forms and manifestations . . . , since it [liberalism] is the very prostitution of freedom," for it does not seek truth, good, and beauty, "but only freedom for freedom's sake, even the freedom of the devil."[35]

As this list of subjects of controversies shows, Darwin was associated with Draper, with Protestantism, with materialism, with spiritualism, or with Bible criticism. It was not a scientific issue; it was a doctrinal one. Actually, each issue of *La Controversia* included a section under the heading "scientific movement," which contained news on technological and scientific developments such as "electricity applied to medical diagnosis,"[36] "photography of stars,"[37] or a "new very small steam engine,"[38] but mostly news on constructions of railways, telegraphy, and other technological improvements from around the world. On these matters, there was no room for controversies or conflicts. The use of the term "science" is here significant and not one common among Catholic intellectuals and clergymen at the time: the editors and writers of *La Controversia* were adopting "science" to mean technological and practical developments, an area where there were no claims to *absolute* truth. Thus, they would not have perceived any conflict between "science" and "religion."

Articles on evolution and Darwinism in *La Controversia*, as in many other Catholic generalist periodicals, were written by philosophers, theologians, or amateur naturalists, like the lawyer and writer José María Menendez de la Pola (1866–1900) or the priest José Robles Lacourtiade (1853–1921). They also included many foreign translations, mostly from French Catholic journals. And besides the fixity or variability of the species and the singularity of the human species, there was an emphasis on relating transformism to liberalism and free speech and with pointing at the moral dimensions of Darwinism.[39]

On the latter, the most significant contributions appeared in another journal, *La Ciencia Cristiana*, and were composed by the early feminist, freethinker, and activist writer Emilia Pardo Bazán (1851–1921), who would eventually turn into one of the best-known, and at times most controversial, female writers of her generation.[40] In 1877, she published seven articles, "Scientific Reflections against Darwinism," in which she tried to popularize the theory that "some call evolution, others transformism, and the rest Darwinism," and which "is rather foreign to most people in our homeland."[41] This terminological confusion was more clearly stated when Pardo Bazán compared Darwinism to the law of gravitation. The former, she said, is just "a hypothesis, or rather as many hypotheses and systems as so-called Darwinists" there are.[42] The main problem was, thus, that while gravitation never abandoned the realm of physics, Darwinism "is more ambitious, invades and tries to explain everything, the nebulae and the planets, the cell and man, the organism of infusorians and social organization."[43]

Like in many anti-Darwinist writings in Catholic media, her discourse drew from the writings of the Italian Jesuit Angelo Secchi and the French naturalist Jean Louis Armand de Quatrefages, so as to stress that her rejection of Darwinism was based on epistemological and scientific grounds, not on theological prejudices. But she also repudiated the moral and racist implications of Darwinism. In the third installment, she warned against the "progressive improvement" supposedly coming from the struggle for existence, the outcome of which would be the "gradual disappearance of the inferior and weakest individuals and species."[44] This, she said, "would engender major pessimism" in Nature as well as in anthropology and in society. Pardo Bazán also described Darwinists as intrinsically racist since, in reducing men to pure matter, they were establishing a hierarchy between superior and inferior human races based on their biological development. But, she said, "even if we were to admit the wrong assertion that savage races are physically inferior, we would not be authorized to extend this theory to the intellectual sphere."[45] No matter how underdeveloped they might be, she argued, all peoples "have literature, traditions and music," which make them fully human. Moreover, the moral debauchery that some European explorers patronizingly reported about so-called primitive peoples were to be found in our society as much as in theirs.[46]

La Ciencia Cristiana, where these articles appeared, was created in 1877 by the traditionalist lawyer, journalist, and professor of metaphysics in Madrid Juan Manuel Ortí y Lara (1826–1904), with the explicit goal, among others, of opposing the Constitution of 1876 for being too liberal. It was also an instrument in the promotion of neo-Thomism, of which Ortí y Lara was one of its major champions in the country. By Christian science, or "true science," this tradition meant knowledge obtained from faith and from reason, both in deep harmony, as opposed to the results of a reason removed from God. Kant had, according to Ortí y Lara, systematized the three "abominable revolutions" that Luther, Descartes, and Rousseau had implemented, the "three subversive manifestations of modern thought" that had their origin in the worst of capital sins: pride, the pride of wanting to make do without God. Without the light and truth of God, "all sciences are vain and fallacious, grotesque delusions engendered by pride."[47] Not surprisingly, the professor of metaphysics asserted, quoting Plato, that "essence is the centre and foundation of any science worthy of the name,"[48] while Kantian rationalism could only provide figments of the imagination.

The tone of this fortnightly periodical was very philosophical, with long articles on theology, metaphysics, and law, as well as comments on papal documents or clerical appointments. And it had a clear political agenda. As a matter of fact, and because of its political influence, the liberal government led by Práxedes Mateo Sagasta (1825–1903), a civil engineer and Freemason, forced its closure for a few months in 1882.[49] Contrary to *La Controversia*, it did not have an explicit section on the natural sciences, and articles on evolution, like the ones by Pardo Bazán, were always critical but mainly addressing philosophical, political, and moral aspects of what was loosely defined as Darwinism. Practitioners of geology, biology, or any other natural science never published papers in *La Ciencia Cristiana*.

The comparison of *La Ciencia Cristiana* and *La Controversia* also helps us understand an aspect often neglected in the historiography of Spanish science-and-religion debates. The so-called Catholic front was not homogeneous.

Many were the groups, political parties, and schools of thought that opposed liberalism, but they also opposed each other. Thus, many articles in *La Ciencia Cristiana* criticized people like Menéndez Pelayo or the "mestizos" at large, namely those Catholics suspicious of compromising too much with the ideals of modernity and who did not commit to neo-Thomism. Perhaps the best example on the most reactionary side was the highly influential book *Liberalism Is a Sin*, written by the priest Félix Sardá y Salvany (1841–1916), a regular contributor to *La Ciencia Cristiana* on political and doctrinal matters.[50]

SATIRICAL PRESS AND CARTOONS

The philosophical and theological arguments for and against evolution were presented side by side with satire, jokes, and cartoons. The improvement and availability of cheaper and better printing techniques helped reduce prices and make journals, newspapers, and printed bulletins more accessible to the general public. Politics was the most common subject for satire before, during, and after the *Sexenio*, and clericalism and anticlericalism were a prominent part of politics. Corruption, fashion, and everyday matters such as hygiene, plagues, and droughts were also discussed in satire. The natural sciences in general and evolution in particular were only seldom directly referred to but, when they were, they often involved political and educational questions.[51]

By the early 1860s, evolutionary ideas were already present in popular culture, as shown in a few cartoons published in 1863 in the generalist weekly *El Museo Universal*, famous for its engravings, in which humans are portrayed to evolve from bulls, pigs, cats, or geese.[52] But these were loose references to a general notion of evolution. During the restoration, the presence of Darwinism in the popular press increased, and so did satire on the subject, especially in reference to the origins of man. Thus, for example, *La Hormiga de Oro*, a popular Catholic journal edited in Barcelona, published in 1884 a long poem describing a conference given "by a gorilla or a chimpanzee, that on this point my memory failed," in which the "gentlemanly ape" exposed his theory; namely, that all primates have evolved from an inferior species: humans. And proof of that was that apes have a more protective skin, longer arms so that they can touch their feet, and useful calluses in "relevant parts," "so as to make it easier to sit down." But moreover, men were inferior because they made the "logical mistake" of believing "the stupid nonsense of the existence of God." The parody finished with the claim that "God is nothing, man is minor, but the monkey is indeed God."[53]

Eight years later, on the occasion of the tenth anniversary of his death, the same periodical published a mock biography of Darwin, starting with a million atoms randomly revolving in interstellar space a few million years ago, which, guided by natural selection and the struggle for existence, ended up

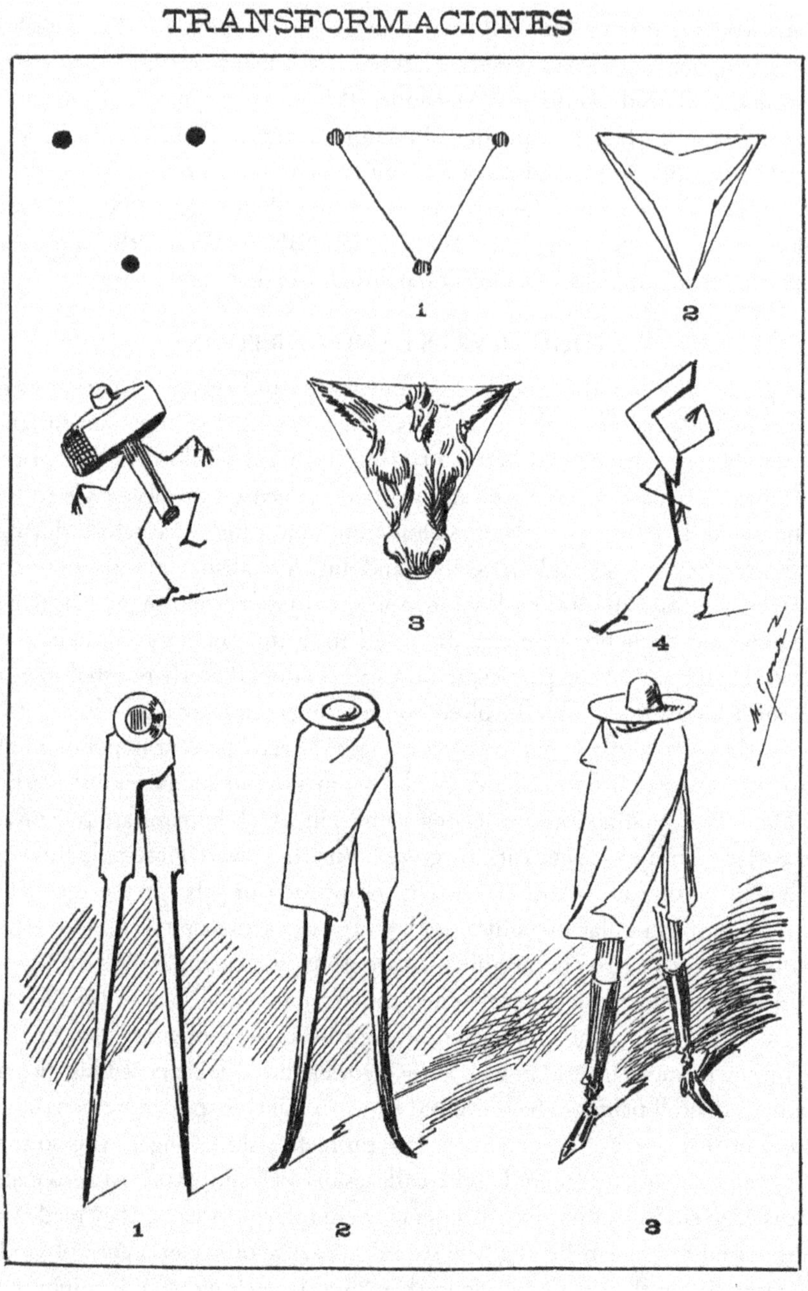

Figure 5.1: "Transformations." The evolution of masonic symbols into a donkey and a human. In *La Chispa: Seminario Católico Casi Humorístico Ilustrado* 24 (1890): 5.

giving birth to the English naturalist. Having his father assign him the task of exploring his genealogical tree, to see if they were of noble origins, Darwin ended up with the disappointing discovery that, in fact, they came from a

METAMÓRFOSIS.

Figure 5.2: "Metamorphoses." The evolution of a moneran into the hat of a revolutionary. *La Chispa: Seminario Católico Casi Humorístico Ilustrado* 28 (1891): 5.

monkey. "Poor father!" the author exclaimed. "But the poor kid was not alone to blame. According to some biographer, Darwin came up with his famous evolutionary theory . . . the day he looked in the mirror!"[54] In spite of the mockery, the story was accompanied by a serious drawing of Darwin.

The analogy between his head and that of an ape was a common trait in anti-Darwinist images. The weekly *La Chispa: Semanario Católico Casi Humorístico Ilustrado* offered two such images. One in July 1890 stressed the apelike traits of a Darwinist, wearing opaque glasses while reading one of Darwin's books. The caption affirmed that, as a blind man, every time he looked in the mirror Darwin's theory was validated.[55] A few months later, in January 1891, another apelike character paraphrased the famous Muslim statement, saying, "There is no other monkey than Darwin and I am his prophet."[56]

More interesting are two other images in *La Chispa* around the same dates, the first of which (figure 5.1) depicts the transformation of the three masonic symbols, the triangle, the hammer, and the compass, into such images as an ass's head and a man wrapped in a cape to suggest secrecy. In the second one (figure 5.2), a moneran metamorphoses into the cap commonly used by revolutionaries.[57] The conservative and traditionalist editorial line of this illustrated journal self-fashioned as "quasi-humorous" was manifest in its first installment: "No to progress of reason that has led us to rationalism; no to exacerbating the senses that pushes us into materialism; no to . . . bestiality that forgets our soul . . . and empowers skepticism and kills greatness."[58]

Núm. 2. Barcelona 28 de Mars de 1879. Any 1.er

LO BURINOT

Preus de suscripció.

Un número, dos cuartos per tot Espanya.
Números atrassats, mitj ral.
Per suscripció, trimestre 5 rals.
Estranger y Ultramar, 5 rals y l' import dels sellos.
No 's servirá cap demanda sino es pagá per endevant

Punts de suscripció.

Llibreria de Subirana, Portaferrissa 16
Taller de encuadernacions de Joseph Mañá, Euras, 4.
Estamperia de Ramon Solá, Nova de S. Francesch n.º 5.
Per assumtos de redacció dirijirse a D. Ramon Barnós, Cirés, 8, Imprenta.

Periodich il·lustrat, satírich de bon geni y millor humor, y gastant ínfulas de literari

SORTIRÁ UNA VEGADA A LA SETMANA

LA CIENCIA D' ANTANY Y D' ENGUANY

Ataneo llibre de micos estudiant lo progrés per tornarse homos. | Ataneo llibre d' homes estudiant lo progrés per tornarse micos.

Figure 5.3: A debate of monkeys discussing the evolution of men, and a debate of men discussing the evolution of apes. *Lo Burinot* 2 (1879): 1.

Another conservative and Catholic journal also explicitly self-fashioned as satirical, the Catalan *Lo Burinot: Periodich Il·lustrat, de Bon Geni y Millor Humor, y Gastant Ínfulas de Literari*, had a number of cartoons in which modern science and modern education were related to the activity of animals, mainly monkeys and donkeys. Figure 5.3 shows an old scientific symposium compared to a modern one. In the first, a group of monkeys "study progress so as to

Figure 5.4: A donkey is the teacher at a state school and Darwin is one of the "absurdities" being taught in them. *Lo Burinot* 51 (1880): 4.

become men"; in the second, a group of men are at the athenaeum holding the skull of a primate, "studying progress so as turn into apes."[59]

Figure 5.4 introduces us to some of the most disputed social and political issues in the 1870s and 1880s: education and freedom of speech. In the immediate years after the restoration, religious institutions recovered their virtual monopoly in primary education. But the first liberal government appointed by the king and led by Práxedes Mateo Sagasta re-introduced and promoted secular schools. In the face of that, *Lo Burinot* depicted a classroom full of boys in which a donkey is teaching "absurdities, tales and falsehoods." Four posters on the walls refer to the four pillars of that kind of education according to the cartoonist, one of which is "Darwin's system" with a threatening, scary monkey.[60]

Not surprisingly, however, one can find more satirical periodicals led by republicans and liberals.[61] The Catalan republican editor Innoncenci López Berganossi (1829–1895), who had taken part in the revolution of 1868, created two successful such periodicals: *La Campana de Gràcia* and *L'Esquella de la*

Figure 5.5: Darwin and Spencer are being rejected by donkeys at the Atheneum in Barcelona. At the entrance, an upside-down sign has the word "Progress" on it. *Almanach de la Campana de Gràcia* (1878), p. 4.

Torratxa, both of which eventually became very influential in early twentieth-century left-wing Catalan politics. Priests were often depicted as donkeys and their followers either as donkeys or as sheep. Figure 5.5 refers to arguments about free speech and the banning of some conferences at the Athenaeum of Barcelona in 1877: a number of donkeys expel the goddess Athena carrying a book with Darwin and Spencer on the cover. The upside-down plaque "progress" and the text below ("modern philosophy and the 'wisdom' of the Athenaeum") identify Darwin and Spencer with modernity and progress.[62]

Another such periodical was the weekly *El Motín: Periódico Satírico Semanal*, appearing for the first time in 1881 and lasting until the death of its creator, the journalist and republican activist José Nakens (1841–1926). The pamphlet was only three pages long, one of which was occupied by a full-page cartoon, and the subjects were mainly republicanism, anti-conservatism, and anticlericalism. By the end of the nineteenth century it eventually reached a print run of 20,000 copies. The first issue was very clear as for its goals: "War on Conservatives!"[63] The extremism of *El Motín* meant that, at least in its first decade, republicans like Salmerón or Francesc Pi y Maragall (1824–1901), who had agreed to a liberal government within the monarchic regime, were also subject to criticism. A cartoon from 1887 (figure 5.6) shows an interesting mixture of anticlericalism, republicanism, and the trope of evolutionism. In it, the new liberal president, Sagasta, becomes the "high priest of the religion of

Figure 5.6: The "great pontifex of the religion of evolution [President Sagasta] and his deacons [Salmerón and Pi y Maragall]" worship King Alfonso XII. *El Motín* 41 (1887): 2.

evolution" and worships King Alfonso XII, helped by Salmerón and Pi y Maragall as altar boys.[64] Evolution here simply means non-revolutionary politics, in analogy to the gradualism of Darwinism.

These examples are significant for, at least, a couple of reasons. In most of them Darwin and evolution are icons for political or educational controversies, not for naturalistic or scientific ones. They do not appear when representing famines, plagues, or health-related issues. The second one is the small number of instances in which they appear. Less than a dozen examples in the Catholic *La Hormiga de Oro* or in the anticlerical *El Motín*, where priests and monks were a constant target of mockery, speak of the meager role that the evolution-and-religion conflict played in the lower end of popular culture if compared to other political and social issues.

EPILOGUE: REGENERATIONISM AFTER 1898

I started this chapter with a reference to what I call the anti-Draper contest. To be sure, the call was not that narrow. The aim was to produce the best book arguing that there was a necessary, intrinsic harmony between (the Catholic) faith and science, and not simply to debunk the historical argument of Draper. Books were, and still are, one of the privileged instruments to disseminate new ideas. But, as we have seen, the increase in the availability of cheaper paper and printing techniques, as well as the (at times limited) freedom of the press, allowed for other media to spread information, opinion, and ideologies in less academic but more incisive ways. Generalist periodicals and satire pamphlets and journals were part of this new culture of printing. But, of course, truth, accuracy, or nuanced arguments were not always necessary traits of these new genres and media. Thus, in a typical overly simplistic strategy, many Catholic publicists identified Darwinism and evolution at large with all possible individual, social, and political sins, while many reformists found the church, as a general concept personified in priests and monks, as the main enemy to beat in their political crusades. And "evolution" and a loose notion of "science" came in handy for their campaigns.

There is still much work to be done on the relationships between science and religion in Spain. Spanish science-and-religion folk culture, as well as many historians of science, still share in the categories that emerged in political and clerical versus anticlerical battles of the second half of the nineteenth century. Indeed, the history of Spain in the twentieth century only helped consolidate such categories on many fronts. A common example is Odon de Buen (1863–1945), naturalist, founder of Spanish oceanography, and a major actor in the popularization of evolution at the turn of the century. De Buen died in exile in Mexico, escaping from Franco's regime, which helped portray him as a martyr of science prosecuted by religion and fascism. But, as historian Agustí Nieto-Galán has shown, from his chair in natural history at the university of Barcelona, which he held from 1889, de Buen taught a version of Darwinism that was the flip side of his republican activism.[65] The problems he had during his tenure in Barcelona, especially a dispute in 1895 after his works were placed in the Index of Forbidden Books, may too simplistically be described as an example of the conflict between science and religion. But, as has happened in other instances, elements of reform, conservatism, revolution, education, and so forth were part of the controversies in the biography of Odon de Buen.

The unproblematic attribution of modern science to self-fashioned "progressives," or the identification of "religion" with a homogeneous movement that extinguishes progress, freedom, and science, misses the nuances and complexities of Spanish science, education, and the role of Catholic institutions in the social fabric of the country. To give an example, the image of a Catholic Church uniformly stagnated in the past does not go down well with the emergence of new religious orders and lay Catholic movements devoted to education, to health care, or to the conditions of the working class, like the work of the aforementioned José Salamero, "the father of the poor."

I have described some of the complexities of the periodical press in relation to science, religion, and evolution in the early decades of the Bourbonic restoration. The year 1898 is a momentous date in Spanish history. The last remnants of the Spanish empire, Cuba and the Philippines, were lost to the Americans, triggering a collective sense of decadence, wounded national pride, and a quest for reform. The spirit of the *Polémica* resurged, with groups and ideologies blaming each other for the "disaster." As I have elsewhere shown, this was also an occasion for a more constructive intervention by religious orders, especially the Jesuits, in taking an active role in the professional popularization of science with, first, the generalist *Razón y Fe* ("reason and faith") created in 1901 and, later, the specifically popular science *Ibérica*, appearing first in 1913.[66] The latter was run by Jesuit scientists, not philosophers, theologians, or clerical authorities, which helped portray a less threatening or political face of the modern natural sciences. What science was, which institutions had to run it, and its optimal relation to faith was still disputed; but the editors' purpose of both journals was to portray the Company of Jesus as a contributor to modern science, to the reconstruction of the nation, and to national pride, while generally shying away from contested and politicized areas such as evolution.

6

Argentine Positivism on Evolution and Religion in the Late Nineteenth Century

Ignacio Silva

Positivist positions traveled from Europe to most countries in Latin America either via the ideas of the French founder of positivism, Auguste Comte (1798–1857), or through the thought of English philosopher Herbert Spencer (1820–1903), particularly through his ideas of progress. Regardless of the effort that Spencer undertook to distinguish himself from Comte's doctrine, Latin Americans usually mixed the two under the label "positivism." By the end of the nineteenth century almost every non-religious prominent intellectual in the country would have passed as a positivist of one or another kind, embracing the ideal of progress as something essential to their thought. Similarly, almost any religious scholar would have used the term "positivist" to refer to those in the other camp.

Latin American positivism was, ultimately, a set of ideas immersed in the continent's political environment, and as such, the narratives offered about the relations between science and religion were influenced by these political settings. Apart from a few publications, these nineteenth-century relationships between science and religion in Latin America have not received much scholarly attention. Most of this research usually either presents some historical study of the intersection between science and religion in discrete national episodes or explores the reception of Darwinism in a specific Latin

American country, with some passing comments on the dynamics between evolutionary thought and religion. The work of Argentine historian Miguel de Asúa in describing some stories of relations between science and religion in Argentina's history is an example of the former strategy. De Asúa's work often features detailed telling of significant episodes in the development of Argentina's political environment. His work on Jesuitic missions and natural philosophy or his more programmatic pieces on nineteenth-century Argentine thought are excellent examples of this strategy.[1] The volume that Glick, Puig-Samper, and Ruiz Gutiérrez edited, *The Reception of Darwinism in the Iberian World: Spain, Spanish America and Brazil*, is a clear example of the latter strategy,[2] as is Lightman's edited volume *Global Spencerism*, which includes several chapters devoted to the reception of Spencer's philosophy in Latin American countries.[3] Many of the chapters in these volumes include some references to the debates between evolutionary science and religion, though this is not their central focus.

This chapter offers a transnational history of the impact that British evolution and French positivism had on Argentine intellectuals. I focus on the case of Argentina's positivists and their Catholic counterparts, allowing for the historical actors to present their own views of the relations between evolutionary science and religion. Later on, I will present some Latin American counterparts in an attempt to paint the broader picture. My main thesis is that, at least during the final decades of the nineteenth century, there were nuanced and ingenious positions on both sides of this debate in Argentina, unlike some local historiographical narratives tend to display. Actual responses to evolution in Argentina raised questions about national identity, education, and the social role of religion and became inextricably entangled with the growth of positivism. These responses can be divided into three main types: atheistic positivism, in the figure of Florentino Ameghino; irenic positivism, exemplified in the thought of Pedro Scalabrini and Alfredo Ferreira; and Catholic anti-positivism, present in José M. Estrada and Manuel D. Pizarro. I briefly show how similar patterns can be found in some of the other Latin American countries, such as Mexico, Brazil, and Chile.

ARGENTINE POSITIVISTS AND CATHOLICS ON EVOLUTION AND RELIGION

Argentine positivism flourished in the last two decades of the nineteenth century and was embraced by notable figures in the natural sciences, politics, philosophy, and education. Some particularly important positivists include Domingo Faustino Sarmiento (1811–1888), Florentino Ameghino (1853–1911), Pedro Scalabrini (1848–1916), and Alfredo Ferreira (1863–1938). As in many other Latin American countries, the way in which Argentine positivists received and assimilated Darwin's theory was framed by an environment in which the national state was being reorganized. On Darwin's role in Argentina, Héctor Palma has recently affirmed that it was so important that "it weaves into a general context of advance and consolidations of the sciences, the organization of the State and a positivist mentality, not only in scientific issues, but also in political and cultural matters."[4] As such, nineteenth-century discussions about the relevance of Darwin's theory of evolution, many a time understood through the prism of Spencer's philosophy,[5] were generally dictated by debates surrounding the progress of Argentine society through scientific education toward models of societies coming from the northern nations, in particular the United States. Argentine intellectuals, thus, offered their understandings of the relationships between evolutionary biology and religious ideas within this context of thinking about the Argentine nation as moving toward progress, and positivist thought was no exception in this trend. In sum, as Argentine historian Marcelo Montserrat expressed it, "evolutionism . . . became the key element of the mental gear of our local version of positivism, which pervaded it with militant biological progressivism both in style and content."[6]

One of the earliest encounters between evolutionary and religious thoughts in Argentina occurred as early as 1862, three years after the first publication of *On the Origin of Species*. In a short essay, a young and passionate José Manuel Estrada, who would later become a fervent outspoken and influential Catholic lawyer, strongly complained about Lamarck's transformism, defended by some intellectuals in Buenos Aires. He referred to the uniqueness of human beings compared to other animals in his objection, but without any reference to Darwin's work: "Man, so infinitely excellent over the brute, cannot be its progression . . . for what is not spiritual cannot produce the spiritual, what is not free cannot produce freedom, what is not intelligent cannot produce intelligence, what is mortal cannot produce the immortal; for the soul . . . is either a lie, or the *spiraculum vitae* from Genesis."[7] Estrada would continue his complaint, asking, "Do you believe in progressive races? Do you believe in the pre-Adamic man? Then you do not believe in the soul, you believe in a mortal brute without destiny, in a being lacking individual self-consciousness, without any notion of absolute justice . . . but you do not believe in God! You are an atheist."[8] Estrada would come back to the forefront of the debate between Catholicism and positivist evolutionary thought during parliamentary discussions about secular education in 1883–1884.

Domingo Faustino Sarmiento, president of Argentina between 1868 and 1874 and one of the most influential positivist characters of his time (even if not a philosopher himself), was a fervent evolutionist, referring to Darwin's theory as an Argentine product. He wrote in his diary of his trip to New York on August 12, 1868: "Darwin's theory is Argentine, and I am committed to nationalize it." In 1882, one month after Darwin's death, the Argentine Medical Circle organized an event to commemorate Darwin's legacy, inviting three guest speakers, of which Sarmiento was the first. In his lecture, he affirmed, "I, gentlemen, support the doctrine of evolution as a development of the spirit, because I need to rest on a harmonious and beautiful principle, so as to silence doubt, which is the torment of the soul."[9] Regardless of Sarmiento's actual understanding of Darwin's theory, which has been shown not to be that of a specialist,[10] he adopted the ideal of evolution and joined it to the ideal of progress, so much in vogue within intellectual and political circles at the time.

THE GREAT ATHEIST NATURALIST

This amalgamation of evolutionary and societal progress ideas was likewise present in the thinking of Florentino Ameghino, also a speaker at the Argentine Medical Circle's event honoring Darwin and one of the most important nineteenth-century Argentine naturalists and positivist figures (see figure 6.1). Ameghino was well known for his theory of the origin of the human species in the Argentine Pampas[11] and for his posthumously published doctrine of the four principles of reality: "The Cosmos is the set of four infinites: the *infinite space* occupied by *infinite matter* in *infinite movement* in the successive phases of *infinite time*."[12] In this posthumous work, *Los Cuatro Infinitos* (1919), Ameghino expressed his views on the idea of God in positivist terms relating it to his four infinites: "The idea of God is a primitive idea, simple, infantile, daughter of fear that the unknown and ignorance engender. . . . An idea born with man since the savage state, it has been slowly modified as man became civilized and cultivated his intelligence, until making such idea a purely metaphysical notion . . . , using this expression in its most vulgar sense, which renders everything that is not understood into a metaphysical notion . . . this notion of God dissipates in front of the much greater notion, as well as real and positive, of the *eternity* of *infinite matter* in *infinite movement* filling the *infinite space*."[13]

Ameghino's reputation was so significant in Argentina at the time that, as de Asúa asserts, "his image as the personification of evolutionary science was utilized by the positivistic establishment as a battering ram to storm what they saw as the crumbling fortress of metaphysical and religious deceptions."[14] Ameghino was part of these controversies between evolutionary thought and religion: in 1884, under the pseudonym Dr. Serafín Esteco, "former member of the Concolic Apostacolic Club" (a fictitious club, clearly), he published a satirical letter in the newspaper *La Crónica*.[15] In this letter, Ameghino recounts what an imaginary French colleague told him when infiltrating a Catholic Congress in Luján—a small town in the province of Buenos Aires where the Catholic Church has venerated a clay-made image of the Virgin Mary since

Figure 6.1: Florentino Ameghino. Source: *[Retrato de Florentino Ameghino, ca. 1900] [material gráfico], Caffaro, José [fotógrafo]*, Biblioteca Nacional Argentina. Public domain.

the seventeenth century—to investigate clay-made antiques. On top of the irreverence expressed in the letter toward religious imagery, there is a passage in which religious people are phrenologically observed and said to be regressing to their evolutionary origins as apes. The text, copying part of the letter of his fictitious French colleague, is rather revealing of some positivist attitudes toward popular religiosity at the time: "During that so called peregrination, I had the occasion to observe a large number of flat, rough, narrow, and angular foreheads, all evidence of sick, obtuse, and infantile brains, such as those observed in inferior races, caused by a stoppage of development, a kind of inversion to the ape." After a few accusations by members of the hierarchy of the local Catholic Church and attempts of denial by his followers, Ameghino acknowledged authorship of the letter, accepting "having used the pseudonym of Dr. Serafín Esteco in a humoristic article that was published in *La Crónica*," thus confirming his sentiment toward religious belief in general and the Catholic Church in particular.[16] In fact, when speaking of human evolution in an unfinished text from 1908, Ameghino acknowledges the possibility of a regressive evolutionary *development*: "If many different kinds can acquire a similar morphological likeness due to a parallel or convergent evolution, other [kinds] can acquire the same likeness through a regressive evolution or bestialization."[17] Over twenty years before this text, by 1880, Ameghino had already fully adopted evolutionary science, highlighting that "its founder," Darwin, had produced a true scientific revolution when he published *On the Origin of Species* in 1859. Reflecting on this revolution, Ameghino commented on the future of this theory, stating that it would be shortly "considered as a definitive truth conquered by science."[18]

Ameghino was well aware of the rejection of evolutionary theory on the part of some "orthodox zoologists," and he wrote against this position. As he explains in his *Filogenia* of 1884 (probably with the thought of those like Estrada in mind), "transformism or the theory of evolution is usually strongly attacked in zoology, for it is considered to be against revelation."[19] After all, "orthodox naturalists read in it [Genesis] that God created each species with all the features that they currently have."[20] Affirming that "Genesis should never be quoted in a scientific work," he claimed that "paleontology does not confirm the sacred text." Instead, "she shows us the different species of fish, of mammals or birds succeeding one to the other in time and not appearing all at the same time. Thus, in opposition to the orthodox idea of the creation of species by a supernatural power with all their current features, the idea of filiation was born and, hence, [the idea of] descent of one species from another, by means of modification."[21] Ameghino even contrasted these ideas with the by then popular doctrine of successive creations: "The theory of evolution teaches that the diverse communities of beings called species . . . do not owe their appearing to successive and independent creations; they are rather the slow and progressive modifications of other communities that preceded them."[22]

Contrasting with these attacks against religion, Ameghino did offer some possible solutions to his counterparts across his works; solutions that were quite popular in other regions of the world, particularly in Asa Gray's

Darwiniana (1876). Considering the doctrine of successive creations, in an irenic passage of his 1875 *Diario de Un Naturalista*, Ameghino wrote,

> It would be more natural that, being God so powerful, he needed not resort to those creations so continued, and that from the moment in which his goal was to produce different kinds, it would have been enough to give certain features to the first animals through which they would have modified, producing, with the passing of the centuries, kinds different from the primitive, and thus successively, without him having to be troubled creating new animals in each geological period; and under this consideration, Darwinism is in more concord with the so-told omnipotence of God, than the theory of successive creations.[23]

Were there an omnipotent God as the Christian claimed, Ameghino argued, this God could have endowed the original species with some features that would make them develop into new species. As Asa Gray did in the United States, Ameghino was simply offering a way in which to bring together evolutionary and Christian thoughts.

This concordism reappears a decade later in his *Filogenia*, in which Ameghino takes up an analogy between evolution and the nebular hypothesis in astronomy that was in vogue at the time for solving the conflict between orthodox naturalists and evolutionary theories.[24] This analogy evokes, among many, ideas one can find in the thought of Gray, who, when reviewing Darwin's *Origin* in the early 1860s, also found in astronomical hypotheses a path to facilitate a more open mind for the reception of Darwin's theory. Analyzing Pierre-Simon Laplace's nebular hypothesis for the formation of planets and stars, Ameghino suggests that everyone agrees that the "planets and the stars went through a long evolution . . . no less astounding and no less different from their current state than the different transformations through which, according to evolutionary theory, animals underwent." Since "no one denies the possibility of such theory, nor anyone holds it in contradiction with revelation," Ameghino argues, then "why would it then be when applied to the formation of species?"[25]

I am not aware of any other passage in which Ameghino rehearsed arguments of this sort, though it is remarkable that they belong to works published with a decade in between. There are, indeed, many other passages in his works in which he openly states that the science of his time (evolutionary theory, geology, chemistry, or any other) disproves religious or theological belief, giving credit to the typical stance that puts Ameghino as one of the most prominent characters in Argentina's stronghold against religious belief and institutions. One does wonder, however, what motivated Ameghino to put forth these arguments on the relation between science and religion in these two influential works.

Upon his death in 1911, Ameghino received a grand civic positivist funeral, filled with allegories to positivist ideals: the stage included banners referring to Labor, Justice, Fortune, and Progress, together with Roman statues symbolizing Science, Humanity, and History.[26] One of the most important Argentine positivist intellectuals from the early twentieth century, José Ingenieros, offered a speech at the occasion titled "The Modern Sanctity," in which he reinterpreted the Christian idea of sainthood, suggesting that heroes in the times of war, apostles during moral crises, and wise men in the times of modern civilization were the saints for each period. Ameghino was a saint of modern civilization, for these modern saints "discover new laws of natures or teach or dominate some of its forces."[27]

IRENIC POSITIVISM AND SECULAR RELIGION

Ameghino's civic funeral recalls the ideals of the secular religion of positivism, which by the end of the nineteenth century became the goal for the local Comtean school, of which the figures of Pedro Scalabrini (1848–1916) and his disciple Alfredo Ferreira (1863–1938) are the most influential (so much so that when the latter of the two died in 1938, one of his disciples claimed, "This cannot be! Without Ferreira, what will become of the doctrine of Comte, Leffitte and Corrá in our country?"[28]). Scalabrini was an Italian scholar and devoted schoolteacher who led the positivist school of thought in Paraná, the center of positivism in the country, situated about 350 km northwest from Buenos Aires. Educated in law, he soon turned to the study of natural history, in particular paleontology, collaborating with Ameghino (and collecting over seventy thousand fossils).[29]

Scalabrini was the greatest disseminator of Comte's positivism in the country, to which he added evolutionary principles with a strong Spencerian component.[30] Of a less fighting nature than those from Buenos Aires, Scalabrini, following the spirit of Comte's philosophy, always called for dialogue and understanding between opposing views, having particular consideration to those views that were "less progressed." This might be the reason for him including Thomas Aquinas, as the "creator of Catholic Philosophy," in the list of the greatest luminaries (creators, discoverers, and inventors) of Western thought, among which he also counted Aristotle, Thales, Plato, Descartes, Krause, Lagrange, Gaudry, Lamarck, Darwin, and many others, all of whom were "spiritual beacons of the human family that brightly represent the growing and ever perfecting civilization."[31]

Scalabrini showed his bivalent thoughts on revealed religions when speaking, in 1888, about Darwin's religious belief: "Notwithstanding his apparent adhesion to the Biblical beliefs of the protestants, [Darwin] was a true independent thinker who knew how to conciliate scientific freedom with the respect that religious beliefs of other people deserve, respectable and even venerable when they are sincere, but that they ultimately represent a delayed stage in the evolution of human thought."[32] About a decade after this statement, Scalabrini openly declared that he was a Catholic, in the sense that he believed that the Catholic faith was a revealed religion that was meant to turn into a demonstrated religion through the progressive evolution of human thought. Ultimately, for him, "there is no doubt that Catholicism is superior to the other religions."[33]

In his widely read 1888 work *Materialismo, Darwinismo, Positivismo: Diferencias y Semejanzas*, Scalabrini finds that "positivism is the unavoidable result of anthropological evolution."[34] Positivism, for Scalabrini, accounts for the "scientific, artistic, industrial, philosophical, political, and religious" and excludes the "chimerical, idle, indecisive, vague, negative, and absolute." In stating the superiority of positivism to materialism, for which he understands a brute reductionism of every natural event to the laws of physics, Scalabrini affirms that materialism "is incomplete because it does not elevate but only one part of knowledge," reducing "the most noble in man to the most gross." Scalabrini makes use of a major problem in materialism, namely its "contempt against all religions," to show the value of revealed religions: "[Religions] have served and continue to serve, according to times and places, among which excels the Catholic religion."[35]

In terms of Darwinism, which for Scalabrini stands for a basic understanding of the theory of evolution through adaptation and natural selection, he explains that positivism includes much of its insights (as well as those of materialism), embracing both the faith on scientific progress and the ideals of humanity present in Comte's thought. Offering a simple modification of Darwinism, Scalabrini affirms that it can be included within positivist philosophy: "Darwinian selection does not need that purported struggle for existence so that it be a demonstrated truth; labor for existence is enough. . . . Modified in this way, Darwinism is better affirmed and can be incorporated to positivism."[36] As a way of conclusion, Scalabrini suggests a brief definition of positivism that includes materialism and Darwinism: "Positivism is a philosophical system founded in science, i.e., in facts and in the laws of nature, and in this it identifies with materialism and Darwinism, which are part of positivism."[37]

In fact, Scalabrini boldly affirms that materialism must be "a precursor and collaborator of Darwinism, as the latter is of positivism,"[38] the final goal of which is the instauration of Comte's religion of humanity: "The final and most eminent construction of positivism is the demonstrated religion, which is defined with Comte as the union of all convergent beings for industrial, artistic and scientific work, which would rule the regimen, the rite, and the dogma of the new religion."[39] So for Scalabrini, materialism and Darwinism lead to religion—not to a revealed religion such as Catholic Christianity but to Comte's demonstrated positive religion. After all, "opposing materialism in the name of the Bible . . . or in the name of the canons of the Vatican Council, is to confront a certain defeat."[40] In any case, for Scalabrini, this new religion looks very much like the Catholic religion.

This goal was to be promoted through society via the education of the youth, starting with natural history, which, as the "essential and fundamental base of education, would not miss the opportunity to awake feelings of admiration, respect and submission in students towards nature, immutable in time, immense in space, omnipotent in its multiple creations that reveal goodness, intelligence, activity, order, progress, unity of plan and of execution, harmony and stability."[41] Nature, in this positivist take, would assume many of the attributes that were once predicated to the Christian God.

In this spirit, Scalabrini attempted a conciliation of materialism, theology, metaphysics, and positivism in his 1896 paper "Conciliación Relativa," affirming that, historically understood (under the lens of Comte's philosophy of history), there is no real opposition between theology and materialism or between metaphysics and positivism, given that they all pertain to different stages in the evolution of human society. What was useful once, however, needed not be useful at a different historical stage.[42] Thus, Scalabrini believed that science and religion actually worked together in two different ways: First, the theological and metaphysical enterprise of humanity acted as the necessary antecedent for positive thought based on scientific grounds to come about in history, and thus *revealed* religion was instrumental for it. Second, scientific thought, in the forms of materialism and Darwinism, served as the precursor of the ultimate demonstrated positive religion. In the end, Scalabrini believed that Comte's positivism had clearer chances than other philosophies of the time (Spencer's in particular) to produce a healthier relationship between science and religion, in the sense of a human and secular religion, which would act as a guarantor of social order. This secular religion would, for Scalabrini, transform the Christian precept of love into the positivist moral as something "pure, disinterested, and Christian." Of course, the pre-humanist religion did not fare well with Scalabrini, not due to its inherent conflict with a scientific worldview but rather due to its pre-positive state. In any case, as Scalabrini elegantly puts it, religious beliefs deserved due respect.

Interestingly, Scalabrini's disciple (and foremost Argentine positivist by the end of the nineteenth century) Alfredo Ferreira (lawyer, educator, and founder of the Argentine Positivist Committee in 1924) went further with this idea, affirming in an 1895 article of the newly launched journal *La Escuela*

Positiva, "Positivism is not atheist. . . . It does not deny the existence of God, idea so swift that the human mind cannot comprehend. . . . Human intelligence is so limited that it is impossible for it to form an idea of what a billion is, much more of the meaning of absolute justice, absolute mercy, etc."[43]

Following on these ideas, Ferreira offered a different way in which Catholicism and evolutionary science would relate. For him, the Catholic religion would incorporate evolutionary thought into its scheme, since Catholicism was, ultimately, progressive and evolutionary thought was not against the idea of God: "Catholicism will eventually accept it [evolution] in the future with its progressive spirit . . . , because this idea is opposed in nothing to the existence of God." In fact, in a passage that resembles Ameghino's irenic arguments, Ferreira affirms that God "in his infinite wisdom and power could have dictated laws over matter so that it would produce beings and things in virtue of such transformations. . . . In this, positivism wants to harmonize religion with science."[44]

THE CATHOLIC STRONGHOLD

Regardless of the ways in which Scalabrini and Ferreira attempted to link their positivist take on evolutionary theory to religious views, Argentine Catholics rejected this philosophy as atheistic and contrary to human progress. José Manuel Estrada was one of the main Catholic advocates in the 1883–1884 parliamentary debate over education laws fervently opposing positivist ideals (see figure 6.2). His main concern was that, as he claimed, positivism led to negating the reality of morality and ethics in human beings. He affirmed in 1883, in one of a series of essays he wrote during this debate, that "modern science . . . , with the positivist school, understands by morality the fatalist system in human actions. It sees in man an organism and nothing else. It sees in man's thoughts and sentiments, physical and chemical phenomena."[45] He was, to an extent, correct. After all, Alfredo Ferreira would later affirm this very idea in his 1914 "Una Ética Química": "The moral content of life . . . is formed by desires and hopes, efforts and struggle, disappointments, sufferings, triumphs. It is possible to explain these phenomena with physico-chemical analyses," boldly claiming that "we are chemical machines," asking, thus, "How could we be ruled by morals?"[46] Hence, Estrada concludes, "the positivist thesis involves the cardinal negation of any principle of ethics"[47] and, thus, the root of Christian morality.[48]

Estrada also finds this lack of foundations for morality in the close relation between positivism and evolutionary thought. After speaking of the "humiliating hypotheses" that Darwin posed and picking up on his earlier criticism of transformism, Estrada claims, "There is no experimental antecedent that can give credit neither to the theory of the convertibility of species nor to their

Figure 6.2: José Manuel Estrada. Source: Biblioteca del Congreso de la Nación. Public domain.

progression."[49] Estrada is here contrasting the view that natural species can progress (or evolve, which was more or less the same for debates at the time) with the idea that every human individual can progress in their moral life, a duty that everyone had for the simple fact of being human—hence Estrada's rejection of the positivist secular religion of humanity, because, as Scalabrini held, this positivist religion found its fundament on evolutionary principles: "The dogma of the spirit and the rules of conscience have conceded their place to naturalist systems, which the most advanced and discrete liberal school call today the 'Religion of the Science.' This religion has at its base the theory of transformism, in vogue among positivists and atheists, which is completed with the 'law of natural selection,' developed in a 'struggle for life.'"[50]

Springing from his background education in law and the Catholic tradition of the humanities, this sort of statement manifests Estrada's worries about the implementation of a positivist education based on the natural sciences, including theories of evolution. He held that this educational system tended to "destroy the study of the classics"[51] and leave no place to philosophy: "An education system calculated in the positivist sense cannot be dominated by philosophy. . . . The natural and exact sciences prevail in it."[52] In the end, Estrada believes that this lack of philosophy (by which he actually means Catholic metaphysical thought) and preponderance of the natural sciences in positivism lead to some personal discontent: "The atheist and the materialist feel it [the discontent] no less stingingly. Why do they wearily seek the laws, unless for the angst and bleakness that leave in their souls the quick passing and fading off of facts and phenomena? Why do they proclaim the eternity of matter, unless for the dismay that leaves the ephemeral manifestation of a law, unrelated to a generating and stable principle in the contemplation of which could rest the mind disturbed by the whirlwind of things hastening from nothingness to nothingness?"[53] In contrast to Scalabrini, Estrada offers a critical assessment of the natural positive sciences: they are sterile in the pursuit of the highest spiritual goals, because they "do not exercise but only one part of intelligence, being necessary that all of human faculties be harmonically strengthen and developed for the goals of education."[54]

Notwithstanding this pejorative take on the natural sciences, Estrada is clear on their value: "We cannot doubt the suitability of initiating the youth in the secrets of nature and in the investigation of its laws, particularly when in their being fixed, in their harmony and splendor, the soul seeks the likeness of the Lord Author of all things."[55] By 1887, Estrada was well aware of the richness that the sciences had brought to humanity, particularly during the nineteenth century, and especially of the benefits of an education in science: "I do not dismiss the illustration that has been called scientific, and nobody is blind to the fact that it is imperative to give it a place of honor, because without it there would be no complete education, as neither a man of this time could neglect the glorious conquests of the century in the dominion of the material world."[56] Most importantly, Estrada believed that "scientific information . . . [is] a condition and a complement of the vast program of education . . . because it perfects the agent in the moral duties."[57]

His rejection of positivist science was rooted in his idea that there were some external principles to science that, when overimposed to it, turned it to be anti-Christian. In fact, without these principles, science was, indirectly, Christian, as he explained by 1888: "Science, inasmuch as it is the knowledge of facts and laws that rule them, is, in fact, on its own, neither Christian nor anti-Christian. I should say that it is always indirectly Christian, for it manifests the truth of the natural order. . . . But there is a current trend of ideas by which men are driven to deny that there are knowable truths, beings, laws, or principles beyond those pertaining to the visible nature. This way of conceiving things is anti-Christian, and science that limits the sphere of thought in this way is anti-Christian." He explains that this trend is exhibited when new hypotheses are added to well-researched facts and laws of science, hypotheses which "are not part of science, because they are false, and they seed uncertainty and extravagance."[58] These hypotheses, Estrada believed, are brought into science by the ideology of materialism, which "penetrates the sciences and throws innumerable scholars into the abyss of misery and pride."[59] Out of this anti-Christian science springs the "insane pretense of creating conflicts between science and faith."[60]

On the contrary, Estrada finds that science without these alien principles "ennobles those who discover God behind cosmic beauty" but "shrinks those who allow themselves to become sick by materialism."[61] In the end, Estrada would affirm that only a science based on philosophical considerations opened to the issues of the origin and end of human beings and the natural world would count as true science. These philosophical principles, by necessity, would have to include the notions of God, natural religion, and revealed religion. Not considering these notions would mean "mutilating science," while engaging with them implies "entering into the religious realm."[62] As such, Estrada believed that true science, even evolutionary science I would suggest, meant opening the doors to religious beliefs.

Fellow lawyer, journalist, and politician Manuel D. Pizarro (1841–1909) escalated the tone of the controversy against positivists ideas of science, particularly in the parliamentary debates concerning a secular government also present during the 1880s. For Pizarro, speaking of positivism (of an English ilk) meant speaking of Darwinism, materialism, and, ultimately, the liberal ideals present in the thought of, for instance, Sarmiento. For him, "those who with Darwin reject the divine genesis of humans and want to make of him nothing more than a perfected beast that proceeds from the ape and is barely superior to it in the zoological scale, that is, the adepts to the positivist or materialist school, believe in the existence of a purely civil government."[63] These positivists, Pizarro would argue, instead of promoting the progress of human society, by aligning themselves with materialist science create a regressive movement:

"In science, they would believe to be the arbiters of truth . . . understanding that science is something material and human, and not the spirit of God in the works of creation . . . believing them to be its [science's] authors, they consider themselves gods: they will invent a rationalist or pantheist theology and will proclaim the dogma of the nature-god or the reason-goddess, infiltrating this first error in philosophy into all the branches of knowledge . . . to obscure truth and retard, not impulse, the progress of nations."[64]

Objecting forcefully to the materialist take on positivist science, Pizarro affirmed that civilization would regress to "the spirit of ancient civilizations, in a descendent movement from the heights of Christian philosophy." For him, "if human intelligence persists in locking itself up in the circle of purely sensible observations . . . , philosophy will become poor. . . . Then, there would be no science, but sciences; no truth, but truths; no religion, but religions. . . . Religion, truth, science, would lose their unity and absolute character, to become simple terms in a relation, transient and fleeting as everything that is contingent."[65]

Pizarro's take on Darwin's theory is of the same character: evolutionary theory is "impious and cruel," and he argued that in the theory of evolution through struggle "human life is nothing more than a mere combination of forces of nature."[66] In evolution, Pizarro claimed, morality, human freedom, and conscience are sacrificed. Similarly to Estrada, Pizarro claimed that morality for this positivist Darwinian mentality was "a manifestation of matter and nothing else; a progressive evolution of organisms: an evolution from animality to intellectuality and to human sociability."[67] In an argument reminiscent of the Platonic ideal, Pizarro held that if one eliminates God—as most positivists tended to do at the time—"the eternal and immutable archetypes of the true, the good, the just and the beautiful are also eliminated," meaning that human beings "could never affirm neither scientific truth nor artistic beauty." With this Platonic argument Pizarro wanted to show how "intimate relations exist between science and religion."[68]

In an attempt to show the radical difference between Catholic and evolutionary thought, Pizarro borrowed and resignified the essential terms of Darwin's theory: "[The Gospel] is the whole *law* . . . which is sanctification, justice, and freedom in the world; that is, the *evolution* of the souls toward uncreated holiness, justice, and freedom. This is the *true struggle for existence*; where existence is the eternal life, and struggle is that of charity, love, fraternity. . . . This is the *law of evolution and mystical selection*."[69] In the end, for Pizarro, the ultimate preoccupation of human beings was eternal salvation, and science was worthy of pursuit as long as it was aligned with this goal.

There were not many Christian scientists involved in these debates, hence my choice of two lawyers to portray the Catholic response to positivism.

Perhaps the first naturalist to engage in these debates was the Jesuit priest José María Blanco (1878–1957), an anti-evolutionist serious amateur anthropologist. After the turn of the century, from 1916 onward, he directly objected to Ameghino's science. His strategy, rather than attacking the philosophical basis of evolutionary thought (as Estrada or Pizarro did), was to argue against Ameghino's thesis about the origin of human beings in the Pampas. Adopting a strong empiricist method, he objected to the hypothetical nature of Ameghino's ideas. In attempting to show that Ameghino's thesis was mistaken, his main goal was to argue that science was compatible with Christianity.[70]

SOME LATIN AMERICAN COUNTERPARTS

Positivist considerations of the relations between science and religion across Latin America were frequent and usually echoed the concerns guiding the debates in Argentina: the ideal of progress of the nations through the scientific education of the people offered by a secular state. Mexico is perhaps where positivism first arrived in Latin America, particularly via the works of physician and philosopher Gabino Barreda (1818–1881), who studied in Comte's Paris and upon his return to Mexico by 1853 educated the first generation of local positivists and led his country's educational reform during the mid-nineteenth century. He assumed Comte's motto of "order and progress" for his reform, as well as Comte's grand narrative of the history of humanity. In 1867, he expressed in his *Oración Cívica* that "science, progressing and growing like a weak child, first tried to augment its strength . . . until slowly, as it grew stronger, would enter into combat with preoccupations and superstition."[71] For Barreda, positivist science was the means to explain all parts of nature and human life. In fact, for him science made morals and religion to "have become part of its dominion."[72]

In his *Oración*, where Barreda traces the history of Mexico under the lens of Comte's law of the three stages of humanity—presenting, perhaps, the first sociological analysis of Mexican society—he rejects religion mostly for its links with Spanish colonial power rather than for its opposition to scientific knowledge. In a parallel image to the political wars between church and state that had just finished in Mexico at the time of writing his *Oración*, Barreda invokes the narrative of warfare between science and superstition (most of which denotes the Catholic theology), out of which science would result "triumphant and victorious."[73] Nevertheless, to understand this narrative fully, one must acknowledge that the Catholic religion was for Barreda the "key stone of the edifice of the Conquest,"[74] indicating that the image of warfare between science and religion was more of a proxy for the conflict between the ideas of the relationship between church and state in Mexico, as it was to happen later in the Argentine parliamentary debates of the 1880s. In fact, this opposition between science and religion did not spring from his considerations of evolutionary theories, since he did not accept Darwin's theory of evolution, for it had not, at least by then, satisfied the strict "conditions required by the scientific method." He would rather take the theory as a "hypothesis the demonstration of which is yet to be done."[75] Again, the conflict between science and religion in Barreda's thought was part of the positivist emancipatory movement of the Mexican people from its colonial ties, an idea that finds echoes in the thought of the political generation of Sarmiento in Argentina.

Barreda's main concern was the emancipation of the Mexican people from all ties with their Spanish colonial past: political, economic, and religious. In this context, a positivist education would promote such emancipation, since education, for Barreda, meant liberation from past bonds rather than indoctrination into new or old ones. Barreda conceived that even if political and economic emancipation had been achieved with the revolutions against Spain in the first half of the nineteenth century, religious emancipation was still to be attained, because the ties that the Catholic Church had with Europe were far too strong and continued to influence the lives of the Mexican people. Mirroring the argument that Scalabrini and Ferreira would offer a few decades later in Argentina, Barreda held that education was the tool through which to progress toward the scientific or positive stage of humanity in Mexico, leaving behind the less progressed religious and metaphysical stages of Comte's philosophy.

However strong his rejection of religion was, as one of Comte's best Latin American disciples, Barreda had also some positive reflections of Catholic theology, which influenced Mexican positivists beyond his time. In the Mexican journal *La Revista Positiva*, at the turn of the century, the editor lengthily quoted Barreda's ideas on theology in a 1904 editorial titled *La Síntesis de Augusto Comte*: "Positivist philosophy abhors neither theologians nor theology. . . . She [positivist philosophy] sees them [theologians] as slow in the march of humanity and seeks to pave and light the way of progress and emancipation . . . ; she is the only one that has offered [to theology] full and impartial justice. The positivist philosophy . . . is the only one that has acknowledged and proclaimed the immense services that theology has offered to humanity."[76] These lines are reminiscent of Scalabrini's ideas about Catholic theology and its services to humanity in its progressive development toward the positive scientific stage.

Positivist ideas in Brazil were also shaped by political and sociological considerations, in particular for their rejection of slavery from the Brazilian empire as a step toward the progress of the country. Perhaps the most renowned Brazilian positivists were philosophers Miguel Lemos (1854–1917) and Raimundo Teixeira Mendes (1855–1927), both of whom became acquainted with Comte's ideas in Paris. These brothers-in-law, following the teachings of Comte's successor, Pierre Laffitte, founded the still-standing Brazilian Positivist Church in Rio de Janeiro in 1881. This church became the foremost apostolate for the religion of humanity in Latin America, promoting the abolition of slavery and the unity of all human beings under the positivist religion. Their influence can still be seen today in Brazil's flag including Comte's motto, "order and progress" (credited to Teixeira Mendes). Many of those involved in the coup against the monarchy that proclaimed the Brazilian republic in 1889 were positivist followers of Lemos and Mendes.

Lemos and Mendes were deeply involved in debates about monogenism and polygenism, crucial for Brazilian positivists. They adhered to monogenism, since they interpreted it as promoting the abolition of slavery (which ultimately happened by 1888). A prominent defender of monogenism, and one of the most influential naturalists of his time, was Ladislau Netto (1838–1894), director of the Brazilian National Museum between 1876 and 1893 and self-proclaimed Darwinian. In a lecture offered in Buenos Aires at the Argentina Scientific Society in 1882, Netto expressed his belief that Charles Darwin's theory, even if feared by those who still embraced scholastic thought (i.e., Catholics), did not conflict with religion at all: "[Darwin's] new doctrine . . . offers the most eloquent evidence of its [reason's] perfectibility, because its influence has been felt in every branch of human knowledge . . . without arising any conflict with religion."[77] In fact, on that occasion Netto spoke of Darwin's reading of the *book of nature*, from which he "will bring the fulgurant irradiation of the rays of truth . . . which no learned spirit can, in good faith, reject."[78] The theory of evolution was so brightly true that it could not contradict the ideas of a God creator. As Brazilian historian of science Regina C. Ellero Gualtieri describes, even by 1878 Netto was of the idea that God oriented the transformative processes of life, holding that it was not necessary to relinquish the belief in a creator orienting the process of evolution.[79] The influence and reception in Argentina of Netto's lecture and ideas on the theory of evolution and its relation to religious ideas is still a matter of research, but it is certainly due a mention given the rarity of his views at the time in Argentina.

Chilean positivists thinkers at the turn of the century also shared some of Scalabrini's considerations, in particular those about the positivist religion and its relation to scientific thought. Juan Enrique Lagarrigue, the founder with his brother Jorge of the Chilean Positivist Church in 1883 (which would later become the Chilean Positivist Society), wrote a letter to Scalabrini's disciple Alfredo Ferreira in 1900, in which he expressed some of these views: "[Positivism] is still often mistaken with simple science, the sterility of which disgusts the swift hearts. Saint Paul's admirable sentence will never cease to be precise: science inflates but charity edifies. Still, positivism does not exclude science, it rather subordinates it to love. Conceived in that way, she is always fully religious."[80] There were close connections between positivists on both sides of the Andes, sharing the idea that positivist philosophy was to be distinguished from modern science, for the fact that positivism included the impulse of growing in virtue.

CONCLUSION

Positivism in Argentina of a more scientific ilk, such as that of Florentino Ameghino, predominantly followed Spencerian ideas of progress and conceived religion as something of the past, which was to be left behind as science continued its path ahead. Positivists more concerned with educating the nation, like Scalabrini and Ferreira, followed Comte's thought more closely and professed the necessity of a rational religion of humanity, which was the ultimate goal of anthropological evolution. Still, most Argentine positivists shared the basic understanding that modern science, particularly under the lens of evolutionary thought, was the right path toward the development of a progressive society, away from revealed religions and metaphysical deliberations. Perhaps the only exception was that of Ferreira's claim that positivism was not atheistic, though more research needs to be done to place this assertion within his thought as a whole.

Catholic responses, at least in the voices of Estrada and Pizarro, were adamant in their rejection of positivist ideas, for their ideals both of historical progress and of an evolutionary basis of morality. For them, forgetting the teaching of traditional (Christian) philosophy so as to leave space for a scientific society was a mark not of progress but of regress in the development of human history. This rejection, however, did not imply a rejection of science per se, but rather suggested a rejection of the materialist lens through which modern science was considered.

What is clear in the Argentine case is that these debates between positivist scientists and educators and Catholic advocates were a proxy for greater discussions that were taking place in the country—in particular, discussions about what kind of education the Argentine society required: an education based on a scientific curriculum, which would, for some positivist thinkers, lead to the growing of the virtues, particularly that of charity; or an education based on the humanities and the classics, which would not neglect the sciences but would place them in a subordinate relation to philosophy and theology, as the Catholic party expected. Ultimately, the national identity of the country was at stake: a secular society or a society based on the Catholic tradition of its people.

As a preliminary conclusion, conflict narratives in public discourses about science and religion acted as proxies for other conflictive issues, for instance, regarding the authority to decide the type of education for the Argentine youth. These issues involved the change toward a secular education and a separation of the secular state and church. The conflict narrative to which historians of Argentine thought often allude becomes, thus, insufficient to understand what was going on during the final decades of the nineteenth century in Argentina. In fact, when actors are left to speak for themselves and transcend these greater issues, such as in the case of Ameghino or Estrada, one finds hints of different understandings of the relations between science and religion. Ameghino, evoking ideas present in Asa Gray's advocacy for Darwin in the US debates, explored analogies between the way astronomy related harmoniously with scripture and how evolutionary biology could relate in a similar manner to theological truths, or even what it meant for an omnipotent God to create episodically. Estrada, on his part, argued that a non-ideological science, disrobed from principles that were alien to it, had no quarrel with the Christian religion.

More research needs to be pursued for other Latin American countries such as Mexico, Brazil, or Chile in order to understand fully positivist positions regarding the relationship between science and religion. In Mexico, Barreda's rejection of religion is clearly not due to his understanding of evolutionary science, while in Chile, Lagarrigue's emphasis on the religion of humanity shows a lesser appreciation of science as such than his positivist colleagues in other countries. Finally, Ladislau Netto's statements that Darwin's theory does not oppose religion and that God does guide the evolutionary process require due consideration against the positivist majority trend in Brazil and Argentina.

7

Prehistoric and Primeval Pasts

Antique Chronology and Civilizational Progress in Semicolonial Egypt

Meira Gold

The emergence of prehistory in the mid-nineteenth century profoundly changed the British perception of the human past. The historicization of humans was engrossed in the Victorian sciences of deep and shallow time, evolutionism, and race. Human antiquity also blurred allegedly contradictory scientific and religious explanations for human origins. Charles Lyell (1797–1875) and his followers fought to reconcile geological evidence with Genesis, while many others upheld the traditional Judeo-Christian time frame of six thousand years.[1] These "conflicting chronologies," as Lyell referred to the debates, extended into the contentious realm of disciplinary authority. Practitioners from the natural and human sciences, including geologists, paleontologists, archaeologists, ethnologists, philologists, anthropologists, and astronomers, offered competing conceptions for the age and development of human civilization. Colonial encounters and scientific surveys throughout the British empire further diversified understandings of remote human history, utilizing new data, materials, and bodies.[2]

With unprecedented British imperial presence in Egypt from the 1840s, Egyptology similarly became a "border discipline among others that took shape in the crucible of empire."[3] Colonial scientists and chronologists alike sought to substantiate their claims in the Nile valley. The alluvial basins of Egypt and Mesopotamia were increasingly regarded as primeval lands containing a unique combination of textual, ethnological, geological, and archaeological evidences for human prehistory. The Orientalist search for the "cradle of [Western] civilization" in the Middle East, as in the Indian subcontinent, was therefore closely tied to Victorian evolution and religion.[4] The special status of pharaonic Egypt in the biblical imagination grew exponentially during the nineteenth century. Yet British Egyptology only fueled disagreements over chronological calculations and the reliability of different textual sources for studying the past, as the subject was not yet distinct from prehistory.[5] Human antiquity moreover became entangled with Orientalist propaganda that claimed pharaonic civilization as ancestors of the "West" while promoting a sharp opposition with present-day Egypt as part of the degenerate Muslim world, or the "East." The supposed decline of Islamic civilization became a crucial tenet of the colonial platform, legitimized by the belief that modern Egyptians were antipathetic to the relics of the country's ancient past.[6] Taking a transnational and pluralistic approach to the growth of prehistory, this chapter shifts focus to the semicolonial Egyptian context and addresses how local writers negotiated these debates.

Egyptian writers in the 1860s took up questions of human antiquity in novel ways. Christian and Muslim evolutionists turned to arguments for the "social" progress of the human races through material and technical advancements, adapting the writings of Charles Darwin and Herbert Spencer to legitimate political reform. Civilizational decline and progress, just like evolution, were seen as universal laws of nature.[7] Conversely, Muslim primordial chronologists focused on the antediluvian period between Adam's fall and the flood. Biblical, classical, and Qur'anic sources were scoured in the literary tradition of *tarikh*, the developmental sequencing of remote historical events.[8] The poet Mustafa Salama al-Naggari (d. 1870) mocked chronologists' disputed calculations: "There is no reliable text and no correct reasoning concerning the age of the world. This is why their theories differ so much and their works contradict each other."[9] Muslim technocrats trained in Europe in the first half of the century simultaneously became aware of the moralizing power of Egyptology. The intellectuals Rifa'a al-Tahtawi (1801–1873) and Ali Mubarak (1823–1893), among others, produced novel interpretations of Egypt's primeval past by appropriating European scholarship and reconciling it with Arabic literature. The two authors promoted the prestige of ancient Egyptian civilization but opposed the Orientalist notion of Islamic decline. Many writers of the *Nahda*, or Arabic renaissance, drew a progressive chronology from ancient to modern Egypt—notwithstanding the early Ottoman conquest. They promoted a new Egyptian modernity that shared ancestry with the pharaohs, crediting the Egyptian Khedivate for reviving the scientific-technological achievements of the ancients.[10]

In some cases, Egyptian counter-tempos were forms of anti-colonial resistance.[11] Temporality was also negotiated with local sociopolitical-religious agendas. As Daniel Stolz shows, late-Ottoman Egypt was a "chronologically plural society" where several religious communities marked the passage of time according to their own calendars. Muslim communities simultaneously used the Islamic lunar calendar (*hijri*) to date historical texts and observe holidays, the thirteen-month Coptic solar calendar to track agricultural time, and the Nile's annual inundations, which dictated the three seasons of flood-based agriculture.[12] Traditional al-Azhar astronomers tactically converted multiple chronologies and calendars, while younger Franco-oriented Khedival astronomer-surveyors like Mahmud al-Falaki (1815–1885) joined colleague Mubarak in promoting pharaonic sciences. Consequently, the relationship between the Ottoman Egyptian government, Egyptian scientists, and agriculturalists in this semicolonial context cannot simply be reduced to "a binary of imperialism and resistance."[13]

Negotiating the various Egyptian and European approaches to human antiquity was the Europeanized, Catholic, Turkish-Armenian, Cairo-based civil engineer and reformer Joseph Hekekyan (1807–1876). (See Figure 7.1.) Hekekyan was an unabashed Anglophile who spent his early scientific education in Britain before entering Ottoman Egyptian governmental service. He gradually embraced the British mantras of industrial progress, free trade, liberalism and freedom of worship, moral education, abolition of slavery, and antiquities conservation.[14] His 1850s geo-archaeological excavations for Lyell's father-in-law, the Scottish geologist Leonard Horner (1785–1864), extended human civilization in Egypt to over 13,000 years.[15] Building on this research, Hekekyan remotely published a dense book for British audiences titled *A Treatise on the Chronology of Siriadic Monuments* (1863) with one main goal: "to throw open a new field for chronological and historical inquiry."[16]

Hekekyan consolidated European theories of primeval measurements in Egyptian monuments while incorporating hermetic teachings into his chronological sequences. Hermeticism was the philosophical and religious system built around the antediluvian divine sage Hermes Trismegistus, a Hellenistic-era amalgamation of two gods of wisdom, Thoth and Hermes. Hermetic lore located the birth of science in the primeval Nile valley, preserved through the mystical Egyptian priesthood. It flourished through the Ptolemaic Egyptian priest Manetho (third century BCE) and numerous Greco-Roman writers before shaping Islamic scholarship and European Orientalism.[17] Steeped in Egyptian Freemasonry, Hekekyan's book demonstrated the advanced metrological knowledge preserved in pharaonic monuments and their universal compatibility with Greek, Hebrew, Arabic, and global chronologies. The monuments were material evidence of the scientific ingenuity of "the remotest generations of mankind," and he argued that the ancients who built them were far more skilled than their modern descendants.[18] Unlike his Muslim contemporaries, Hekekyan believed the superior ancient Egyptian race had directly "passed to Europe the sacred fire of science and civilization."[19] To extend his personal prestige in Britain, his treatise drew equally on the Victorian human antiquity debates and questions about primeval chronology and civilizational progress in Egypt.

Siriadic Monuments was a unique local reflection on the problems of remote Egyptian antiquity that arose around midcentury—a topic that has received little scholarly attention from historians of science despite a growing interest in deep history. Pratik Chakrabarti has argued that the nineteenth-century sciences of prehistory, archaeology, geology, and evolution need to be better aligned with histories of imperialism. Historians must tackle Eurocentric universalizing tendencies by pushing for multiple, overlapping, and coexisting narratives.[20] While in Europe the distinction between history and prehistory was crystallizing through disciplinary specialties, these components of antiquity overlapped in colonial India, where there was an "essential interplay" between geology, ethnology, and mythology.[21] The nineteenth-century search for prehistory was heterogeneous, as was the meaning of empire itself, as Stolz has noted for Ottoman territories. Late-Ottoman science was characterized by competition among several factions of the elite, while non-elites practiced emerging sciences "as a means of resistance to state power."[22]

Hekekyan brought a distinctively composite approach to human prehistory precisely because he was an elite member of a minority faction in Egypt who used his scientific and multilingual skill set to his own advantage. It has been previously noted that Hekekyan's ethnicity, education, and religion alienated him from the evolving bureaucratic elite in his adopted Egypt. He did not fit neatly into any of the dominant communities of power and influence.[23] His isolation has led historians to largely exclude Hekekyan from accounts of Egyptian Egyptologists in the imperial era.[24] However, it is precisely Hekekyan's hybridity that makes him a compelling case study. Rather than disqualify his research activities as un-Egyptian, it is more salient to illuminate his role as a cultural mediator, who straddled multiple audiences as well as scientific and textual practices. From this perspective (and despite their respective Catholic and Protestant faiths), Hekekyan is comparable to Christian Syrian émigrés Yaqub Sarruf, Faris Nimr, and Shahin Makarius, who campaigned for religious and scientific tolerance in their Cairo-based periodical *Al-Muqtataf* while authoring tracts on Freemasonry.[25] The emergence of Egyptian prehistory

Figure 7.1: Photograph of Joseph Hekekyan Bey (1807–1875), date unknown. From Ardashes Kardashian, *Niuter Ekibdosi Hayots badmoutyan hamar* [Materials for the history of the Armenians of Egypt], vol. 2 (Venice: Armenian Publishing House of St. Lazarus, 1986), 241. Courtesy of AGBU Nubar Library, Paris.

must be considered alongside these transnational figures, who straddled literary genres as much as political, scientific, and religious loyalties. Not only was Hekekyan an important go-between linking British, Egyptian, and Ottoman actors, but his body of work also provides an opportunity to better integrate histories of science, archaeology, and Middle Eastern studies.[26]

Building particularly on Marwa Elshakry's and Stolz's explorations of science in semicolonial Egypt, I argue that nineteenth-century Egyptian treatises on primeval chronology were crucial landscapes for debating human origins. These proto-pharaonic histories were tied to urgent debates about evolutionary progress, decline, and stagnation vis-à-vis the rise and fall of civilizations, particularly Islam and Europe. Hekekyan became an intermediary between the entangled nascent sciences of Egyptology and prehistory and a new Egyptian modernity that sought a direct lineage to the grandeur of ancient Egyptian civilization. While Hekekyan's research was directly taken up by European geologists, chronologists, and pyramidologists, his book was also in tacit conversation with Muslim contemporaries who offered Arabic-language alternatives to European notions of "prehistory." Their antique chronologies, often linked to evolutionary themes, had profound implications for Christian and Islamic beliefs. Yet Hekekyan failed to be accepted in either elite community due to biases on both sides. This chapter situates Hekekyan as a scientific minority in his Egyptian context to demonstrate the plurality of agendas across the colonial spectrum that shaped prehistory. In the process, I illuminate the shared transnational, religious, and scientific contexts for human antiquity at midcentury and the political tensions that fueled debates.

AN ARMENIAN ANGLOPHILE IN CAIRO

Nineteenth-century Egypt was unique as the locus of multiple, overlapping empires.[27] An influx of European capital exploited the country's strategic trade location, cotton cash crops, coal, cheap labor, and long history. The Ottoman Egyptian government under Muhammad Ali Pasha was granted administrative autonomy from the Ottoman Porte after the Second Turko-Egyptian War (1839–1841), in exchange for being recognized as the hereditary pasha of Egypt. British merchants flooded the country as Egypt effectively became a producer of long-staple cotton for the British textile industry. By midcentury, British and French engineers were employed throughout the Ottoman Egyptian bureaucracy, giving them an advantage in the administration of antiquities and public works, including the overland trade route to India and the Suez Canal. The viceroys increasingly borrowed money from European investors for Egypt's infrastructural development, until Ismail Pasha announced the country's bankruptcy in 1876. Yet the Egyptian Khedivate remained Ottoman in terms of loyalties and can be regarded as "colonized colonizers."[28] Under the oppression of multiple colonial encroachments, many Muslim Egyptian elites rose to cultural, political, and scientific prestige. Their sympathies varied as well. The Egyptian nationalist movement encompassed a variety of agendas whereby "anti-colonialist" could mean anti-Ottoman, French, British, or any such combination.

Hekekyan was one of several Europeanized Armenians who were educated abroad and recruited into the Ottoman Egyptian bureaucracy during the reign of Muhammad Ali. They were multilingual and held high-ranking government positions. They worked as translators, in education, and in foreign affairs and commerce. Their close ties with European consuls and fluency in European languages made them valuable mediators. They were an ethnic-minority faction—Armenian authority was due not to popularity but to viceregal patronage. To extend their personal power and influence in the 1830s through the 1850s, Armenian officials strategically advocated on behalf of European, rather than Turkish or Egyptian, interests.[29] These technocrats tended nonetheless to remain loyal to the viceroy, reaching their peak political influence in the 1880s.[30] In the decades preceding the Egyptian crisis known as the ʻUrabi revolt (1879–1882) and the British occupation, Armenian Egyptian technocrats straddled sociopolitical interests in advantageous ways.

Hekekyan was born in Istanbul and raised Armenian Catholic. His father was among the first generation of Armenians recruited to the viceregal court by Muhammad Ali. In 1818, at the age of ten, Hekekyan was sent to complete his preparatory training in Britain at Clapham Academy and the Catholic Stonyhurst College. The pasha covered his educational expenses with the expectation that he would come to Egypt to apply his expertise to the new modernizing projects underway. During his first seven years in Britain, Hekekyan was exposed to subjects as broad as natural philosophy, chemistry, geology, minerology, arithmetic, geometry, geography, architecture, English and French, handwriting, and drawing. He undertook an additional five-year apprenticeship in theoretical and practical mechanics at Bramah's Engineering Factory in London. Hekekyan studied all the subjects necessary for a career in civil engineering in Egypt: steam engines, machinery, hydraulics, surveying, and irrigation. He toured cotton mills in the industrial north, noted the mismanagement of the working class, and attended lectures in mechanics institutes. His student notebooks are filled with sketches of nature, canals and bridges, and sections of pumps and steam engines.[31] These experiences had an enduring impact on him.

Upon his arrival in Cairo in 1830, young Hekekyan was the only Armenian technocrat in Muhammad Ali's court to know both English and Arabic, making him a valuable translator. He was also fluent in Armenian, French, Latin, Greek, and Persian, quickly reacquainted himself with his native Turkish tongue, and later added Italian and German. The viceroy took advantage of Hekekyan's expertise in civil engineering, and he was quickly put to work in industrial and educational reform. Hekekyan was appointed supervisor of several cotton mills, oversaw the construction of roads, bridges, buildings, and model villages, led coal-mining expeditions, and briefly served as president of the board of health.[32]

As an educationalist, Hekekyan trained students from the Qasr al-Ayni engineering college in geometry and mechanics and was tasked with managing the engineering schools in Bulaq and al-Qanatir al-Khayriyya. From 1834 he served as co-director of the Polytechnic Institute in Cairo, or Muhandiskhanah, along with the Saint-Simonist Charles Lambert, and designed courses in hydraulics and machine theory.[33] He left in 1837 to establish the School of Arts and Crafts, the Madrasa al-Amaliyat. In 1849, Hekekyan was sent to assess student progress at the Egyptian Military School in Paris. He suggested several reforms to the students' integration in Parisian life and extended his proposal to the reorganization of primary education in Egypt: "European civilization and power press upon us from all sides; sooner or later it will be necessary to open our barriers to them, or they will force the barriers open."[34] The proposal reflected his anglicized vision of an Egyptian modernity through the moral and disciplinary education of the next generation of leaders. Yet Hekekyan despised the condescending Christian missionary attitude in England and argued that future Egyptian administrators required a firm basis in the history and faith of Islam.[35]

Hekekyan's later career, from 1851 to 1875, came during a period of sociopolitical instability in Egypt. This period saw, among other things, the "cotton boom," a debt crisis, and contentious changes to Khedival autocracy. The Armenian influence had waned, as native Egyptian technocrats began to replace their once-privileged expertise. Scholarship on the Muslim Egyptian technocrats during this period has highlighted the ways Western science was adapted to fit their agendas.[36] Hekekyan's moral attitudes toward science and religion provide important additional insights into the changing sociopolitical milieu of this period, during which scientific practitioners were forced to choose between (or juggle) loyalties toward the Egyptian Khedive, European officials, and anti-colonial intellectuals.

Hekekyan was dismissed by Abbas Pasha from government service in 1851 and was placed on a quarter-pay pension. His retirement was supposedly due to his chronic ophthalmia. But the decision was likely prompted by his vocal opposition to Abbas Pasha's autocratic power. Growing tensions between Armenian bureaucrats intensified after Hekekyan's relative Artin Bey was exiled from Egypt. Worried for his family's safety, Hekekyan wrote to the former British consul Samuel Briggs asking for "British protection."[37] His focus on Egyptian human antiquity began after receiving an appropriately timed letter from the Scottish geologist Leonard Horner. The letter inquired whether Hekekyan might direct excavations on Horner's behalf to determine the annual sedimentary increase of the Nile at sites of archaeological interest.

Their long-distance collaboration reached its peak between 1851 and 1858. Hekekyan directed four years of excavations mainly at the ancient sites of Heliopolis and Memphis. The cost was offset by funds from both the Royal Society of London and Abbas Pasha (r. 1848–1854). Their work resulted in two publications by Horner for the Royal Society's *Philosophical Transactions*, the latter of which claimed that "civilized" humans had lived in Egypt for 13,371 years.[38] Horner's publication provoked a variety of responses in Britain, ranging from geological enthusiasm and Egyptological skepticism to outright rejection by biblical apologists. Their collaboration highlights Egypt's important role in the human antiquity debates of the late 1850s through early 1860s. It was the first time that geological stratigraphy was adapted to archaeology—by Hekekyan no less—and left Hekekyan with several important supporters in England, including Leonard and Anne Horner, Charles and Mary Lyell, and the other Horner daughters.[39]

Hekekyan's fieldwork for Horner lasted only until 1854, at which point he escalated his own investigation into Egypt's Nilometers and other records of the Nile's inundations in antiquity. This side project ended in 1858 when Said Pasha cut off his funding. Hekekyan speculated that this had much to do with him being an Armenian technocrat, a status that previously allowed him European protection and privileged access to Muhammad Ali and, to a lesser extent, Abbas Pasha. Internal squabbles among the Armenians in Said Pasha's court, followed by Ismail Pasha's Franco allegiances, left Hekekyan more isolated from his Ottoman Egyptian colleagues than previously.

GEOLOGY, CHRONOLOGY, OR SPECULATION?

In 1858, Hekekyan sent Horner a "little cahier of 48 pages" on the Roda Nilometer. He envisioned this "preliminary chronological study" would initiate "a new field of inquiry."[40] The cahier was ultimately not included in Horner's 1858 manuscript and instead formed the basis of Hekekyan's separate "archaeological" memoir. The reason for its exclusion was that Hekekyan's drafts increasingly focused on Freemasonry, which Horner deemed unsuitable for his scientific audience. In one journal entry, Hekekyan recalled an encounter with a local sheikh guarding the Nilometer in 1851, who provided "suspicious" calculations: "I could have wished the Sheikh to have afforded me further information, but he declined to comply with my request, affirming that he was engaged by a sacred promise, like all his predecessors and ancestors, to reveal the secrets of the Nilometer to no one but to that of the reigning caliph."[41] These Muslim "hereditary Nile observers" adopted multiple scales of measurement "in order that its construction should baffle every attempt to penetrate into the mysteries of the Nilometer." Hekekyan substantiated this claim by referring to Herodotus, who wrote that the first Pharaoh Menes and his Egyptian priests "turned their attention to knowledge of the heavenly bodies and *their configurations*." In case "their science should anytime be lost among men, and *what they had previously acquired* should perish," they constructed several monuments and engraved "upon each of them their discourses" to preserve their knowledge.[42]

Assertions of secrecy permeated Hekekyan's description of the Heliopolis obelisk as well. "The obelisk is the philosopher's stone importing golden information," he wrote, adding that the ruling pharaoh ordered the manipulation of its data so that no one unworthy could decipher its valuable information.[43] Hekekyan speculated that only "civilized" nations could comprehend and appreciate this knowledge. The oldest monuments predated the first dynasty, he thought, and could only have been constructed with an advanced knowledge of geology. "But geology as a science can only be appreciated by nations who have obtained a high state of wealth and civilization," so it may be inferred that the ancient Egyptians "had attained to a certain state civilization and the culture of science—not inferior to ours of the present day."[44]

Horner replied diplomatically that he could not include Hekekyan's conclusions in his paper because it would be limited to geology. Hekekyan's chronological memoir "should be a separate jurisdiction," because "it belongs more to archaeology than to geology proper." Hekekyan, Horner believed, did not fully comprehend the expectations of geologists in the Royal Society. To excite British geologists, "whose subject has now expanded over so vast a surface," and yet "remains within its fixed narrow limits, one must condense, avoid wandering the alluring flowering fields of speculation that has not the firm basis of facts rigidly belonging to the science."[45] Indeed, Horner's papers focused mainly on the geological sedimentation in the Nile valley. It is interesting that Horner believed his discussion of human antiquity was safely in the domain of geology, even though his main argument revolved around fragments of ancient pottery and burnt bricks. Horner did not consider this to be "archaeological"—by which he likely meant chronological and speculative.

Horner further dismissed Hekekyan's "chronological speculations" on the grounds that geology had nothing to do with Bible history. "The most recent period in the geological history of Egypt is immeasurably remote from all the legends of Genesis." He added, "You may find something in the monuments of Egypt that may be able to throw light on the latter in Genesis about a deluge—but the Nile was flowing and depositing mud probably for thousands of years before the existence of the human race."[46] Horner agreed to publish Hekekyan's archaeological research at a later date, but he warned Hekekyan that he did not expect he would do his chronology justice, "for in truth there are sections of them I cannot understand."[47]

As consolation, Horner proposed to edit and oversee the publication of

Hekekyan's forthcoming memoir in Britain. He offered to be Hekekyan's "medium of communication" to the Syro-Egyptian Society or the Royal Society in London. By 1861, Hekekyan's treatise was nearly complete. He declined Horner's offer, explaining, "I always felt it extremely necessary that I should be present in person to explain it." Hekekyan campaigned to be the Egyptian commissioner for the 1862 Universal Exposition and come to London at the pasha's expense, though he lost the position to Auguste Mariette, the newly appointed French director of the Antiquities Service. Hekekyan nonetheless visited London with his family in July 1862 for six weeks, meeting the Horners and Lyells. He brought with him a copy of his memoir in the hopes that it might be printed in London while he was there.

His book ultimately took two more years to publish because it was incomprehensible. Horner secured the editing services of a Mr. Foster and had Taylor and Francis print, bind, and distribute the book. Foster was delayed, as he too found "difficulties both with regards to discrepancies in [the] figures and from not understanding part of [the] nomenclature."[48] He complained that correcting the proofs had been "very heavy."[49] Finally, 250 copies were printed in October 1863. Hekekyan gave Horner a list of the people and institutions to whom he wished a copy to be sent. Horner sent 177 copies to individuals in Britain and ten copies "to most public bodies," while Taylor and Francis retained just five copies for their own dispersal. Horner further sent thirty-four copies to Hekekyan, one of which Hekekyan personally presented to Said Pasha. Thirty-four additional copies remained in London for future distribution.[50]

Despite Horner's characterization of Hekekyan's work as non-geological, the engineer's geological methods were informed exclusively by Horner and Lyell. Hekekyan received a copy of *Principle of Geology* in 1857 and told Horner it was "the only work I have on the science." He took Lyell's "sections on the Nile" as his guide to stratigraphy.[51] The irony of their collaboration was that Horner and Hekekyan initially shared similar objectives in Nile sedimentation; yet they ended up on opposite sides of the same chronological debate. Horner used geo-archaeological evidence to argue against biblical chronology, a conclusion that Lyell initially supported in *Geological Evidence for the Antiquity of Man* (1863). Hekekyan conversely concluded, "The discoveries made in recent times need not shake our faith in the chronology of the Bible. The Bible chronology is now demonstrated to be perfectly correct and to be believed."[52]

THE LOST SCIENCE OF SIRIADIC MONUMENTS

Siriadic Monuments provided an esoteric interpretation of Egyptian antiquity that made Horner's accusations of "chronological speculation" understandable. Hekekyan coined the term "astrogeology" to refer to the ancient

Egyptian application of geology and astronomy to Nile irrigation. He called this astrogeological science *khemy*, "the science of sciences," after *kmt*, the hieroglyphic term for black, fertile soil and for Egypt itself.[53] He argued that *khemy* was used to construct the oldest hydrological monuments in Egypt to navigate the annual Nile inundations. This lost priestly knowledge was preserved in the pharaonic monuments, termed "Siriadic" because their measurements corresponded to the revolutions of the Dog Star, Sirius (Egyptian *Spdt*, Greek *Sothis*). He added that the Egyptian priests were "experts in recording their learning" in a "masonic nature," utilizing the universal language of geometry, which was "capable of being understood by all nations in all times."[54] His most novel argument related hydrology, geology, and astronomy. Analysis of all three together could confirm the exact date a monument was erected. This could be determined by considering its geometrical properties alongside its "hydromasonic depth." The latter was the depth at which each monument was "hidrymatized," namely, constructed at a particular elevation using "nilometric cubits" and adjusted to the maximum Nile flood level at that spot.[55] Using these monuments as fixed points, the priests could analyze whether the Nile was increasing or decreasing—which was paramount to agricultural prosperity.

Hekekyan was particularly inspired by the French engineers of the 1798 Napoleonic expedition, who used scientific instruments to recover what they believed to be the metrological precision preserved in Egyptian monuments. Like the concurrent British imperial surveys in India, these were ambitious Orientalist projects to recover the lost sciences of ancient weights and measures and politically repurpose data toward the legitimization of national standards. It was crucial for colonizers to establish "primeval measurements" that linked "European standards with oriental territories."[56] Heated debates focused on the true value of the ancient Egyptian *cubit*: the basic linear unit defined as the length of the forearm from elbow to extended fingertips, subdivided into *palms*, each equaling four *digits* (fingerwidth).[57] European surveyors disagreed about recovery methods and techniques. Hekekyan evoked Pierre-Simon Girard's 1799 hydraulic measurements of the Elephantine Nilometer, posited to be "a primordially precise hydraulic metrological record." Edme-François Jomard asserted that Egyptian metrology was better understood through astronomical surveys of the Great Pyramid, as the builders were well versed in "practical geometry and practical astronomy."[58] Hekekyan resolved these "cubit wars" by showing that pharaonic monuments were built with geometrical, astronomical, and hydrological knowledge. The savants had also been embroiled in chronological disagreements with Egyptologist Jean-François Champollion over dating the Dendera Zodiac. The two ostensibly incompatible approaches, between the exact sciences and erudition, created

APPENDIX.

A Table of the principal Levels of the Nile in the Roda Observatory.

[Taken from the public Registers.]

[N.B. The Sheik of the Niloscope informed me, in 1852, that the Nile of the previous year rose 24 Cubits 11 Dactyles. He did not explain in a satisfactory way why he had quoted it 9 Dactyles. It is right that he should not divulge the secrets of the Niloscope before the proper time.]		Lowest Ebb of Nile.		The Nile Level in the Summer Solstice.		The level at the opening of the Cairo canal.		The Nile Level in the Salib or Autumnal Equinox.		Highest level reached by the river.	
		C.	D.	C.	D.	C.	D.	C.	D.	C.	D.
L. Horner's researches in Memphis	A.D. 1854.*	6	16	6	17	15	19	23	18	23	23
	1853.*	6	3	6	6	18	11	24	1	24	9
	1852.	7	0	7	13	16	8	20	22	21	8
Heliopolis.	1851.*	6	1	6	4	17	15	23	20	24	9
	1850.	5	11	5	11	14	18	21	6	21	20
	1849.	5	11	5	11	16	4	23	12	24	5
	1848.	...		5	11	16	0	23	10	23	23
Operations of the International Commission to determine the relative levels of the two seas and the Nile.	1847.*	...		5	16	16	7	22	17	22	23
	1846.*	...		6	21	15	21	23	1	23	23
	1845.	...		6	5	16	8	20	6	20	15
Lepsius discovered the Semneh Instrument	1844.*	...		6	7	16	6	22	3	22	3
	1843.	...		7	5	16	0	22	6	22	6
	1842.	...		8	0	16	6	23	14	23	14
	1841.	...		5	14	16	4	23	18	24	0
	1840.	...		7	16	16	10	23	18	23	18
Abdel Medgid Khan ascended the throne, 1 July 1839.	1839.*	...		5	13	16	2	19	18	19	23

Figure 7.2: Table showing the various levels of the Nile by the Roda Observatory between 1839 and 1854. Hekekyan calculated these in cubits (C) and dactyles (D) to correspond to Nile levels measured at other locations. From Joseph Hekekyan, *A Treatise on the Chronology of Siriadic Monuments* (London: Taylor and Francis, 1863), appendix. Courtesy of Cambridge University Library.

long-lasting feuds between "scientific" and "philological" approaches to Egyptology.[59] Hekekyan promoted the utility of textual sources while presenting his calculations as a continuation of the savants' approach, thus showing the limits to purely philologically based chronologies.

Hekekyan also derived data from the 1847 international commission to determine the levels of the Nile relative to the Red and Mediterranean Seas. It was a pet project of the Saint-Simonist Prosper Enfantin, who established the Société d'Etude de l'Isthme de Suez in 1846 to assess the viability of constructing the Suez Canal as a trade route. The surveys were carried out by three groups of European civil engineers and resulted in the 1856 commissioner's report by the French diplomat Ferdinand de Lesseps.[60] Hekekyan appropriated their calculations to confirm his astrogeological theories (see figure 7.2). Egyptian priests, he argued, had also carefully studied the elevation and depression of the country's soil in relation to the Nile and sea levels.

Hekekyan thus drew on new hydrological calculations to revive geodesic claims previously made by French engineers. These monuments "masonified data" commensurable with the dimensions of the earth and Nile levels. However, he coined new metrological terminology for the Nilometric cubit. Hekekyan employed *noctas* and *dactyles* that precisely recorded annual accumulations of Nile silt. Fifteen noctas formed one dactyle, four dactyles formed one palm, and six palms formed one cubit. Noctas were simultaneously spatial and temporal units: they measured silt deposits stratigraphically, and in the process, traced those deposits back through time. Accordingly, "360 noctas make a standard Nile cubit and 360 Nile years make a cubit measure of time."[61] This sacred number corresponded to the Egyptian civil calendar, which consisted of three agricultural seasons organized around the river's annual inundation. Egyptian priests added five epagomenal days to the new year that coincided with the return of Sirius in the night sky.[62] Accounting for the ancient solar and astronomical calendars, Hekekyan showed that symbolic numbers guided the dimensions and placement of ancient Egyptian monuments for the benefit of future generations.

The location of the Great Pyramid, for example, was carefully chosen in relation to the Roda Nilometer, or the "Memphis Astrogeological Observatory," which Hekekyan believed was the oldest monument in the world. The summit of the Pyramid was hidrymatized in 4863 BCE, exactly 118,140 noctas above the highest Nile level that year (see figure 7.3). The Pyramid "masonifies geographical and geodesical records of great historical value" and, like all other pharaonic monuments, was built to withstand the ravages of time. Hekekyan predicted that like geological fossils, the Pyramid "will again emerge to light in a state as perfect as that of the diminutive seashell now found embedded in the summits of high mountain ranges." He added that the Pyramid "will in proper time reveal to the human race the science of their constructors, and teach them useful lessons for their guidance, and should they have forgotten it, teach them the existence of a Great Maker of the makers of these curious monuments."[63]

Siriadic monuments provided compelling evidence of primordial wisdom, and moreover, the metrological recovery of this lost astrogeological knowledge supplied the means to accurately date pharaonic history. Once Hekekyan fixed the date of a monument according to scientific calculations, he argued, it

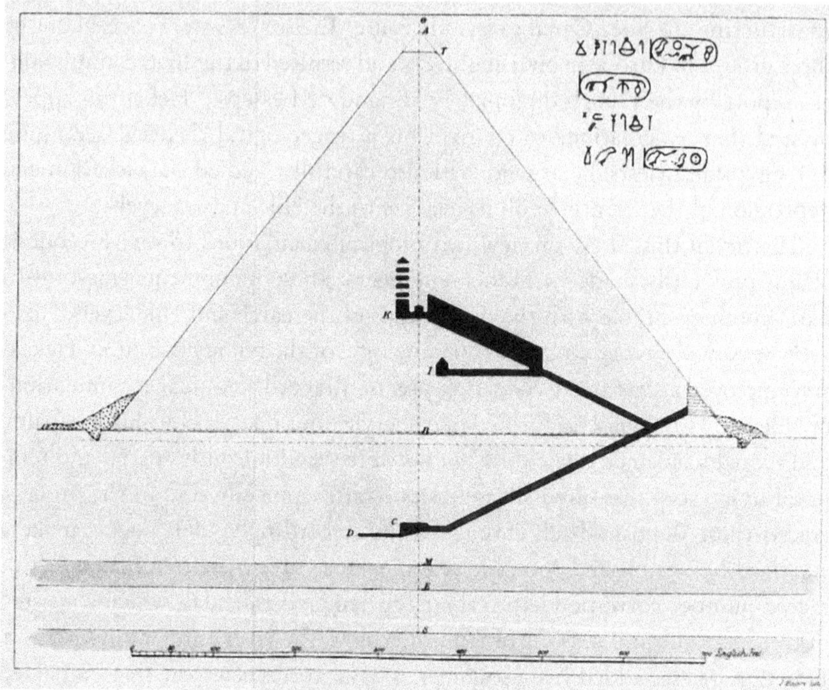

Figure 7.3: Hekekyan's sketch of the Great Pyramid showing its "hydrogeological elevation," or the coordinates of its base and height in relation to the Nile and sea levels. The measurements were commensurable with "Sothic periods." From Joseph Hekekyan, *A Treatise on the Chronology of Siriadic Monuments* (London: Taylor and Francis, 1863), foldout between pp. 48–49. Courtesy of Cambridge University Library.

could substantiate Manetho's *Aegyptiaca*, the book responsible for chronologically reconstructing periods of pharaonic rule into royal dynasties. Manetho was the head "Nile observer and recorder" who supervised the irrigation of canals, oversaw all matters of Egyptian agriculture, and maintained the mysteries of the Nilometers through secret numbers.[64] Hekekyan argued that Manetho's pharaonic dynasties were precise records of astrogeological Nile observations. Manethonean Nile-lists dated back to the primeval "Dynasty of the Demigods" and corresponded to the prediluvian genealogies of Genesis. Hekekyan's book thus found a primeval origin for numerous metrological sciences while harmonizing ancient Egyptian, Hebrew, Babylonian, Greek, Arabic, and Hermetic sources.

This reconciliation of textual sources was important in an era when chronologists were analyzing ancient calendrical systems and converting them to modern timekeeping standards. European Egyptologists resolved discrepancies between the ancient agricultural seasons, the solar and lunar cycles, and the "sothic cycles," which were widely acknowledged to be periods of 1,460 years commencing with the heliacal rising of Sirius. British Egyptologist Reginald Stuart Poole, for example, turned to astronomical data supplied by George Airy to fix the dates when the ancient Egyptians had recorded their observations of the star.[65] Hekekyan confirmed that Masonic measurements corresponded exactly to the movements of Sirius, because the start of the Sothic cycle indicated the annual inundation of the Nile. While at times disjointed, his comparisons of lunar years to solar years, Sothic periods to geometric measurements, and Nilotic cubits to noctas illustrate his stake in these debates.

Notably, *Siriadic Monuments* discredited Muslim Egyptian chronologists who saw the superiority of ancient Egyptian civilization progressing into the Islamic period. The modern Egyptian "has forgotten the science and the language of his forefathers who built the pyramids."[66] The Egyptian priests created a hydrological system precisely because they had foreseen that "it was possible for nations to fall down from the pinnacle of civilization to the lowest depths of barbarism." Egyptian inferiority was not due to any universal law of development or decline, as "nations have no infancy, manhood, old age, nor decrepitude." Rather, the monuments teach us that nations can "flourish in ever-increasing strength so long as they do not deviate from the path of progress."[67]

The exception were the secretive Nile observers, the Muslim descendants of the ancient Egyptian priesthood. He identified them as Sabians (*sabi'un*)—a mysterious pre-Islamic religious community mentioned three times in the Qur'an and identified, alongside monotheistic Christians and Jews as *ahl al-Kitab* ("people of the book"). Their identity puzzled Muslim and Orientalist scholars. Early and Medieval Islamic exegetes conflicted over whether they were magicians, survivors of an ancient Jewish-Christian sect, or a pagan community of star worshippers from Harran.[68] Drawing on this confusion, Hekekyan claimed astrogeology could "be understood only by Sabeans and those who are initiated in the mysteries of the monumental sciences." Their Egyptian descendants "belonged to the same great confraternity of the Sabean school who during thousands of years have received the sacred torch of science from the hands of their predecessors."[69]

Hekekyan therefore disparaged the Arabic hermetic texts that informed the primordial chronologies of his Muslim contemporaries. "The absurd fables and impossible combinations . . . having their origin in a science applied to the uses of common life, are merely the confused images and echoes of realities which have been bequeathed to us by our ancestors." He was specifically referring to Islamic alchemists who regarded Hermes Trismegistus as the founder of their science and sought out treasures, hidden spells, and ancient temples that contained magical secrets.[70] Hekekyan's use of "to us by our ancestors" is particularly telling. He dissociated himself from Muslim colleagues but presented himself as ethnically "Egyptian" for the benefit of his British readers, thus deliberately self-fashioning an ambiguous identity. This choice may have been informed from the poor reception of Horner's paper. Several reviewers objected, on racial grounds, to the reliability of Hekekyan and the Egyptian laborers in his charge.[71] His treatise sought to correct this misidentification by demonstrating his pro-British sympathies, while also positioning himself as a native spokesperson for Egyptian civilization.

To this end, his book employed numerous strategies designed for his British readership. He produced numerical tables with measurements of the Nile's elevation in ancient Egyptian cubits, English inches, and French meters. Elsewhere he compared the lengths of Egyptian dynasties to the reigns of English monarchs and, significantly, the Christian civilizing mission to the ancient Egyptian priests. Benevolent Westerners who sought to enlighten "benighted nations" were a "manifestation of the same religion which made the Egyptians of the pyramidic epoch feel it to be their duty to masonify, for the benefit of those who were to come after them . . . the knowledge accumulated in their monumental history."[72] An important conclusion can be drawn from his equivalence of ancient Egyptian wisdom with European superiority: He was inserting himself into this Orientalist lineage.

This comparison was particularly evident in an imaginary didactic dialogue prefixed to his book, set in Heliopolis ca. 500 BCE. This was the standard literary convention of the Greek Hermetica. A dialogue between master (usually Thoth-Hermes) and disciple normally inverted the wisdom of the dominant Greek culture with that of the indigenous Egyptian priesthood.[73] The conversation involved an Egyptian priest and a Greek student who traveled to Egypt to study astrogeology.[74] The Egyptian priest represented Manetho, to whom Hekekyan likened himself. The priest tells the student about an excavation in Memphis forty years earlier that "was found studded with remnants of the work of man" and "an abundance of baked clay fragments" dating back 8,280 years. Hekekyan was directly referring to his excavations with Horner in the 1850s. The student lamented that he does not "know how these things will be received in Athens, where it is the general belief that the world itself was created only 4,004 years before the fall of Troy"—a reference to James Ussher's creation date of 4004 BCE.[75] He was alluding to an anachronistic conflict between supporters of human antiquity and defenders of scriptural chronology. The priest's system was superior as it was based on Lyellian observable scientific phenomena. "We hold that system the most rational which, confining us to the knowledge of the present as a starting point, allows us to descend to the past, with the aid of history and monumental guides and traditions, as far as we can." He added, "We believe that the universe and the human race were created; but we assign no time to the Creator—no date to his works."[76] Hekekyan presented an apologist primordial history, allowing for creationism while moving beyond the traditional biblical time frame.

Hekekyan simultaneously drew on genres of writing that would have evaded British readers. His dialogue was also in the Islamic tradition of *'Ilm al-Kalam*, which included "conversation, discussion, controversy, and defense" as a literary tool. *Kalam* often took the form of dialectical disputations between various schools of thought. For contemporary Muslim writers, *kalam* was "intended to meet the challenges to faith posed by the modern sciences," frequently taking the form of religious apologia.[77] Hekekyan used the genre to defend primordial chronology against scientific debates in Britain. But because of his negative experiences studying in England, and despite his Catholicism, Hekekyan too believed that faith in Islam, not Christian civilizing, was the key to modernizing Egypt. His text therefore highlights the degree to which he was utilizing Islamic theorizing to make cases for both Judeo-Christian and Muslim religiosity.

EGYPTIAN EGYPTOLOGY AND PRIMEVAL CHRONOLOGY AT MIDCENTURY

Hekekyan's motivations make sense when situated among the growing body of Egyptian intellectuals who studied ancient Egypt to facilitate their political motives. There was a range of chronological sources available to these nineteenth-century writers. Most utilized biblical and classical sources but relied more heavily on the Qur'an and medieval Arabic literature—genres less accessible to their Western counterparts. Until the early nineteenth century, Muslim writers often described their hostility toward the pharaohs; ancient Egypt was merely the pre-Islamic period of ignorance, polytheism, and idolatry.[78] The *Nahda* writers collectively sought to correct this belief by pointing to the superiority of ancient Egyptian science and its commensurability with Islamic beliefs. There was in fact a great deal of Islamic literature on ancient Egypt that assessed antiquities positively as confirmation of God's transcendence, as objects for moral education and admiration, and as evidence of pharaonic magical wisdom.[79] Particularly influential was the Arabic hermetic tradition, which identified Hermes Trismegistus with the Qur'anic Idris and the biblical prophet Enoch. The Egyptian Hermes, known as Hermes the Copt (Al-Ya'quibi) was a post-diluvian traveling philosopher perceived to be the source and preserver of scientific knowledge. Corresponding legends by Ṣa'id (d. 1068), Abd al-Latif al-Baghdadi (d. 1231/2), Jamal al-Din al-Idrisi (ca. 1238), and Al-Maqrizi (d. 1422), among others, marveled at the mystical and technical knowledge preserved in the Giza Pyramids.[80]

The 1860s witnessed a growth in Arabic-language books on Egyptology, most commissioned by Ismail Pasha for the technical education of young Egyptian bureaucrats. The Muslim writer Abdallah Abu al-Suud translated Auguste Mariette's textbook on ancient Egyptian history in 1865 and his Arabic guidebook of the new museum in Bulaq in 1869. Both translations served the *Nahda* by teaching Egyptians "about how their forefathers lived," thus inserting a new lineage between the pharaonic past and modern Egypt.[81] Even more influential were the prolific writings and translations of Rifa'a al-Tahtawi.[82] His 1838 book *Bidayat al-Qudama* (*The Origin of the Ancients*) was the first history of ancient Egypt in Arabic, translated from mainly French sources. He is often praised as a founder of the *Nahda* and responsible for spreading Egyptian awareness and interest in Egyptology.

Al-Tahtawi was a native Muslim Egyptian, educated at al-Azhar before joining the first student mission to Paris in 1826. He and Hekekyan were called to Egypt within one year of each other and appointed to leading positions in educational reform, al-Tahtawi as director of the new School of Languages in Ezbekiyeh and the Translation Bureau in 1835. They worked together on several projects, including Muhammad Ali's abortive national museum in 1835.[83] Both fell out of favor with Abbas Pasha. Al-Tahtawi was exiled to the Sudan from 1850 to 1854, and upon his return, their career paths diverged, as al-Tahtawi was favored by both Said and Ismail Pasha. Yet both produced primordial chronologies of Egypt in the 1860s, albeit for very different audiences.

Al-Tahtawi produced simultaneously "sacred and secular histories" through a novel combination of sources.[84] *Bidayat* summarized Arabic and European scholarship on ancient civilizations, beginning with a Bible-based primordial chronology of Adam and his descendants and proceeding to Egyptian chronology and Greek mythology.[85] In 1868, al-Tahtawi published *Anwar tawfiq al jalil fi akhbar Misr* (Glorious light on the story of Egypt), an encyclopedic survey covering the pharaonic period to the Islamic conquest, drawing on Egyptological scholarship while still allowing for hermetic lore.[86] *Anwar* began with lengthy discussions of the Nile, its annual floods, nilometers, and agriculture, before proceeding chronologically through Manetho's dynasties. He associated Noah's grandson Misrayim with the pharaoh Menes, thus moving his chronology further away from strictly biblical and Qur'anic sources and aligning it with classical accounts. *Siriadic Monuments* and *Anwar* were similar in their assumptions about the religion of the ancient Egyptian priesthood. Al-Tahtawi similarly identified the ancient Egyptians as Sabians, which was an interpretation designed for his Muslim readers.[87]

Al-Tahtawi's book praised Khedive Ismail and justified his reign as "the renewer of the Islamic community."[88] It directly influenced Sheikh Mustafa

Salama al-Naggari's treatise *The Garden of Ismail's Praise* in the late 1860s, another Arabic primordial chronology, which fused "hybrid fragments of Muslim, Christian and Judaic narratives."[89] Al-Naggari, as Adam Mestyan argues, represented an alternative *Nahda* to al-Tahtawi. Though both straddled several Islamic literary genres, including the chronological tradition of *tarikh* and *'aga'ib* (wonders), which discussed various legends and tales about magical buildings, sculptures, and peoples.[90]

These three texts written in the 1860s—by Hekekyan, al-Tahtawi, and al-Naggari—shared a focus on pre-flood chronography. Their books were also, to varying degrees, apologist treatises for Bible-based primeval history. More significantly, all three writers faced the problem of disputed calculations out of Europe. The geological evidence for human antiquity was disruptive for all Bible-based chronologists, whether Judaic, Christian, or Muslim. In Egypt, primordial chronologies became common "hermeneutic devices" to fill in the undocumented gap between creation and the flood.[91] By writing entirely for a British audience and omitting medieval literature, Hekekyan was in more direct conversation with the Victorian debates. His chronology offered a complementary thesis of Islamic decline by expressing a direct lineage from ancient Egypt to ancient Greece to Western modernity.

The proliferation of Arabic-language primordial chronologies should have also been familiar to European Egyptologists through the mixed membership of several institutions that emerged in the 1860s and 1870s. Some existed previously—Hekekyan co-founded the Société Égyptienne in Cairo in 1836, which became a rendezvous for European travelers.[92] L'Institut Égyptien was established in Alexandria in 1859. Membership was dominated by French residents, though several Egyptians and Armenian Egyptians were founders.[93] The institute operated in proximity with the Egyptian Antiquities Service (est. 1858), the Egyptian Museum in Bulaq (opened 1863), and the Khedival Geographical Society in Cairo (est. 1875). These institutions facilitated collaborations between Muslim Egyptian and European practitioners.[94] Hekekyan was not alone in engaging Western-dominated scholarship through these venues.

HEKEKYAN AND SYMBOLIC PYRAMIDOLOGY

Reviewers in British periodicals were baffled by Hekekyan's book. The *Athenaeum* noted it was "meant only for the initiated," though even learned Freemasons were warned about "its peculiar phraseology" and "the new words employed." The main problem was that Hekekyan wrote "too much from the standpoint of a civil engineer."[95] The *Westminster Review* was more condescending: "With one statement in it which we feel pretty certain we do understand, we as certainly disagree." His approach was dubbed "pseudo-erudition" and would do nothing "to revive our languishing faith in the reality of Egyptological science."[96]

Hekekyan never publicly confirmed that he was a Freemason, although this was not unusual for membership in the clandestine lodges. Indeed, local Egyptian Freemasonry grew substantially from the 1840s through the 1870s as Muslim Egyptians joined the fraternal lodges originally established by Europeans. Freemasonry played a prominent role in the establishment of the so-called Egyptian "National Party" from 1879.[97] Masonic lodges became "common meeting-grounds" for the Egyptian elite, serving as an unofficial political platform before the heyday of mass print culture. They used these ready-made structures to disseminate radical views, gain favor with European diplomats, and demonstrate their allegiance to the Khedive, who was often the honorary (if not actual) Grand Master of the National Grand Orient of Egypt (est. 1870s).[98] Confirmed members of the Anglo-influenced Kawkab al-Sharq (Star of the East) in Cairo included the political reformers Jamal al-Din al-Afghani and his protégé Muhammad Abduh, who joined the Masonic movement for "political and social purposes."[99] Other Freemasons included the Christian Syrian émigrés Faris Nimr and Shahin Makarius, editors of the leading Cairo-based Arabic newspaper *Al-Muqtataf*.[100]

These secret societies played equally important roles in evolutionary and religious debate, particularly between pharaonic civilization and Islam. Freemasons believed in a monotheistic God as the "Grand Architect of the Universe," alongside the study of the past through symbolic stonework and secret rituals. While Freemasonry was allegedly not a replacement for religion, the Masonic stance on liberalism and religious tolerance likely made the lodges more attractive to initiates Hekekyan, al-Afghani, and Abduh—all of whom stressed the importance of religious unity. Although Hekekyan was Catholic, he emphasized commonality: "All religions have fundamental principles of morality in no manner differing from one another."[101] For both Afghani and Abduh, religion was the "primary motor of civilizational progress."[102] Abduh argued that science and Islam were uniquely compatible because of the latter's emphasis on independent reasoning and liberal free thought. Eastern civilizational progress would be found through the unification of science and religion, and like Hekekyan, Abduh believed that pedagogical reform in Egypt must include instruction in both. The evidence for the harmony between science and religion was found in the Qur'an: "Men are created from one soul, which is humanity . . . and all men are brothers in humanity," adding "which is why it matters little if they claim their father is Adam or a monkey or something else."[103] Abduh's evolutionary history of religions proposed it was possible to reconcile all the followers of revelation. Freemasonry was likely an attractive platform for Abduh as a religiously tolerant institution for bringing together multiple faiths.

Similarly, articles by Nimr and Makarius in *Al-Muqtataf* sought to recontextualize the taint of paganism associated with ancient Egyptian religion for Muslim readers. One 1883 article on polytheism (*shirk*) and monotheism (*tawhid*) stressed that while the ancients believed in a pantheon of gods, the superior Egyptian priests—those forefathers of Freemasonry—were more progressive: "Their wise men were monotheists, believing in one God, the Single Creator of all that is in heaven and on earth."[104] Like Hekekyan and Abduh, who faced similar allegations of pro-British allegiances, the editors of *Al-Muqtataf* saw Masonic sociability as a vehicle for political radicalism to overthrow the Ottoman sultanate. At the same time, their newspaper spread their beliefs in individual liberalism by promoting the evolutionary cosmic "unity of science" in the writings of Herbert Spencer.[105] Egyptian Freemasonry was undoubtedly a social platform for the dissemination of religious and scientific propaganda, in addition to political activism.

Masonic literature commonly saw similarities between followers and the secret initiations of the ancient Egyptians priests and generally believed that the Great Pyramid stored the priests' forgotten knowledge. Symbolic Masonry was familiar to European and Egyptian readers alike through pyramidology. As Hekekyan was negotiating the publication of his book in Britain, the viceregal astronomer Mahmud al-Falaki was also investigating the relationship between the Pyramids and Sirius. The Muslim astronomer-surveyor became a crucial figure among the technocrats in the viceroy's service (reflected in his moniker al-Falaki, "the astronomer").[106] Mahmud overlapped with Hekekyan at the engineering school, which the former attended as a student from 1833. He completed his studies in Paris and was recalled to Egypt in 1859 by Said Pasha to undertake a new astronomical map of Egypt. That year Mahmud also became the lone Egyptian co-founder of the Institut d'Égypte. Hekekyan had by then fallen out of favor with Said Pasha, who encouraged French collaborations with Egyptian scientists. It was therefore Falaki, not Hekekyan, that Said commissioned during the spring equinox of 1862 to "go to the Pyramids to determine its orientation and try to draw some conclusions."[107]

Mahmud published multiple papers for European societies in the 1860s in French, two of which focused on the Great Pyramid and ancient Alexandria.[108] Hekekyan's *Siriadic Monuments* was written in the late 1850s and published in 1863, while Mahmud's *L'Âge et le but des pyramides lus dans Sirius* was written in 1862, published that year as a stand-alone paper and as a monograph in 1865. Hence Mahmud wrote his book after Hekekyan but published it first. Their books were remarkably similar. Both were written for the

European elite. Both drew on the measurements of the Napoleonic engineers and European scholarship on Egyptology and philology. Both supplemented their metrological calculations with chronological and textual analysis. Both focused on the constructions of pharaonic monuments in relation to Sirius. Both authors used their technical surveying skills to rewrite an Orientalist version of ancient Egyptian history that promoted pharaonic civilization as distinct from Ottoman, Arab, or Muslim societies.

Like his Muslim colleagues, Mahmud incorporated Arabic medieval scholarship to make the topic more palatable. He offered a henotheistic interpretation of ancient Egyptian religion, minimizing its pagan associations by elevating only the jackal-headed gods associated with the star Sirius, who had "played the most important role in Egyptian antiquity." Arab lore, which attributed the construction of the Pyramid to Hermes, was therefore correct, and "one can infer that the Pyramidal form was really dedicated to Hermes who is nothing other than Sothis."[109] This orientation aligned Mahmud with the Egyptian historiography that traced its lineage to ancient Egypt. His book "claimed the pyramids as a part of Egyptian history at a time when these most recognizable monuments on the Nile had to be *made* Egyptian."[110] Hekekyan and Mahmud were likely aware of each other's projects, yet neither author cited the other. Citations signaled loyalties to European scientific communities. Hekekyan was consistently pro-British, whereas Mahmud's career flourished under the Franco-oriented reigns of Said and Ismail Pasha.

While Mahmud won favor with the Khedive and French engineers like Jomard, Hekekyan was favored by British pyramid theorists sympathetic to his objectives. Scottish astronomer Charles Piazzi Smyth popularized the geodesic theory that the dimensions of the Great Pyramid were divinely inspired and corresponded to the dimensions of the Earth. Smyth linked these conclusions to the competing moralities of the French and British empires. The pyramid cubit closely resembled Anglo-Protestant measurements and not the atheistic French metric system, legitimizing British control over Egyptian territory.[111] Smyth cited Hekekyan several times in *Life and Work at the Great Pyramid* (1867). He spoke highly of Hekekyan as a "subtle and ingenious theorist, with a good knowledge of mathematics, and of extensive as well as varied and quaint reading." While he admitted that Hekekyan's book did not have scientific support, he was given the benefit of the doubt. Smyth was more condescending to Mahmud, whose work was a failed example of "symbolic astronomy."[112] Smyth chastised him for citing only French astronomers but excluding British pyramid theorists, particularly Howard Vyse's 1837 pyramid survey. Smyth ultimately disregarded the book as suffering from the "tastes, predilections, and perhaps patriotic fancies, rather mistakenly interpreted, of the author" and accused Mahmud of catering his work to please his scientific mentors at the Imperial Observatory in Paris.[113] He was not entirely wrong, but Smyth's allegations of prejudice were nonetheless ironic.

Mahmud's pyramidology influenced the prolific reformer Ali Mubarak, Ismail and Tawfiq Pasha's Minister of Education and Public Works. Mubarak also studied at the Qasr al-'Ayni engineering school and was among the Egyptian students sent on the 1844 mission to Paris. Mubarak's governmental career from 1849 marked an important turning point. He was the first Muslim Egyptian trained in the European sciences to achieve a prominent role in the Ottoman Egyptian bureaucracy that had previously been dominated by Turkish and ethnic-minority elites like Hekekyan.[114] Several of Mubarak's publications stressed the need for forward-looking Egyptians to possess knowledge of the Egyptian past. His monumental *Al-Khitat al-Tawfiqiyya*, written in 1877, was an encyclopedic topography of modern and ancient Egypt. Chapters on ancient sites like Memphis drew equally on European scholarship and the Arabic hermetic interpretation that the Pyramid was purpose-built to preserve knowledge during the biblical flood.

Mubarak's engineering background also led him to take special interest in the measurements and placement of the Pyramid. In *Al-Khitat* and other works, he promoted Jomard's original geodesic hypothesis. The Pyramids were proof that the Egyptians had mastered engineering and geometry, that their system was the basis for all subsequent standards of measurement, and that Egypt was indeed the birthplace of advanced civilization. Mubarak, however, did not share the "divine inspiration" attributed to the ancient builders by the likes of Smyth. He similarly disregarded Hekekyan's thesis of the Pyramid's astrogeological dimensions as an accurate record of primordial chronology. Pharaonic achievements were instead elevated as a benchmark by which to judge the recent Khedival successes in land cultivation and irrigation and as an industrial model for the future progress of the Egyptian state.[115] For Mubarak, as for his colleague Mahmud, cartographical-historical representations of ancient Egypt brought a territorial reorientation that served the technical demands of the modernizing nation: "Egypt as a defined space-with-a-past."[116]

Hekekyan's Masonic sociability provided opportunities to preach his theories beyond the limited circulation of his book. In 1864, he shared "a few conversations" about the Pyramids with A. Dufeu, a longtime resident in Egypt and French member of L'Institut Égyptien. He regarded Hekekyan as "a just and serious appreciator of the seriousness and the high wisdom of the builders of these marvelous buildings."[117] Dufeu dedicated to Hekekyan his book *Quatre Pyramides de Gizeh* (1873), which reproduced the conclusions in *Siriadic Monuments* with additional theories. Unsurprisingly, Smyth wrote a scathing review of the book. "Though the author is on every few pages declaring that his theory is entirely new, and that he is therefore a hero to publish it, I am sorry to say that it is not more new than it is true; for the greater part of it was invented, written, and printed . . . by Hekekyan Bey." Smyth's objections seemed to focus on the author's French identity; his results differed entirely "from those of the English scientific and *sacred* theory." Dufeu's additions to Hekekyan's theory were "precisely the most flagrantly erroneous parts of the whole book."[118]

Hekekyan's arguments were shared more widely among the lay public in Britain and the United States. Late-Victorian occultists revived the belief that ancient Egypt was the primeval source of esoteric wisdom.[119] Pyramid theorizing was popular among theosophical theorists like Helena Blavatsky and Masonic authorities, like George Oliver and Albert Mackey, who claimed Freemasons were lineal descendants of the Egyptian Priesthood.[120] Freemasons shared Hekekyan's opinion that the Great Pyramid was designed for initiation. The British pyramid popularizer James Bronwick contended in his 1877 book that the Pyramid was likely a Masonic lodge: "Its secret passages, its dark solitudes, its mysterious chambers" could have made it "a fitting place for initiations into those sacred mysteries."[121] One anonymous correspondent lectured Smyth, who strongly rejected any relationship between the Pyramid and Freemasonry: "Had you been a Freemason or studied the rites and ceremonies of the ancient mysteries, you would have had no difficulty in understanding what the pyramids were built for." The only certainty, for Freemasons, was that the Great Pyramid was a site of Masonic initiation. "The initiated *easily* understand [these] allusions; the 'profane' do not."[122]

As the turn of the twentieth century approached, Western Egyptologists increasingly followed British archaeologist Flinders Petrie in purifying the Great Pyramid from connotations of hermetic lore, mysticism, and symbolic Masonry.[123] They distanced themselves from the earlier theories of the French engineers and British pyramidologists and wholly ignored the work of Egyptian primordial chronologists, translators, engineers, and astronomers, to rebrand colonial archaeology with scientific legitimacy.[124] Egyptian Egyptologists, conversely, would promote the Great Pyramid as a symbol of the *Nahda* and celebrate the pioneering efforts of Rifa'a al-Tahtawi, Mahmud al-Falaki, and Ali Mubarak in the "pharaonism" movement of the 1920s.[125] Just as he was alienated from scientific debate in his lifetime, Hekekyan too was excluded from both these intellectual lineages.

CONCLUSION

This chapter traces how Joseph Hekekyan mediated between European theorists of human antiquity and Egyptian primordial chronologies and symbolic pyramidologies of the emergent *Nahda*. This more nuanced picture reinforces the necessity of moving beyond the simplistic binary of imperialism and resistance when discussing science in semicolonial Egypt. Hekekyan was a privileged figure in a transnational story whose experiences highlight the pluralistic adaptations and development of prehistory in colonial settings. The episode likewise demonstrates the complex global dimensions of the European human antiquity debates—specifically Egyptian antiquity—and the extent to which they were taken up by diverse writers in both Britain and Egypt. Hekekyan's approach ultimately shared more in common with al-Tahtawi, al-Nagarri, al-Falaki, and Mubarak. These Muslim Egyptian intellectuals wrote to inspire nationalist reformations and rightfully took their places in the historiography of Egyptian Egyptology. Hekekyan, on the other hand, with his British loyalties, fell into relative obscurity.

The Orientalist dichotomy between Western scientific and Eastern text-based approaches to human origins has ensured that "prehistoric" and "primordial" chronologies are typically treated separately. Yet this is a misleading equivalency. Hekekyan and his Egyptian colleagues offered alternative explanations for Egyptian civilization, suggesting the need for historians to rethink the homogenizing divisions between science and ancient history in the nineteenth century and the Eurocentric assumptions they perpetuate. Taking a transnational approach to disentangle these narratives provides insight into the pluriverse of local actors, religious and scientific perspectives, and the politics of constructing the Egyptian past.

8

Evolution, Religion, and Racial Politics in New Zealand, 1814–1930

John Stenhouse

In *Dealing with Darwin* (2014), historian David N. Livingstone argued that "as scientific theory is encountered in different locations, it is translated into different idioms, fashioned into different forms, and harnessed for different purposes." As the *Origin of Species* (1859) circulated around the globe, people read and applied Darwin's theory in ways he could not control. Presbyterian communities in Edinburgh, Toronto, and Princeton, Livingstone showed, gave the *Origin* a more receptive hearing than their co-religionists in Belfast, mainly because there the Irish physicist John Tyndall, a leading Darwinian, aggressively attacked Christian theology in his 1874 presidential address to the British Association for the Advancement of Science. Despite their common Scottish Calvinist heritage, Presbyterians in different cities dealt with Darwin in surprisingly diverse ways. Local contexts shaped what they made evolution mean.[1]

Like most historians of science and religion, Livingstone accepted John Brooke's influential 1991 critique of the two metanarratives dominant since the mid-nineteenth century: the warfare thesis of J.W. Draper and A.D. White depicting science and religion as eternal enemies; and, emerging in reaction to it, a less influential harmony thesis, emphasizing congruence and cooperation. Both general theses, Brooke argued, distorted the complexity of the past by ignoring all the evidence that did not fit.[2] Historian Ronald L. Numbers observed that although Brooke's "turn to complexity" convinced most academic historians of science, it left many "emotionally and intellectually unsatisfied." To find a way beyond the old metanarratives and more recent complexity-oriented micro-studies, he identified as important five mid-scale trends developing from the late eighteenth century. The first, naturalization, saw practitioners of science invoke only natural processes and laws in scientific explanations. Privatization, the second, saw God-talk steadily excluded from science. Secularization—the loss of faith among scientists—constituted a third, though sketchier, pattern. The fourth, globalization, saw public discussion about science and religion spread around the world. Radicalization, the fifth, made extremists on both sides more visible than moderates in between.[3] Applauding Numbers's attempt to simplify complexity, Peter Harrison suggested that by integrating histories of science and religion with general theories of modernization and secularization, historians might be able to construct mid-scale narratives interesting to a wide range of readers.[4]

This chapter, exploring how New Zealanders read Darwin after 1859, aims to integrate Livingstone's emphasis on place with Harrison's on secularization. My site for investigation is the southernmost colony of the British empire from 1814, when the first Anglican missionaries arrived, until about 1930. I begin by sketching the polarized religious, racial, and political context into which Darwin's *Origin of Species* arrived in the early 1860s, the start of the most consequential phase of the New Zealand wars. Fought between British and colonial troops and their Māori allies and several major North Island tribes, these lasted until 1872. A settler-dominated government confiscated millions of acres of Māori land from "friendly" as well as insurgent tribes to crush the "rebellion" and deter any repeat.[5] These events continue to trouble the nation.

Colonial New Zealand constitutes an ideal site in which to combine Livingstone's interest in place with Harrison's in secularization. A brief summary of my main points may help orient readers. Between 1860 and 1890, many, probably most, Pākehā (that is, whites in Māori) journalists, settler politicians, and scientific leaders read the *Origin* to support the government—against Christian critics of both races—by portraying racial conflict, British victory, and the defeat and probable extinction of the "inferior" Māori as normal and natural evolutionary processes. Although a few pressed evolution into the service of atheism or materialism, the most numerous and influential commentators were Protestant laymen, often Anglican, who pressed evolution interpreted theistically into the service of the state. Operating in an overwhelmingly British and Protestant colonial culture, these secular theologians moved not only scientific, political, and intellectual but also moral and religious authority away from the old bicultural and transnational Anglican clerical and missionary establishment and toward the state, the laity, and a bustling settler democracy. Ultimately, lay settlers, politicians, and men of science, not clergy, missionaries, or Christian Māori, decided what kinds of science, religion, and racial politics would dominate the public sphere. The public culture that emerged after the wars was more secular, pragmatic, utilitarian, and hostile to so-called religious "fanatics" trying to "interfere" in public affairs, much as Harrison would lead us to expect.

This chapter writes above the nationalist framework that has dominated much modern historical writing in New Zealand, as elsewhere. Ever since the left-leaning nationalist politician-turned-historian William Pember Reeves published *The Long White Cloud* in 1898, many, probably most, historians of New Zealand have focused on the gradual emergence of an independent, progressive, and democratic nation-state as the natural organizing framework for writing its history. Without denying the important insights this approach has generated, here I take a transnational approach, highlighting how important transnational connections were to many of the individuals and groups discussed below. In a 1999 essay, for example, I argued that New Zealand men of science embraced evolutionary theory more quickly and easily than in settler colonies such as Canada and Australia. While I stand by that claim here, I did not then appreciate just how much transnational links with British Darwinians shaped the receptivity of colonial evolutionists. Darwin and Joseph Dalton Hooker visited during the 1830s and over the following decades enrolled New Zealand men of science such as William Colenso, Walter Mantell, James Hector, William Travers, F.W. Hutton, and Walter Buller into their transnational information networks. Rising to power in British science during the 1860s and 1870s, these British Darwinians provided vital patronage and support for colonial scientific leaders. Embracing evolutionary theory early and enthusiastically, this colony's scientific community was not as self-contained and independent as I once assumed.

MISSIONS, MĀORI CHRISTIANITIES, AND THE MAKING OF A TRANSNATIONAL ANGLICAN ESTABLISHMENT

The first section of this chapter outlines the emergence before 1859 of a bicultural colonial Anglican Church well connected to the English establishment. Historian Tony Ballantyne has rightly argued that the Anglicans of the Church Missionary Society (CMS), arriving in 1814, soon became entangled in Māori worlds. The intertribal musket wars, spreading and intensifying during the 1820s, killed and displaced thousands, while introduced diseases took a further toll. By the time the wars petered out in 1837, roughly a fifth of a total population of 100,000 had died. Around 1830, growing numbers of Māori set out to remain in *te ao mārama* (the world of life and light) by reconfiguring their spiritual lives as inventively as they transformed their material culture by using iron fishhooks, axes, and muskets. Younger chiefs and liberated slaves took the lead in spreading *te rongo pai* (the good news) among communities that had never seen a white missionary. Thousands began attending mission services, reading or listening to *Te Paipera Tapu* (the Māori Bible) and *Te Rawiri* (the Anglican prayer book), freeing slaves, embracing literacy and agriculture, and resolving disputes non-violently. By 1845, at least 40 percent identified as "Mihinare" ("missionary," i.e., Anglican), 15 percent as "weteriana" ("Wesleyan," i.e., Methodist) and 5 percent as "katorika" (Roman Catholic).[6] The CMS, working around the globe, considered New Zealand by far its most successful mission field.[7]

Determined to protect flourishing but young and vulnerable Māori churches, a loose coalition of missionaries, clergy, and humanitarian officials, often Anglican, defended Māori rights and welfare more consistently than any other Pākehā between 1830 and the 1860s. Their racial politics were inspired partly by the leading British authority on the "science of man" (anthropology), the Bristol physician James Cowles Prichard (1786–1848). An Anglican Evangelical, abolitionist, and supporter of missions, Prichard disagreed with those who exaggerated racial distinctions. The Scottish polygenist Lord Kames had argued in *Sketches of the History of Man* (1774) that such distinctions, too deep to have been caused solely by differing environments or cultures, reflected the separate creation of different races in distinct regions of the globe. In *Researches into the Physical History of Mankind*, first published in 1813, Prichard drew on linguistic, archaeological, historical, and anatomical evidence to argue for the unity of humankind, the common origins of races, the relative insignificance of racial differences, and the reliability of a non-literal monogenist (single origin) reading of the Genesis account of the creation of humanity. All humans suffered from the same diseases; all could interbreed and produce fertile offspring; and the myths and legends of different cultures displayed many similarities.[8]

After parliament abolished slavery in 1833, abolitionist politicians led by Thomas Fowell Buxton, an Anglican Evangelical with Quaker connections, turned their attention to Aboriginal peoples in or near British settlements. Taking evidence from colonial officials, traders, naval officers, missionaries, and indigenous informants, Buxton's Select Committee produced a report, written and shaped by Buxton women such as Anna Gurney and Priscilla Buxton, showing that British expansion had visited "calamity on the heathen and savage nations" by introducing disease, drink, violence, and crime.[9] In 1837, Buxton and his Quaker colleague Thomas Hodgkin, a physician friendly with Prichard, founded the Aborigines Protection Society (APS) to disseminate information, mobilize public opinion, and lobby politicians. Dominated by

Quakers and Evangelicals, it had close ties to the Protestant missionary societies.[10] Its motto, "Ab Uno Sanguine" ("Of One Blood"), came from the Apostle Peter's sermon in the Acts of the Apostles (Acts 17:26). Echoing the language of Genesis 2–3, Peter declared that "God hath made of one blood all nations of men." The APS argued that if Britons followed the Golden Rule ("do unto others . . ."), then Aborigines' "existence might be preserved and their situation ameliorated."[11] In a paper, "On the Extinction of the Races of Man," read to the British Association for the Advancement of Science (BAAS) in 1839, Prichard warned that wherever "Europeans have settled, their arrival has been the harbinger of extermination to the native tribes." If colonists in south Africa, Australasia, and North America claimed "the whole of the unpeopled regions of the earth" as "British ground," the annihilation of "aboriginal nations of most parts of the world" would occur unless government acted to prevent it.[12]

Sharing such concerns, and alarmed by the destruction of Tasmanian Aboriginals, the Protestant missionary societies in 1837–1838 opposed the English political economist E.G. Wakefield's attempt to win a parliamentary charter allowing his New Zealand Association (later Company) to colonize. Skeptical of Wakefield's assurances to the contrary, the societies argued that colonization was likely to damage if not destroy Māori society. Furious, Wakefield savaged the missionaries as dogs in the manger, selfishly excluding fellow Britons from their "Theocracy" and "Levitical Republic."[13] In 1839, William Wakefield, Edward's brother, claimed to have bought much of the lower North and upper South Island from local chiefs. The Colonial Office reluctantly stepped in.

In February 1840, CMS missionaries helped negotiate te tiriti o Waitangi (the Treaty of Waitangi) between the Crown and Māori chiefs and translated it into Māori. In return for ceding sovereignty, the English version guaranteed Māori all the rights and privileges of British subjects and "full, exclusive and undisturbed possession of their lands, forests, fisheries and other treasured possessions." The second article required Māori to sell land to the Crown, not private buyers. According to historian Claudia Orange, "almost all the main [Māori] speakers voiced a preference to retain the services of the missionaries and [British Resident] Busby, which they argued were sufficient to Māori needs." CMS missionaries carried the treaty around the country, encouraging over 500 chiefs to sign.[14] In 1842 the CMS urged the Colonial Office to appoint Robert FitzRoy, Darwin's captain on HMS *Beagle*, second governor of New Zealand. Poorly resourced and temperamental, FitzRoy quickly earned a reputation among settler-critics as a missionary-ridden, Māori-loving religious fanatic.[15]

Bishop George Selwyn, educated at Eton and Cambridge, arrived in 1842 as the first (and only) Anglican bishop of New Zealand. Learning to speak

Māori en route, Selwyn embraced the traditional role of Anglican bishops as the conscience of the state. Well connected to England's Anglican establishment, he believed that his church incurred serious political responsibility when its missionaries urged Māori to sign the treaty. Political realism as well as pastoral concern for a biracial flock motivated Selwyn's concern to defend Māori rights. Assertive chiefs, often Mihinare, quickly taught him that they would not put up with serious injustice. He knew that a race war would hurt both peoples and burgeoning Māori churches. From 1842 until he returned to England in 1868, Selwyn led what critics would call the "philo-Māori" or, later, the "Church party." Initially supported by Methodist as well as Anglican missionaries, this coalition regularly defended Māori rights against the government, as the following incident illustrates.[16]

In 1846, Earl Grey, Whig secretary of state for colonies, wrote to the new governor, George Grey, suggesting that Māori had rights only to lands they currently occupied or used. All other areas should be considered "waste lands," which the government could take over and sell. Senior Anglican and Methodist missionaries condemned this proposal as breaching the treaty and betraying Māori and missionaries alike. Selwyn warned Earl Grey that he intended to use "all legal and constitutional measures" in his power to "inform the Natives of New Zealand of their rights and privileges as British subjects, and to assist them in asserting and maintaining them."[17] Writing to Governor Grey, CMS missionary Robert Maunsell placed Māori at the center of the landscape: "England has her unoccupied territories, and no doubt Earl Grey has his unsubdued lands, ranged over by nothing else than the deer or the pheasant. So also has the New Zealander [Māori] his bird preserves, his 'runs' (the grand sources of supply to our colonial markets), his useful timbers, his valuable plants, his fisheries, and localities sacred in his regards as having been the abodes of his forefathers, the scenes of their triumphs, and the resting places of their bones." Missionary protests helped secure the Crown's recognition of Māori title to all of the lands of New Zealand in 1848.[18]

The staunchest supporters of Company colonization, and the strongest critics of philo-Māori Anglicans, were Philosophic Radicals on the secular left of Britain's diverse religio-political spectrum. Embracing the utilitarian philosophy of Jeremy Bentham and James and John Stuart Mill, most Radicals supported Wakefield's vision of systematic colonization to enhance investment opportunities for capitalists, create jobs for workers, relieve Britain of surplus population, and promote the greatest happiness of the greatest number.[19] In an 1845 Commons speech, the Radical MP for Bath, John "Tear 'em" Roebuck (1802–1879), a lawyer friendly with Mill and Chartist leaders, dismissed the treaty as "not worth the paper on which it was written," insisting that Māori "only occupied small pieces of land" and had "no notion of boundaries" and "no traditions respecting the ownership of land." It was "an undoubted fact," said Roebuck, "that all uncivilized races of savages ultimately perished wherever civilized man came." Since "the brown men would entirely disappear" from New Zealand, it was simply "idle to talk of putting them in possession of the soil." In Roebuck's outlook, the doomed Māori, Radical politics, utilitarian philosophy, and the interests of Company colonizers happily coalesced. If Roebuck's "undoubted fact" was true, then Māori posed no obstacle to colonization.[20]

Some Company officials disseminated the "doomed Māori" theory equally eagerly. Born into a middle-class Anglican family, William Fox studied at Oxford before reading law in London. Impressed by Wakefield, he sailed to Auckland in 1842 on a Company ship. He soon fell out with Chief Justice William Martin, Selwyn's friend, for refusing to swear an oath that he had never done anything to make him unfit to be a lawyer. Angry at being unable to practice, Fox went to war against Auckland's government and missionary establishment. Dismissing the treaty as "shallow, flimsy sophistry," Fox claimed that Māori only had rights to those lands they inhabited and cultivated. Succeeding William Wakefield as the Company's principal agent in 1848, Fox sailed to England in 1851 to persuade the Colonial Office to grant full self-government. In *The Six Colonies of New Zealand*, published in London that year, Fox claimed that "statistics compel us" to conclude that "the annihilation of the New Zealanders, as a race, will occur in about fifty years at longest." Spending the colony's entire income on "hospitals" and "medical science" would make no difference because the Māori's "ultimate fate is certain . . . extinction." An active Low Church Anglican, Fox would serve four terms as premier of New Zealand.[21]

The "doomed native" theory, flourishing in and beyond Company circles, galvanized formidable opposition.[22] In 1851, Wiremu Tamihana Tarapipipi Te Waharoa, a powerful Ngāti Haua Christian chief soon to lead the Kingitanga (Māori King movement), attacked the campaign for settler self-government led by Fox and Isaac Featherstone, superintendent of the Wellington provincial government (another "doomed Māori" advocate). Many Māori felt "disturbed by the evil talk of the newspapers of the bitter doctor [Featherstone] and the man of the Pākehās [Fox], that tried to take our lands from us previously," wrote Tamihana. These "bad men" intend "to oppose the Queen" and "govern New Zealand and the Maori people" themselves. Urging the queen "never to give up to the White Men the Maori Men she promised to protect," Tamihana warned that his people would decline as fast as "the Maories of America" if settlers took full control. Māori "would not like to be given into the hands of the bitter Doctor, or his men that wish to possess our lands." British humanitarians reprinted Tamihana's letter, originally published in Wellington, in the *Colonial Intelligencer, or Aborigines' Friend*, an APS publication circulating around the Atlantic. Literate Christian chiefs entered transnational humanitarian networks.[23]

Alarmed by Roebuck's inflammatory rhetoric, in 1846 Selwyn wrote to his old Eton and Cambridge friend William Gladstone, Tory secretary of state for colonies, that he intended as bishop of New Zealand "to take care that none of those damnable crimes, by which the coloured races of other lands have melted away before that pest which he calls civilization, shall be cloaked under the name of a decree of Providence or an inevitable law of nature." Selwyn would leave the "civilized invader" no such "mysterious dispensations" with which to "veil" British "neglect or violence."[24] Edward Shortland (1812?–1893), educated at Harrow and Cambridge, trained in medicine before emigrating in 1841. From mid-1843 he toured the South Island, collecting information about Māori land tenure and demography, forging relationships with chiefs and tohunga (religious experts), and meeting Selwyn.[25] In *The Southern Districts of New Zealand* (1851), Shortland, an Anglican and pioneer anthropologist, noted that "many" Britons considered "the ultimate extinction of the Aboriginal race" a "matter of certainty." This "fallacy" would "mislead the intending colonist, if he consider it a favourable circumstance," and disadvantage "the native race, if it be thereby regarded as merely sojourning for a time on the earth." Shortland could see "no sufficient reason to anticipate the extinction of the Māori race, except by the possible means of its becoming blended with the European stock," a prospect he favored.[26]

Long-standing conflicts between Anglican "Church party" leaders and Christian Māori, on one hand, and Company colonizers and their settler supporters, on the other, peaked after war broke out in Taranaki in 1860. Previous governors had respected Anglican missionary advice not to buy disputed land. But Governor Gore Browne, pressured by Taranaki settlers and politicians, decided to buy a block of flat and fertile land beside the Waitara river offered by a minor Te Ati Awa chief. Ignoring repeated protests from the senior Te Ati Awa chief, Wiremu Kingi Te Rangitake, Browne bought the land. When Te Rangitake, a Mihinare, told his people peaceably to resist government attempts to survey and occupy the Waitara, Browne sent in troops.

Anglican clergy, missionaries, and Christian Māori, often Mihinare, created a crisis in church-state relations by holding the government responsible. In April, Bishop Selwyn wrote to the colonial secretary "respectfully" protesting the government's decision to declare martial law before Māori had taken arms. Invoking Christian just war principles, Selwyn insisted that the government should not have resorted to "military force" against one of "the most loyal, peaceable and industrious of the New Zealand tribes" before exhausting the resources of "civil power" to resolve the dispute. Since Britain annexed the country "for the protection of the New Zealanders," this "primary object" of colonization must not be sacrificed "to the aggrandisement of the English provinces."[27]

Christian Māori far outnumbered Pākehā critics. Almost 500 Otaki Māori, mostly Ngāti Raukawa and Te Ati Awa Mihinare, wrote to the queen that, despite their living for years "under the peace-securing principles of religion" and faithfully observing her "laws," Governor Browne had "unwarrantably" seized the land and then "fired upon" Her Majesty's "loving subjects," who aimed "not to trample upon the law" but simply to keep lands they did not wish to sell.[28] In Hawke's Bay, Renata Kawepo Tamakihikurangi (?–1888), a senior Ngāti Kahungunu chief and Anglican *kaiwhakaako* (missionary-teacher), told a great *hui* (meeting) at the Pawhakaairo that when one member of the Body of Christ suffers, all suffer; Te Ati Awa deserved support from Christians everywhere. Governor Browne forced "hard and uneatable food" on Te Ati Awa by committing "abominable murder" in Taranaki, which no "palliating words and excuses" from the "lick-platter assembly of the governor"—the colonial parliament—could excuse. The queen must remove this "Governor gun-feeder, powder-feeder, ball-feeder" and send one offering "soft and eatable food" such as "runangas [councils] . . . quiet judgements . . . love and charity, and good works." Octavius Hadfield of the CMS brought Kawepo's views to an international audience by reprinting this speech in his *The Second Year of One of England's Little Wars*, published in London.[29]

CHARLES SOUTHWELL, ATHEIST MISSIONARY

The fiercest foe of the government's mostly Anglican critics was Charles Southwell, founder and editor of the *Auckland Examiner and People's Voice* newspaper. He and a colleague had launched Britain's first openly atheistic newspaper, the *Oracle of Reason*, in Bristol in 1840. A penny weekly designed to popularize atheism, the *Oracle* spread French evolutionary materialism, depicting life evolving blindly from below, to discredit the theistic science of the Anglican establishment. Having evolved from matter, wrote Southwell, humans possessed no immortal soul or divine image conferring any "superiority over the brutes" whatever. If Christians considered this strictly naturalistic view of human origins "impious," an "insult to the creator, in whose image they tell us we are made," so much the better; "clerical obstructives" had long rejected "discoveries in science" that undermined their "prejudices." A lurid attack on the Bible as "The Jew Book," both anti-Jewish and anti-Christian, saw Southwell arrested for blasphemy and jailed, as he had hoped. Martyrdom sent *Oracle* sales soaring.[30]

Emigrating to Auckland in 1856, Southwell presented himself, as ever, as liberating the people from Christian oppression. Determined to make New

Zealand great, he advocated exploiting its natural resources, importing "millions" of settlers and developing Māori land to make it the "brightest jewel" in the "British crown."[31] He repeatedly attacked three obstacles to this glittering vision: Anglican missionaries, humanitarian officials, and Māori, particularly Christian Māori. "Humanity-mongers" constituted the "curse of the colony," he thundered in 1857, by urging settlers to "piece up a unity" with "savages" who enjoyed "too much property and too much law." If only England let the settlers govern themselves, they would quickly solve the colony's land and Māori problems.[32]

Southwell sustained transnational freethought networks. Having regularly encountered "savages" in Auckland, he told readers of British freethought leader G.J. Holyoake's *Reasoner* that Māori might "be exterminated, but never can be civilised." Christian chiefs, "self-confessed cannibals" all, were in the religion business purely for "tobacco and blankets." Power-mad missionaries, through "wrong advocacy of native rights," aimed to "win for themselves the undisputed sovereignty of these fair islands" by "graciously" allowing "the Maoris to govern the Government."[33] During the Taranaki crisis, Southwell urged patriotic settlers to capture "rebel" chief Wiremu Kingi "dead or alive." Ridiculing Anglican missionary Richard Taylor's claim that Māori were embracing Christianity and civilization, Southwell retorted that "no Native can ever be civilised" until "in his coffin"—the only good Māori was a dead one. After Selwyn's protest, Southwell urged the government to impose the "full penalty" of the law on "treason-hatching bishops" for their "high crime." Several readers endorsed this call for the government to execute Selwyn. Southwell died of tuberculosis in August 1860, before reading the *Origin*.[34]

His *Examiner* writings, which settlers knew Māori might read, created consternation when Governor Browne sent a particularly violent article, "Blood for Blood," to the Colonial Office. Settler politicians worried that Southwell's article might alarm Māori enough to spark a full-scale race war and convince the British public that most colonists were ruthless warmongers. Debate erupted in the colonial parliament and the House of Lords.[35]

READING THE *ORIGIN* DURING THE WARS

Although Southwell's language was extreme by contemporary standards, many more respectable settlers read the *Origin* in similar ways when it arrived during the early 1860s. Many, probably most, commentators in the settler press read Darwin to naturalize racial conflict, the triumph of the "fitter" British, and the defeat and probable extinction of the Māori. Many also dismissed as unrealistic the old missionary-humanitarian ideal of building a society in which both peoples flourished.

Some Auckland journalists found the *Origin* particularly illuminating.

"Whatever may be thought of Mr. Darwin's views concerning natural selection and the origin of species," wrote Joseph Giles in Auckland's *Southern Monthly Magazine* in 1863, "no one will be disposed to deny the existence of that struggle for life which he describes, or that a weak and ill-furnished race will necessarily have to give way before one which is strong and highly endowed." Declaring the "fitness of things" the "basis of all rights and obligations," he rejected as "fallacious philanthropy" the missionary-humanitarian view that settlers should "save and elevate" the natives. Enlightened colonists should rather "admire" the "natural laws" effecting "the decline and extinction of the inferior race." Applying "the scientific curiosity of zoologists" to "human society" might seem "frigid," Giles acknowledged, but waiving "our just rights in the attempt to save from extinction a declining race" was pointless: "Our war is essentially a war of colonization and justifiable on that ground." An English-born Anglican, physician, and journalist who interpreted the *Origin* theistically, Giles served from 1861 as a military surgeon in the Waikato and Tauranga wars. His attack on the "fallacious philanthropy" of the Church party shows how deeply the wars divided colonial Anglicans.[36]

In Taranaki, the Richmond-Atkinson "mob" of intermarried English Anglican and Unitarian families read Darwin similarly. Arriving in the early 1850s, they struggled to farm rugged, bush-clad areas while Te Ati Awa retained much of the best land, including a flat and fertile block beside the Waitara river. Determined to obtain more land, members pressured one of their own, William Richmond, a lawyer-politician who became minister of Native Affairs from 1858 to 1860. A politically liberal Unitarian, Richmond encouraged Governor Browne to buy the Waitara, staunchly defended him against Māori and Church party critics, and blamed war on Native-loving Anglicans. Richmond's brother-in-law, Arthur Atkinson, enjoyed riding out to the local *pa* (village) at Mataitawa and talking with its inhabitants. After reading the *Origin* in 1862 and joining the Taranaki Volunteers, a settler militia, he now considered lying in wait to shoot Māori "a sort of scientific duty."[37] By 1862, the mob had captured all three Taranaki seats in the House of Representatives. Harry Atkinson served four terms as premier of New Zealand, Arthur became a leading anthropologist, James Richmond became minister of Native Affairs (1866–1869), and William became a Supreme Court judge.

Although most fighting took place in the North Island, southern commentators read the *Origin* much like their northern counterparts. Noting that the "grave has just closed over the last male aboriginal of Tasmania," the *Otago Witness* observed in 1869 that the "Maoris are going too" and they "know it themselves." Settlers were "powerless to save and civilise these moribund races" from that "mysterious law that bids the man of colour perish, when the white man treads upon his soil." In the "struggle for existence" of "Mr Darwin," the writer concluded, "the best men must win."[38] In Canterbury, the conservative Christchurch *Press* claimed in 1862 that if "there is any truth in the new and popular theory of Darwin on the Development of Species, and on Progress by Natural Selection, (which is another name for the old proverb 'the weakest goes to the wall') we need not be surprised at the gradual extinction of the Maori race in the presence of the Anglo-Saxon."[39] The *Lyttelton Times*, more liberal, quoted Darwin's *Journal of Researches* to illustrate how well the *Origin* illuminated New Zealand's natural history: "As the white man's rat has driven away the native rat, as the European fly drives away our own, as the clover kills our fern, so will the Maori disappear before the white man himself." The *Times* identified the principle that the "hardier and subtler vitality of 'selected,'—i.e. civilized—nature" quickly "beats the luxuriance of wild nature" in the "struggle for life" as "entirely in the spirit of Mr Darwin's great work."[40]

DARWIN AND RACIAL CONFLICT IN AUSTRALASIA

Although Darwin avoided discussing human evolution in the *Origin*, he had been including humans in his theorizing since the late 1830s. Writing in 1859 to his friend Charles Lyell, the geologist, Darwin argued that intellectual ability aided individuals and races in the struggle for existence: "The intellectual powers are as important for the welfare of each being as corporeal structure; if so, I can see no difficulty in the most intellectual individuals of a species being continually selected; & the intellect of the new species thus improved, aided probably by effects of inherited mental exercise. I look at this process as now going on with the races of man; the less intellectual races being exterminated."[41] Writing again to Lyell in September 1860, Darwin commented that the "White man is 'improving off the face of the earth' even races nearly his equals." Darwin was referring to the Taranaki war, well covered by British newspapers. This letter sheds no further light on his attitude to the war or on the tone of this remark.[42]

Elaborating his ideas about human evolution in *The Descent of Man* (1871), Darwin devoted a section of the seventh chapter to the probable extinction of "wild" races, including the Māori, in competition with "civilized races." Elsewhere, however, he warned that to check our sympathy for the weak and sick would result in "the deterioration of the noblest part of our nature. . . . If we were intentionally to neglect the weak and helpless, it could only be for a contingent benefit, with an overwhelming present evil." We must "therefore bear the undoubtedly bad effects of the weak surviving and propagating their kind."[43] In private correspondence, by contrast, he sometimes expressed less ambivalence about racial competition. Writing in 1878 to G.A. Gaskell, an early advocate of birth control, Darwin expressed reservations about the impact of "artificial checks" on world history. "Suppose that such checks had been in action during the last 2 or 3 centuries, or even for a shorter time in Britain, what a difference it would have made in the world, when we consider America, Australia[,] New Zealand and S. Africa! No words can exaggerate the importance in my opinion of our colonization for the future history of the world."[44] In 1881, writing to an Irish political economist, Darwin insisted that natural selection did promote progress. "I could show fight on natural selection having done and doing more for the progress of civilization than you seem inclined to admit," he wrote. "The more civilized so-called Caucasian races have beaten the Turkish hollow in the struggle for existence. Looking to the world at no very distant date, what an endless number of the lower races will have been eliminated by the higher civilized races throughout the world."[45] Darwin's biographer, Janet Browne, rightly argued that "direct experience of imperial expansion" in the Southern Hemisphere encouraged Darwin "to see struggle, war and extinction as inescapable truths of nature."[46] The colonial commentators discussed above were not seriously distorting or misinterpreting Darwin's science. As the following section will show, members of a scientific community closely connected to the British Darwinians and the colonial government read Darwin similarly.

THE COLONIAL SCIENTIFIC COMMUNITY

Following the Royal Society of London (founded 1660), the New Zealand Society—the colony's first significant scientific body, founded in 1851—prohibited "the introduction of politics and polemics" as "totally inconsistent with the fundamental principles of a scientific Society" and decreed that "any member guilty of introducing any subject of such tendency shall cease, ipso fact, to be a member of the Society."[47] The New Zealand Institute (NZI), established by government grant in 1867 to promote science, literature, and the arts, was equally opposed to fruitless and divisive wrangling about politics and religion. Addressing a large and distinguished audience at Wellington's Colonial Museum in 1868 as its first president, Governor George Bowen said he considered it one of his "highest pleasures" and "most important duties" to meet "members of all religious communions . . . all social classes, and . . . all political parties, on the common ground of education, science and literature." In sponsoring science with an annual grant of £500, parliament aimed "to provide guidance and aid for the people of New Zealand in subduing and replenishing the earth," which Bowen called "the 'heroic work' of colonization."[48]

The NZI and its affiliated provincial societies were not quite as open, inclusive, and non-partisan as historians who have read such liberal and ecumenical language uncritically have assumed.[49] A larger proportion of colonial politicians were active members, often officeholders, in the NZI and affiliates

Figure 8.1: Wedding party of Dr. James and Maria Georgiana Hector, née Monro, daughter of the Speaker of the House of Representatives. This photo, taken at Nelson in 1868, illustrates the unusually close connections between science and the New Zealand government during Hector's long career as its chief scientific advisor. Box-014–005, Hocken Collections, Uare Taoka o Hākena, University of Otago.

between 1860 and 1890 than since. During the thirty-six years that James Hector, a Scottish-born physician and geologist, managed the NZI (1867–1903), he sustained vital financial and political support for science by forging "links between New Zealand's tiny band of professional men of science and the colonial political elite."[50] (See figure 8.1.) By the mid-1870s, most government ministers, almost half the Legislative Council (upper House) and over a third of the House of Representatives had joined a scientific society. Over half the members of the House of Representatives active in science were farmers or ex-farmers (79), with 33 identifying as large landholders (pastoralists or runholders). Lawyers were the second-largest occupational group (29), followed by surveyors (8), engineers (6), medical doctors (5), and naturalists (3). The colonial scientific community "largely comprised well-educated and well-off gentlemen, in professional occupations and on the land, who pursued science as a leisured vocation, just as they entered politics as a leisured vocation."[51]

Hector—who married Georgiana Monro, the daughter of the Speaker of the House, and botanized with his farmer-politician father-in-law—forged "an unusually close nexus between science and government" as compared with the Australian colonies.[52] In 1865 parliament appointed Hector director of the government Geological Survey, Colonial Museum, and Colonial Laboratory and, in 1867, manager of the NZI and editor of its *Transactions and Proceedings*. Later, he acquired responsibility for the Meteorological Department, the Colonial Observatory and time service, the Wellington Botanic Garden, and the Patent Office Library. Joining the council and senate of the University of New Zealand in 1871, Hector served as its chancellor from 1885 until 1903.[53]

As the most powerful government scientist of the century, Hector deserves attention. Wellington geologists had urged Premier Frederick Whitaker to create the national Geological Survey, which Hector directed from 1865. Whitaker, an Anglican active in Auckland diocesan affairs, led a group of Auckland lawyers, farmers, businessmen, and politicians that contemporaries called the "war party." Members worked tirelessly to extend war from Taranaki to the Waikato and beyond, to keep Britain paying and providing troops, and to confiscate Māori land to promote economic growth, from which they benefited handsomely. They devised some of the most punitive legislation of the 1860s, confiscating huge areas from "friendly" as well as "rebel" tribes. If Hector, a lifelong Anglican churchgoer who interpreted evolution theistically, had supported the government's Church party critics during the early 1860s, he might never have been appointed government geologist in 1865. In fact, he cooperated cordially with war party politicians and the Richmond-Atkinson mob because he shared their disdain for the Church party and its Māori "pets." Writing as war raged in the Waikato, Hector told Joseph Dalton Hooker in 1865 that "there is nothing for it now but to turn a deaf ear to all the Exeter Hall nonsense" of the Church party until "the bumptiousness of the Native has been reduced" by "very harsh treatment." The government must "let the settlers take reprisals by shooting down a few of the Chiefs who are the real instigators" of "massacres by treachery," and then "the Maories [*sic*] soon have freedom." A "war of extermination," though "dreaded by the Exeter Hall folks," would force the natives to "very soon come to terms."[54] He also disliked anti-evolutionist "parsons." In 1870, he told Hooker that his "talking Darwinism" in Wellington had galvanized opposition from the "parsons & old folks here," who "came out before the great change in the public mind had been wrought" by the *Origin*. Hector had reason to complain. Church party "parsons" such as the Reverends Richard Taylor, Robert Maunsell, Octavius Hadfield, and Bishop Abraham dismissed as overly speculative and unproven the evolutionary theories of Robert Chambers and Darwin during the 1860s and 1870s.[55]

One of Hector's staunchest supporters, William Travers (1819–1903), an Anglo-Irish lawyer, politician, and botanist, served as chief parliamentary spokesman for science until 1903. He drew up the statute that established the New Zealand Institute, served on its board of governors until his death, published around forty articles in its *Transactions*, and sent plants and information to Hooker at Kew Gardens. In 1869, Travers gave a series of well-attended public lectures at the Colonial Museum in Wellington that drew on the *Origin*, whose argument he summarized accurately, to explain how the "vigorous races of Europe" were transforming New Zealand. Many native species were "unable to cope" with European "intruders." The large, aggressive European rat had displaced the smaller Māori rat, the "European honey bee now swarms in our forests," and introduced pigs, cattle, and goats were dominating bush and high country. The displacement of indigenous by introduced "races" operated as surely among humans as among plants and animals. "It is indeed a fact, which does not admit of doubt, which is even presented to us as a law of nature,—as a necessity—," said Travers, "that wherever a white race comes into contact with an indigenous dark race on ground suitable to the former, the latter must disappear within a few generations." Even "the most sensitive philanthropist may learn to look with resignation, if not with complacency, on the extinction of a people which, in the past, had accomplished so imperfectly every object of man's being."[56]

According to Livingstone, Travers's lectures illustrated how Wellington scientific leaders read Darwin to endorse "the runaway triumphs of white colonial settlement." Here I highlight rhetorical dimensions of these lectures in order to illustrate Livingstone's point about "speech spaces"—historians must investigate how things are said as well as how they are heard.[57] Unlike in Australia and Canada, whose first scientific professionals and leaders were too committed to pre-Darwinian theistic science to embrace evolution before the 1880s, most of their New Zealand counterparts—Travers, Hector, F.W. Hutton, Julius Haast, Walter Buller, William Colenso, and A.P.W. Thomas—embraced evolution from the early 1860s and interpreted it theistically. Travers, for example, attributed the laws of nature to its "Great Author." Using humanitarian language, he acknowledged that "we have shown too little regard to [Māori] feelings of pride and nationality," ridiculed "their habits and manners," and reduced "the power and influence of the greater chiefs," with "unfortunate results." Although he considered Māori extinction highly likely, he thought "generous statesmanship" might yet allow Māori an "equal share" in the colonists' "progressive civilization."[58]

Social evolutionists in the Wellington Philosophical Society (WPS) who consigned Māori to extinction callously, omitting any religious and humanitarian language, found trouble. Born in Madras into an East India Company family, Alfred K. Newman (1849–1924) grew up on a farming estate in Hawkes Bay. Studying medicine in Scotland and London from the late 1860s, he became a self-styled evolutionary materialist after reading Darwin, Huxley, Tyndall, and the continental medical materialists Carl Vogt and Jacob Moleschott. Returning in 1875, Newman went into business and married Octavia Featherston, daughter of the powerful Wellington politician Isaac Featherston. Eschewing medical practice, Newman became a shareholder and director of the Gear Meat Freezing and Packing Company, which prospered by buying Māori land sent its way by the Darwinian ornithologist Walter Buller, a lawyer who had worked with Featherston in the 1860s to purchase the vast Rangitikei-Manawatū block.[59]

Newman's early scientific papers in the WPS won him a well-deserved reputation as a speculative metaphysical materialist careless about facts. Travers and Hector led a bevy of critics of Newman's overblown claims.[60] Undaunted, the ambitious young physician read "A Study of the Causes Leading to the Extinction of the Maori" to the WPS in 1882. Drawing on his medical training and the breeding studies of Darwin, A.R. Wallace, and Francis Galton, Newman claimed that the race was heading for extinction *before* Europeans arrived. He asserted, without much evidence, that "many" died from cannibalism and that killing infants and old people was "very common," suicide "exceedingly common," and murders "numerous." Endemic diseases and inbreeding, producing widespread "sterility" and "barrenness," took a further toll. Initially characterizing "imported diseases" such as influenza, measles, typhoid, and leprosy as "very powerful agents" of depopulation, Newman later minimized their impact. Condemning "observers not trained in medicine" for exaggerating the toll taken by venereal diseases—a jab at missionaries—Newman claimed that medical doctors, the "only class of men whose opinion is worth taking," confirmed his "astonishment at the smallness of the evil." Having treated a "large number" of Māori, Newman claimed that he failed "to detect any evils" from syphilis.[61] Unlike Travers, Newman depicted the coming of the colonizer as an unmixed blessing. The English offered modern science, medicine, and an "abundance of easily-got food . . . infinitely more nutritious" than a traditional diet. But Māori made the worst of it. Young women, eating much and exercising little, grew fat and lazy, which led, "as in all the lower animals," to "lessened fertility" and, sometimes, "absolute sterility." Breeding with the "more vigorous fertile white race" would not help, because half-castes all "die young." Their offspring, even more "feeble" still, showed that the "effete" and "worn out" Māori was doomed "before we came here." The "disappearance of the race is scarcely subject for much regret," Newman concluded: "They are dying out in a quick, easy way and are being supplanted by a superior race."[62]

Regularly quoting these now notorious lines, prominent historians have depicted Newman as an influential and representative figure, expressing the scientific racism that burgeoned around the empire after 1859. According to

James Belich, Newman said what "many others thought" about the Māori, articulating the "Social Darwinism" that became "quite prevalent" between 1870 and 1900.[63] Malcolm Nicolson characterized Newman's lecture as "the most detailed medical justification of the settlers' position" ever offered, an attempt to deploy the "full authority of modern medical science" in the service of "European dominance."[64] Yet such scholars failed to investigate how colleagues responded. In fact, Newman's speculations, as usual, took a pounding. Dr. Morgan Grace, a Roman Catholic army surgeon, rejected Newman's claim that the race was doomed and predicted—correctly as it turned out—that the population would grow over the next fifty years. Hector attacked Newman's arguments on both empirical and moral-religious grounds. Newman had underestimated the impact of introduced diseases. Even so, the Māori population was declining neither as quickly nor as universally as Newman claimed. Hector knew of Māori couples in the isolated King Country raising "large families" of perfectly "healthy" children. He also rejected Newman's attempt to invoke a mysterious "inherent tendency to decay." We were "really to blame," insisted Hector. In serving "our own purposes," we forced Māori to abandon "old habits and customs," undermined "the authority of the chiefs," and failed to establish effective "European authority" as a substitute. Echoing Travers, Hector could see "no reason why the race should have decayed if it had been left alone, or gradually assimilated." He had "no use," he concluded, for attempts "to excuse ourselves by any other natural law but that of might." Rejecting as unethical the principle that "might makes right," Hector condemned Newman's paper as an empirically shoddy and morally dubious attempt to whitewash the colonists. Newman had transgressed the fuzzy but potent boundary separating acceptable science from unacceptable polemic in WPS speech space.[65]

This experience probably contributed to Newman's decision later that year to join the revolt against Hector's dominance of government science led by younger men such as F.W. Hutton and G.M. Thomson. But their *New Zealand Journal of Science*, a rival to Hector's *TPNZI* (*Transactions and Proceedings of the New Zealand Institute*), lasted only two years.[66] Learning from his mistakes, Newman carefully avoided overt anti-Māori racism during an almost forty-year parliamentary career beginning in 1884. But he continued to advocate excluding every other non-white race. Hector's criticisms may have helped Newman's career. He was elected president of the WPS in 1885 and became mayor of Wellington in 1909. His family prospered on what had been Māori land, owning several farms around the lower North Island and a successful thoroughbred stud.[67]

Walter Buller (1838–1906) became the colony's leading ornithologist during the 1860s partly by exploiting his connections with Hector and Governor

Grey. Born into a Methodist missionary family, Buller, a fluent Māori speaker, worked from 1855 as a Native interpreter in the Magistrate's Court. Invited to edit the government newspaper *Te Karere Maori* (*The Maori Messenger*), he joined most Methodist missionaries in supporting the government against Church party Anglicans over the Taranaki war. Writing in *Te Karere* in November 1860, Buller warned the "rebels" that their race, "already fast declining, will surely, if exposed to the ravages of war, ere long cease to exist, and the land on which the warrior has spilt his blood will pass into the hands of strangers."[68] From 1862, as resident magistrate in the Manawatū, Buller helped Featherston, Wellington superintendent, buy the vast Manawatū-Rangitikei block over Ngāti Raukawa protests. Traveling to London to work for Featherston, now New Zealand agent-general, Buller published *A History of the Birds of New Zealand* (1873), which won him wide acclaim and the awarding of a CMG in 1875. With Darwin's support, Buller was elected a Fellow of the Royal Society (FRS) in 1879. After qualifying as a barrister, Buller returned to New Zealand in 1874 to specialize in legal work in the Native Land Court, which facilitated the transfer of Māori land to settlers by individualizing native title. Buller sent land to the Gear Meat Company, whose profits enabled Newman, Featherston's son-in-law, to prosper. In parliament, Newman defended Buller's purchase of the disputed Horowhenua block against Māori and Pākehā critics.[69] In an 1884 presidential address to the WPS, Buller, a self-styled "disciple of Darwin," accepted "the fact that the Maori race was dying out very rapidly." Echoing Darwin's *Journal of Researches*, Buller told members that he had "often heard Maoris themselves speculate on their speedy extinction, saying in a melancholy way, that as the Norwegian had destroyed the native rat, and as the indigenous birds and shrubs were being supplanted by the introduced ones, so surely would the Maori disappear before the pakeha." With their having declined from 100,000 Māori in Captain Cook's time to no more than 30,000 today, Buller insisted there could be "no doubt whatever" that "the race was doomed": what "happened in other parts of the world . . . must inevitably happen . . . here." That the "aboriginal race must in time give place to a . . . more civilized one" seemed "to be one of the inscrutable laws of Nature." If "true to our watchword of progress," Buller could "scarcely believe that even the most earnest Philo-Maori would deplore the change." He concluded by recalling what Featherston told him in 1856: "The Maoris . . . are dying out and nothing can save them. Our plain duty as good compassionate colonists is to *smooth down their dying pillow*. Then history will have nothing to reproach us with."[70]

Again, some disagreed. G.M. Waterhouse—son of the general superintendent of Wesleyan Missions in Australia and Polynesia, a Wairarapa runholder, and seventh premier of New Zealand (1872–1873)—questioned Buller's "decided . . . opinion" and, appealing to government census data, suggested that Māori numbers were "rather on the increase." Judge Robert Hart, a member of the Legislative Council, agreed that "the general notion of rapid decrease among the Maoris was premature" because official statistics "pointed to the opposite conclusion." By contrast, Dr. Newman "entirely agreed" with Buller. So too, more surprisingly perhaps, did Hector. He "entirely concurred in the views put forward by Dr Buller" and dismissed the census data appealed to by Waterhouse and Hart as "quite delusive."[71]

Why did Hector, the most powerful government scientist of the century, endorse Buller's formulation of the "doomed Māori" in the WPS in 1884 when he had excoriated Newman's version just two years earlier? Several differences between the two men and their papers' language and contents seem relevant. First, Newman consigned Māori to extinction callously, without "regret." Buller, by contrast, took care to use humanitarian language, urging "good, compassionate colonists" to "smooth down" their "dying pillow." As Livingstone would lead us to expect, the poetics of social evolutionary discourse mattered. Second, Buller and Hector were both FRSs well connected to each other, to powerful British Darwinians, and to colonial politicians such as Travers, Monro, and the Richmond-Atkinson mob. Newman had lesser credentials, connections, credibility, and clout. Finally, Newman pressed science into the service of metaphysical materialism, sometimes in sloppy and speculative ways that made him an easy mark. Buller and Hector, by contrast, were active Anglicans who interpreted evolution theistically, as God's method of creation, and rarely offered critics an easy target.

MĀORI LEADERS CONTEST THE "DOOMED MĀORI"

Māori leaders, refusing to lie down and die, rejected the "doomed Māori," continuing the earlier skeptical tradition of Tamihana, Selwyn, and Shortland. James Carroll (Timi Kara), the son of an Irish father and Ngāti Kahungunu mother, identified as Māori. Entering parliament as member for Eastern Māori in 1887, in 1891 he dismissed "the fashion to regard the Native race as one rapidly becoming extinct" as "sentimental nonsense" that settler governments had long "indulged in" to rationalize "obstructive legislation." Rejecting the Featherston-Buller notion that parliament should simply "smooth down the dying pillow," Carroll argued, on the basis of "careful observation," that "it is a mistaken theory to assume that the Native race will rapidly decrease."[72] Similarly, Hoani Taipua Te Puna-i-rangiriri of Ngāti Raukawa, member for Western Māori between 1884 and 1893, told parliament in 1892, "I do not look upon Dr Newman as an authority at all in Native matters."[73]

From the 1890s, a group of talented high-born Māori men educated at Te Aute College in southern Hawkes Bay rejected the "doomed Māori" in theory and practice alike. Te Aute was founded in 1854 by CMS missionaries and Mihinare committed to equipping Māori for leadership in church, state, and society. Te Aute leaders, calling themselves the Young Māori Party from 1909, embraced social evolutionary thinking without accepting the additional assumption that their race was doomed. In 1892, for example, Āpirana Ngata (1874–1950) of Ngāti Porou, a strong Mihinare, warned that the principle that "the weaker [race] must die out before the stronger," well under way in Australia and North America, threatened Māori with "extinction" unless they took action. Unlike Newman, Ngata identified the "advent of the pakeha," bearing guns, alcohol, tobacco, violence, and disease, as largely responsible. Criticizing superstition, vice, and ignorance among both peoples, Ngata urged leaders to act: "Only let the chiefs . . . rescue the race." Māori must stop mixing with "low, vulgar, uneducated, degenerate pakeha," because "intercourse, vice and immorality" between this "lamp-post adorning class" and the "Maori maiden" had "destroyed the purity of the race, . . . stunted . . . its physique," and "degraded" traditional virtues of "hospitality, liberality, bravery and manliness." The "pigmy race" resulting from such unions constituted "a degenerate cross between the pakeha and the Maori, inheriting the worst qualities of both."[74]

After graduating from Te Aute, Ngata studied political science and law at Canterbury College, graduating LLB in 1896.[75] Working with Carroll, Ngata entered parliament in 1905, working hard to improve Māori health, farming, landholding, and prosperity. He campaigned, successfully, for the appointment of the first Māori Anglican bishop, helped revise the Māori Bible, and spearheaded a renaissance of traditional arts and crafts. But he did not want Māori breeding with Asians. In a 1934 parliamentary speech urging more government support for Māori landholding, he quoted the Māori proverb "divorce the Maori from the land and you make of him a slave." To buttress this demand, he said, "We do not want Maori slaves wandering about New Zealand mating with the Chinese and the Hindus—people who are alright in their own country, but, in white New Zealand and white Australia, are undesirable."[76] From the late 1920s, Ngata supported the campaign of Auckland's Te Ākarana Māori Association to prevent liaisons between young Māori women and Chinese market gardeners and Indian fruit-and-vegetable sellers in and around the city.[77]

Peter Buck (Te Rangi Hiroa, 1877?–1951), of Ngāti Mutunga and Irish descent, valued both of his ancestries. After a brilliant career at Te Aute, he entered the University of Otago Medical School. Founded by Free Church Presbyterians who embraced the Scottish tradition of the democratic intellect, the university admitted women and men of all classes and races from 1871. Tall, handsome, witty, and well-spoken, Buck became the national long jump champion before graduating in medicine in 1904 and earning an MD

Figure 8.2: Sir Peter Henry Buck (Te Rangi Hiroa), 1877?–1951. Of mixed Irish and Māori ancestry, Buck identified as Māori. After graduating in medicine from the University of Otago, he forged a distinguished career as a politician, soldier, medical administrator, professional anthropologist, and public intellectual. MS-1689/001, University of Otago Medical Library: Historical Collection. Hocken Collections, Uare Taoka o Hākena, University of Otago.

in 1910 with a thesis titled "Medicine amongst the Maoris in Ancient and Modern Times." He worked closely with fellow physician Maui Pōmare to reinforce his people's demographic recovery. Pōmare had made the most of transnational church networks to study medicine in Seventh-day Adventist missionary medical institutions in the United States before becoming New Zealand minister of health in 1923, where he improved infant and maternal health among both peoples. Representing Northern Māori in parliament between 1909 and 1914, Buck (see figure 8.2) served as a New Zealand government medical officer in the Cook Islands and Niue. Making himself an authority on Māori and Polynesian anthropology, he became a lifelong friend of Ngata. Receiving a Distinguished Service Order at Gallipoli during the Great War, he succeeded Pōmare as director of the Māori Hygiene Division of the Department of Health in 1921.[78]

In a 1922 paper to the Auckland Institute published under his Māori name, Te Rangi Hiroa criticized previous scientific papers by Newman, Buller, and others that depicted his people as "facing nothing but rapid extinction." The "facts" undermined this "sad prognosis." Reviewing population estimates from Captain Cook's arrival until the latest government census of 1921, Hiroa agreed that depopulation "considerably accelerated" after contact. But he rejected claims by "European writers" that the race was doomed as, probably, "an excuse." Māori enthusiasm for European firearms increased deaths in intertribal warfare, while introduced diseases and the "abuse and misuse of European alcohol, foods and clothing" exacerbated decline. Whereas Pākehā routinely referred to the "Maori wars" of midcentury, Hiroa, turning the tables, claimed that the "unnecessary European wars of the 'forties' and 'sixties'" lost more lives. While virtually all observers agreed about depopulation between Cook and the mid-1890s, recent data showed that trend had reversed. Census statistics since 1906, properly interpreted, showed a sustained increase. Hiroa supported this argument with a table derived from census reports showing a rise in young Māori (under fifteen years), in numbers and as a proportion of the population, from 14,251 in 1891 (34 percent) to 21,071 (40 percent) in 1921.[79]

Hiroa identified "the gradual elimination of the factors that caused decay" as the main cause. Māori "acceptance of Christianity," the "first great change," ended the musket wars. Embracing "the teaching of peace and good will," most "laid aside their arms and thoughts of revenge." It remained "a matter for regret," however, that the "European wars" had "ruptured" this "peace" through "lack of full appreciation of more pacific ways of dealing with the warlike Maori." Quoting the British evolutionary anthropologist R.R. Marett (1866–1943), who identified elements of both "rigidity" and "plasticity" in the "evolutionary process," Hiroa argued that "plasticity" offered his people "an avenue of escape." Māori adapted better to environmental and cultural changes than other Pacific peoples: "Five centuries in a temperate climate toughened his constitution, sharpened his mentality, and altered his material culture." The adaptable Māori left "far behind" islanders living in tropical climates, with "open houses, scanty *tapa* clothing and food without labour." Shedding "the indolence of the tropics," the "more vigorous and virile" Māori "was in a better condition to survive extinction than his more easy-going kinsmen in Polynesia."[80]

CONCLUSION

In New Zealand, moral and religious language made a difference to the acceptability of social evolutionists' racial science into the 1880s. The mildly humanitarian versions of the "doomed Māori" of Travers, Buller, and Hector proved more acceptable than Newman's callous materialist formulation. In the southernmost colony of the British empire, moral and religious language continued to matter when it came to human evolution. The secularization of scientific discourse, and the retreat of religious language into the private sphere, varied by place and topic. Probably these humanitarian versions of the doomed Māori were more acceptable and influential than Newman's precisely because—like Darwin's *Origin* and *Descent*—they did not systematically reject older moral and religious language.

The "doomed Māori," which Tamihana, Bishop Selwyn, and Shortland had attacked as wicked wishful thinking at midcentury, became more popular than ever during and after the wars. Many settlers, politicians, and men of science read Darwin as conferring scientific sanction on racial warfare, British victory, and Māori defeat and probable extinction. In this process, secular theologians—laymen untrained in theology—such as Giles, Atkinson, Fox, Whitaker, Travers, Buller, and Hector played significant roles. When science, religion, politics, and race are placed in a single analytic framework, the war years facilitated larger secularizing processes, as Harrison would lead us to expect. Laymen, often Anglican, wrested not only intellectual, cultural, political, and scientific but also moral and religious authority from the Anglican clergy, missionaries, officials, and Christian Māori previously influential. The former, not the latter, decided what kind of science, religion, and racial politics would dominate the public sphere. If a new generation of Māori Christians consigned the "doomed Māori" to oblivion from the 1890s with energy, intelligence, and wit, considerable damage had already been done.

PART III

Appropriation and Response

9

Evolution and Religion in Transnational Contexts

Britain, Japan, and China, 1859–1920

Fa-ti Fan and Bernard Lightman

As Peter Harrison has shown in his *Territories of Science and Religion*, the modern concept of religion had its birth in the early modern period when religion came to be understood in terms of beliefs and practices. At the same time, the religious fragmentation of Western Christendom was projected onto the rest of the world.[1] Harrison argues that the West imagined that there were four religions: Christianity, Judaism, "Mahometanism," and heathenism. After the early modern period, additions were made to this understanding of comparative religions. "Eastern religions" appeared in the early and mid-nineteenth century. The first recorded use of "Buddhism" was 1801, followed by "Hindooism" (1829), "Taouism" (1838), and "Confucianism" (1862).[2] By the middle of the nineteenth century these terms were firmly established in English discourse, and the religious groups to which they referred became permanent features in how world religions were understood. But world religions could be resistant to concrete categorization. They were in a state of flux, as the global encounter between nations, some driven by the imperialist enterprises of European powers, was a constant source of change. The modern Western category of science was also, not coincidentally, established during the nineteenth century.[3] After midcentury, evolution had an impact on each of the scientific disciplines as well as on every aspect of human culture, including religion.

Questions about the nature of science, evolution, and religion were asked alongside questions about the categories being used to conceptualize nations and cultures. What did it mean to be British, Japanese, or Chinese in a world where science and religion did not provide stability as they were beginning to be understood through an evolutionary lens? What did science and religion mean when intellectuals had far more opportunities in the nineteenth century to travel the world and experience new cultures? British figures like T.H. Huxley and Charles Darwin, especially early in their lives, were not just beneficiaries of the expanding British empire but actively contributed to its expansion. Their voyages around the world on British naval vessels, encountering new cultures, prepared them to see themselves as a new breed of intellectual, whose work depended on imperial networks and global perspectives.[4] Either on their voyages, or perhaps afterward in Britain, they encountered non-Western scholars. For Japanese and Chinese intellectuals, opportunities for international travel did not become possible until the latter part of the century. The experiences of Japanese intellectuals like Minakata Kumagusu,[5] who traveled to the United States and Britain, and Chinese intellectuals, such as Yan Fu and Liang Qichao, who visited Britain and Japan respectively, had a profound impact on them. But even those intellectuals who could not travel could participate in global debates about evolution by tapping into the increasing quantity of affordable books and periodicals on science, many translated into a variety of languages, that resulted from the use of new printing technologies. Transnational and transregional intellectual spaces were created through different means, including travel, the circulation of knowledge, and imperial spaces. Intellectual spaces were no longer confined merely to the country in which an individual lived. In the nineteenth century an exciting intellectual space emerged in which the implications of evolution for religion, and modernity in general, could be discussed. In this new global space, intellectuals actively grappled with the issues generated by debate, rather than passively accepting an evolutionary model originating from Europe.

This chapter therefore argues that the reception models that focus on how evolutionary theory was received in a national context are inadequate. While it is important to understand the national context, it is also necessary to recognize the various trans- and intra-regional spaces of intellectual discourse that went beyond national boundaries. For example, the global spaces created by the European empires played an important role not only in the production of both evolutionary and religious knowledge but also in how the two overlapping cultural fields interacted with each other in both colony and metropole. Several chapters in this volume, including the ones on South Asia, Australia, and New Zealand, demonstrate this point. As religious traditions and practices in South Asia became objects of scholarship in Europe, they were categorized, classified, and labeled as Hinduism, Buddhism, and so on, and the study of these religions so defined was housed under the discipline of Oriental studies. The knowledge thus produced had significant impact in both Europe and Asia. Herbert Spencer and Huxley were not alone among their contemporaries in employing world religions to goad the religious establishment.

Non-Western intellectuals in South Asia, Japan, and China had to reckon with this scholarship, which arrived with the might of Western imperialism as part of Western learning. Yet they also drew on the scholarship as a valuable source for their intellectual enterprises. This is not to recenter Oriental studies from the metropole in our historical inquiry. In fact, the formation of Orientalist studies itself started with the encounter of Europeans with non-Western scholars and their learning in other parts of the world (see, e.g., Simon Schaffer's chapter in this volume). It should be noted that these spaces of intellectual encounter and discourse were not evenly distributed. It was by no means a level playing field as a result of the power differentials between the groups of historical actors who encountered each other on the world stage. Nevertheless, this perspective enables us to observe more clearly the pattern of knowledge in transit on a global scale. Just as in the case of "world religions" produced by Orientalist scholarship, evolution was not a prepackaged theory made in Europe and exported to other parts of the world. And translation, both linguistically and conceptually, was only a small, though important, part of the story. We need a broader, more radical interpretive framework if we wish to better capture the scope, actions, connections, and processes of knowledge production in relation to evolution and religion at the local and the global level.[6]

This chapter will discuss how British, Japanese, and Chinese intellectuals engaged with the issue of the relationship between evolution and religion within a global context. Of all the prominent British evolutionists, T.H. Huxley and Herbert Spencer devoted the most time to exploring the connections between evolution and Chinese and Japanese religious traditions. They integrated those religious traditions into their evolutionary worldviews and used them to bolster the status of evolutionary theory while questioning the modernity of Christianity. Then we will analyze how Japanese and Chinese intellectuals looked to science and evolution to strengthen, reimagine, and modernize the religious traditions with which they were familiar, though not without objections. In both cases, these Chinese and Japanese thinkers were participants in the global discourse of evolution and religion. James Secord asserts that Darwin was a "characteristic figure of the first great age of globalisation."[7] But there were many others in this time, not located in Britain, who can also be seen as characteristic of this age. By adopting this global perspective, we can avoid the somewhat static reception model adopted by previous studies (e.g., Glick's collections) that don't do justice to the dynamic and active role played by non-Western intellectuals who were responding to, and engaging with, the vast changes that surrounded them. Though disruptive to traditional religious belief systems, evolution also provided a new way to understand a rapidly changing world, during a time in which global spaces for intellectual exchange were in the process of being created.

"ORIENTAL EVOLUTIONISTS" IN BRITAIN

Thomas Henry Huxley, "Darwin's Bulldog," and Herbert Spencer, often referred to as the "philosopher of evolution," were well-known British evolutionists who aimed to revolutionize how science was conducted and theorized. Their ideas on this topic contributed to the global circulation of evolutionary ideas. They were part of a group of reform-minded men who arrived on the British scientific scene in the middle of the nineteenth century. Besides Huxley and Spencer, the group included the physicist John Tyndall, the mathematician William Kingdon Clifford, the founder of eugenics Francis Galton, the statistician and biometrician Karl Pearson, the anthropologists John Lubbock and Edward Tylor, the biologist E. Ray Lankester, the doctor Henry Maudsley, and a group of journalists, editors, and writers such as Leslie Stephen, G.H. Lewes, John Morley, Grant Allen, and Edward Clodd. Referred to as "scientific naturalists" by scholars, these men put forward new interpretations of nature, society, and humanity derived from the theories, methods, and categories of empirical science. Scientific naturalists were naturalistic in the sense that they ruled out recourse to causes not present in empirically observed nature, and they were scientific in that they interpreted nature in accordance with three major midcentury scientific theories: the atomic theory of matter, the conservation of energy, and, especially, evolution. They defended Darwin when his *Origin of Species* first appeared in 1859. Huxley, Spencer, and others of this influential group of British scientists aimed to use the debate over Darwin's theory of evolution to secularize nature and restrict British science to well-trained experts.

In 1871, Huxley and Spencer found themselves in the position of controlling the British component of an ambitious publishing project devoted to the dissemination of scientific knowledge in the Western world. The International Scientific Series (ISS) was one of the most famous of all Victorian attempts to codify and popularize scientific knowledge in a systematic fashion to a wide reading public across national boundaries. The central purpose of the ISS was not just to disseminate scientific thought but to circulate a naturalistic, scientific perspective thoroughly informed by the science of evolution. Published over the course of four decades, it included ninety-eight books, some published in five languages—English, Russian, German, French, and Italian. Six publishers from six countries were involved, and contributors were recruited from Britain, North America, and continental Europe. The ISS was the result of the entrepreneurial vision of Edward Youmans, an American disciple of Spencer's who worked at the New York publishing firm of D. Appleton and Company. With the support of Huxley, Spencer, and other British men of science, Youmans was able to create an international publishing project that could be used to increase the circulation of scientific ideas from one country to another.[8] In addition to coordinating the efforts of the British branch of the project for Youmans, both Huxley and Spencer contributed books to the series, Spencer his *Study of Sociology* (1873) and Huxley his *Crayfish* (1880).

Huxley's and Spencer's status as global intellectuals was not due solely to their participation in the International Science Series. Their perception of themselves as global intellectuals would have been due in part to how much they traveled to Europe and even farther afield. Huxley had served as assistant surgeon from 1846 to 1850 on the British naval survey ship the HMS *Rattlesnake*, which took him to Australia and New Zealand. In 1876 he embarked on a lecture tour of the United States. Spencer was less adventurous, though he too visited America in 1882. But they were also global intellectuals because their ideas circulated around the world through the many translations of their books and articles. For example, Huxley's "Evolution and Ethics" became a key evolutionary text in China when it was translated into Chinese in 1896. Some of his books were translated over time into French, German, Chinese, Japanese, and Russian.[9] Spencer's books on psychology, biology, sociology, education, politics, and ethics were widely translated at different points into Russian, Japanese, Chinese, Spanish, French, Italian, Danish, and Swedish.[10]

The attempts of Huxley, Spencer, and their fellow scientific naturalists to redefine science brought them into conflict with the Anglican clergy and the older generation of natural philosophers and natural historians committed to conceiving of science as based on natural theology. But Huxley and Spencer were not anti-religious. Rather, they were hostile toward Christian theologians who attempted to dictate which scientific theories were acceptable. They argued that science should be autonomous and that the legitimacy of scientific theories, such as evolution, be judged on purely naturalistic grounds. Separate from religion, theology was actually a part of the world of science, as theologians made claims based on fact. By contrast, religion, properly understood as emotion and feeling, was entirely valid and could not come into conflict with science if distinguished from theology. In his *First Principles*, Spencer even accepted the existence of a shadowy deity that he referred to as the "Unknowable," which, he claimed, lay at the center of all religions. Huxley and Spencer worked to redefine the meaning of science at a time when the study of world religions in England was supposedly becoming scientific. They appropriated the Orientalist construction of world religions for their own purposes. Scholars have discussed their criticisms of Christian theology but have rarely noticed their interest in Buddhism, Confucianism, and other non-Christian religious traditions.[11] However, Huxley and Spencer integrated non-Christian modes of thought into their articulation of evolutionary theory, so much so that we have referred to them as "Oriental evolutionists." They wrote about Oriental religions to emphasize the backwardness of Christianity, especially its hostility toward modern science. Not only were Oriental religions more congenial to evolutionary forms of thinking, in the minds of Huxley and Spencer, but they were superior because long ago they had already anticipated philosophical evolutionism. They were crucial for understanding the true nature of religion, the evolutionary process that had produced the world religions, and the evolution of human thought in general.

SPENCER AND THE EVOLUTION OF RELIGION

Herbert Spencer, the philosopher of evolution, was indebted to the sociologists and anthropologists who were creating a new scientific study of religion. He began to work on world religions as part of his research on sociology. In the course of preparing his *Principles of Sociology* (first edition in 1876), Spencer collected and classified "large quantities of facts relating to societies of different types, ancient and modern."[12] This work began in October 1867, with the help of several assistants, and was originally intended as the preliminary data for *Principles of Sociology*. But Spencer later decided to publish the information in table form under the title *Descriptive Sociology*. Each of the thirteen volumes, which appeared from 1873 to 1934, contained data on diverse types of societies at different stages of social evolution, using a classification system that Spencer had developed.[13] For example, the volume on China, published in 1910, was compiled by E.T.C. Werner, the consul at Foochow. It contained raw data on Chinese ecclesiastical institutions and religious ideas through the ages. Table 9, covering the Manchu Period (1644–), had a column on religious ideas that recorded "Religion domestic and centered mainly in reverence for ancestors. Largely agnostic (Confucianism)."[14]

The assertion that Chinese religion was based on "reverence for ancestors" in *Descriptive Sociology* is an obvious clue to how Spencer and his assistants approached religion and the evolution of culture. Starting in the early 1860s, a distinct field of scholarly study flourished in Britain devoted to the scientific analysis of religious practices and belief systems, known as the "science of religion" or "comparative religions" or "the history of religions." Contributors came from anthropology, sociology, classics, and Oriental studies. The key figures included James G. Frazer, Jane Ellen Harrison, Andrew Lang, Friedrich Max Müller, William Robertson Smith, and Edward B. Tylor.[15] Tylor and his theory that the origins of all religions in *Primitive Culture* (1871) can be found in ancestor or ghost worship was particularly important to Spencer.[16] During a controversy with the positivist Frederic Harrison in the pages of *Nineteenth Century* in 1884 over the origins and future of religion, Spencer explained how the facts given in his *Descriptive Sociology* demonstrated the accuracy of the ghost theory and disproved Harrison's assertion that in the past humans worshipped inanimate objects. Spencer also referred to Tylor for support, stating that the anthropologist "has probably read more books about uncivilized peoples than any Englishman living or dead." Both China and India, Spencer argued, perfectly illustrated the origins of religion according to the ghost theory. In China, for example, the "State-religion continues down to the present day to be an elaborate ancestor-worship, where each man's chief thought in life is to secure the due making of sacrifices to his ghost after death." Spencer then outlined his larger views on the evolution of religion. Religions begin with the idea of a spirit or ghost; then evolve to the idea of supernatural beings of all kinds; and then evolve into monotheism, in which one major supernatural agent becomes "an Omnipresent Power to which no attributes can be ascertained," which Spencer referred to as the Unknowable. All religions undergo this evolutionary process.[17]

Spencer developed his theory on the evolution of religion in more depth in his section "Ecclesiastical Institutions," which was added to his *Principles of Sociology* in 1885. Here he repeated his point that the origin of all religions in the ancient past was the concept of god that derived from "ghost-propitiation." Spencer found examples of this everywhere. "Mohammedans" practiced ancestor worship, as did ancient peoples who lived "in the forests of North America," in African jungles, in the Andes, in Polynesian islands, and throughout every other geographical location in the world. The widespread evidence for the ghost theory provided "abundant proofs of the natural genesis of religions."[18] Spencer then went into detail on the evolutionary development of the idea of god in China, Japan, and India. In both China and Japan, the eldest male descendants performed the proper sacrifices after the death of a parent and became quasi-priests. Spencer concluded this part of the analysis by declaring that "the ghost-theory, which explains the multitudinous phenomena of religion in general, explains also the genesis of the priestly function." This leads Spencer from the evolution of god into his next theme, the development of ecclesiastical institutions. Of course, they were part of the social evolutionary process. "Among social phenomena," Spencer affirmed, "those presented by Ecclesiastical Institutions illustrate very clearly the general law of evolution." For Spencer this meant that the development of ecclesiastical institutions fell into line with the general evolutionary process that moved from the homogeneous to the heterogeneous. Spencer pointed out that the evolution of science was a parallel process to the evolution of religion. Rather than undercutting religion, the development of science "enlarges the sphere for religious sentiment," as the proper study of nature led to "a consciousness of the relation which we stand to the Unknown Cause." In a subsequent chapter focusing on "man of science and philosophers," Spencer argued that in "Christian Europe, as in the pagan East, the man of science and the philosopher were of priestly origin." But science and philosophy, at first a part of theology, gradually became differentiated, and then science and philosophy were separated from each other.[19]

HUXLEY, THE EVOLUTION OF THEOLOGY, AND ORIENTAL RELIGIONS

Like Spencer's, Huxley's views on science and Oriental religions reflected a profound debt to the new science of religion that began in the 1860s to draw on anthropology, sociology, classics, and Oriental studies. Huxley was deeply involved in the study of anthropology and wrote several key essays that discussed Oriental religions. He served as the president of the Ethnological Society of London from 1868 to 1871. In that role, he delivered the opening address to the society on March 9, 1869, "On the Ethnology and Archaeology of India." Here he announced that the council of the society proposed to "direct public attention to the desirableness of subjecting the physical characters, the languages, the civilisations, the religions, in short, the ethnology, of the various peoples over whom the rule of Britain extends, to systematic investigation." The ethnology of British colonies was important, Huxley argued, to keep the empire strong. Huxley announced a series of meetings "devoted to the ethnology of one or other of the British possessions." The rest of the address was taken up with India, including the "endless shades of diversity" in its religion.[20] Huxley's interest in India, then, began at least as early as 1869.

But the fruits of Huxley's engagement with Oriental religion are to be found in two later essays, "The Evolution of Theology: An Anthropological Study" (1886) and "Evolution and Ethics" (1893). In the first, rather lengthy, essay Huxley began by proposing to analyze the origins, growth, and decline of theologies as an anthropologist. He would treat theology as "a natural product of the operations of the human mind," akin to the development of architecture, music, or painting. Then Huxley began to examine the role of the ghost theory in the development of ancient Eastern religions and in the creeds of the old Israelites as depicted in the Old Testament. Huxley acknowledged that he did not create the ghost theory and refers the reader to three works: J.G. Müller's *Geschichte der amerikanishen Urreligionen* (1855), Tylor's *Primitive Culture*, and Spencer's "recently-published 'Ecclesiastical Institutions.'" The theory, Huxley asserted, is essential to any anthropological understanding of how all religions originated and developed. "It is a matter of fact that," Huxley declared, "whether we direct our attention to the older conditions of civilised societies, in Japan, in China, in Hindustan, in Greece, or in Rome, we find underlying all other theological notions, the belief in ghosts." When he examined Chinese and Japanese theologies, he asserted that ancestor worship attained "its acme in the State-theology of China and the Kami-theology of Japan." The establishment of Christianity begins a "new course of evolution," the steps of which "are obvious" and which end with the spread of "true scientific culture" in the modern period. Theology, Huxley concluded, has from the beginning of time offered religious symbols that are slowly refined over time. Science's role in this process is to occasionally warn of the tendency of dogmatic theology to produce idols that prevent the growth of new religious symbols.[21]

After the publication of "The Evolution of Theology," Huxley began to use comparisons between Oriental religions and Christianity to confound his opponents. In "Mr. Gladstone and Genesis" (1886), Huxley rejected William Ewart Gladstone's claims that the author of Genesis was either inspired by God or gifted with superhuman faculties, and that the Pentateuchal statement could be taken as a scientific document. Huxley argued that Genesis had no more scientific importance than the cosmogonies to be found in the holy books of other ancient civilizations. He pointed to the "evolutionary character of Indian philosophy, and especially of that of Buddhists." Although Huxley admitted that he was not intimately acquainted with Indian philosophy, he claimed to have "taken a great deal of pains to secure that such knowledge as I do possess shall be accurate and trustworthy."[22] In his 1889 essay "Agnosticism," Huxley responded to an insult hurled at him by Henry Wace, Anglican minister and ecclesiastical historian. Wace might consider him to be an "infidel," but Mahommedans would agree in "reciprocating that appellation to Dr. Wace himself." Whereas Wace believed it should be unpleasant for a man to say that he did not believe in Jesus Christ, Huxley replied that it depended on whether or not the man was brought up in a Christian household. "I do not see why it should be 'unpleasant' for a Mahommedan or Buddhist to say so," Huxley declared.[23]

Huxley's views on the relationship between evolution and Eastern religions are developed at length in his Romanes Lecture, titled "Evolution and Ethics" (1893). It is easy to miss just how much Huxley engages with Eastern religions in this well-known essay. Scholars have read this essay primarily to understand Huxley's rejection of Spencer's attempt to derive a philosophy of ethics out of the evolutionary process, largely ignoring the sections on Buddhism. A lot of the interesting material on Oriental religions is buried in the extensive notes at the end of the essay. A major theme in "Evolution and Ethics" is the anticipation of evolutionary theory in Buddhism as an explanation for why it is more compatible with modern science than Christianity. The Buddhist and Hindu notion of transmigration, though it may seem absurd, paralleled the concept of evolution, as both had their "roots in the world of reality" and therefore were scientifically grounded. For through the evolutionary process, the moral and intellectual essence of each person passed to another through heredity. Huxley even suggests that the concept of *karma*, which allowed for the role of will in the modification of the individual, illustrated that Indian philosophers were "strong believers" in the Lamarckian theory "of the hereditary transmission of acquired characters."[24] There was valid scientific evidence to support aspects of Oriental theology.

In the notes Huxley acknowledged the scholarly works that he had drawn on in his discussion of Buddhism. He referred to the second edition of Hermann Oldenberg's *Buddha* (1890) and to several books by Rhys Davids.[25] Thomas W. Rhys Davids was the most influential British scholar of Buddhism in his day.[26] His works were part of a burgeoning interest in Buddhism in Britain in the latter half of the nineteenth century.[27] In one note, Huxley quoted from Davids's *Hibbert Lectures* for 1881 on how Buddhism was unique in the way it emphasized how humans could reach salvation through knowledge of nature rather than knowledge of god.[28] Huxley's embrace of this rather agnostic-like depiction of Buddhism was at odds with the conventional portrayal of Buddhism as atheistic.[29] But, as Donald Lopez Jr. has observed, for Huxley and other Victorians, Buddhism was a tradition that pictured the universe as subject to natural laws with no requirement for an intervening divine being, for dogma, ritual, or even scriptures. This Buddha became popular among anticlerical European scholars.[30] So Huxley was not alone in presenting Buddhism as the religion, as Lopez Jr. puts it, "most suited to serious dialogue with Science, because both postulated the existence of immutable laws that governed the universe."[31]

In his account of the history of religions, Huxley claims that ancient Greek and Indian thought converged on the same evolutionary position, and it is here that modern thought could take its starting point. "Modern thought is making a fresh start from the base whence Indian and Greek philosophy set out," Huxley declared, "and, the human mind being very much what it was six-and-twenty centuries ago, there is no ground for wonder if it presents indications of a tendency to move along the only lines to the same results." In effect, Huxley was asking his readers to erase the entire Christian era from intellectual history, as it inhibited the development of science. Huxley was conscious when writing this essay that he had to abide by the restrictions of the Romanes Lecture: religion and politics were to be avoided. Before giving the lecture at Oxford in May 1893, he wrote to the founder of the annual Romanes Lecture, George Romanes, several times in April to assure him that he had observed the limitations placed upon him. "There is no allusion to politics in my lecture," he wrote to Romanes on April 22, 1893, "nor to any religion except Buddhism, and only to the speculative and ethical side of that." If people did apply anything he said about these matters to modern philosophies, "that is not my affair." Huxley wrote again six days later, saying that his lecture wouldn't be "too much for the Dons' nerves," and joked that his wife had gone over the paper twice "without scenting a heresy."[32] But there is no mention of a role for Christianity in Huxley's vision for the future. Buddhism emerges as the religion most congenial to modern science and as one of the starting points for modern thought to build upon.

Spencer and Huxley explored the theme of Oriental religions to reflect on the understanding of science, evolution, and religion in their own culture and then disseminated their views around the world in their publications, in their lectures, and in the International Scientific Series. Drawing on the new science of religion, especially anthropology, they outlined the origin and evolution of religion, starting with the ghost theory. As religious and scientific reformers, they were disruptive forces, contributing to the redefinition of both science and religion. Their research on Oriental religions was a part of their attempt to reimagine science, religion, and their relationship. They knowingly made use of Oriental religions in order to demonstrate the inability of Christian theology to engage in a constructive way with evolutionary theory as well as science more broadly. The Oriental religions were considered by them to be farther along in the evolutionary process by culminating in an agnostic-like position, and therefore these religions were presented as having more in common with modern science. Exploration of Oriental religions provided a way of highlighting Spencer's and Huxley's religiosity in the face of charges of materialism and atheism. Strikingly, their conception of how religions developed implied that the principles of evolution had not been produced in Britain. Rather they had a long lineage in Oriental religions, theologies, and philosophies.

SPENCER AND EVOLUTION IN JAPAN

The Meiji period in Japan (1868–1912) was an era of continuous change. A new central government replaced the shogunal regime that had continued for more than two centuries, and the country was reunited under the emperor. The government adopted a rapid modernization policy, which was designed to transform Japan into an industrialized nation like those in the West. During this time, the evolutionary theory was brought by Christian missionaries (mostly

Anglo-Saxon Protestants) as a theory that was fully compatible with Christianity, and the assessment of it entered the ongoing debate on what the role of religions should be in the process of modernization.[33]

Categories such as "religion," "philosophy," and "science" were also brought into Japan. As there were no Japanese equivalents, Japanese intellectuals were "puzzled" by these categories.[34] In searching for meaningful definitions, they placed "science" at the opposite side of "religion" and "philosophy." As a result, the modern scholar Isomae Jun'ichi claims that religion was gradually considered more irrational and began to be referred to as private practices in the 1880s and the 1890s.[35]

Among the evolutionary theorists, the works of Herbert Spencer had a significant impact on Japanese intellectuals. Spencer and his works were introduced by Protestant missionaries and visiting scholars in Japan, such as American naturalist Edward Sylvester Morse (1838–1925), and the translations began to appear in 1877. In this process, Spencer was often linked to Darwin, but the works of other evolutionary theorists did not gain the same attention—the popularity of Darwin and Spencer overshadowed the others from Britain, including W.K. Clifford and Huxley.

A number of Japanese intellectuals took part in a discussion on evolution and religion, but this discussion did not take place only within Japan. Because Japanese scholars traveled abroad and some had direct communications with Spencer during the Meiji period, some of the early responses—acceptance, appropriation, or rejection—occurred outside of Japan. Minakata Kumagusu (1867–1941), for example, left Japan and learned of Spencer's work in the United States and the United Kingdom from 1887 to 1900. He read many works authored by Spencer in English and examined them in connection with his religious belief, Shingon Buddhism (a Japanese Buddhist sect derived from Vajrayana Buddhism). The Meiji period also saw the large-scale importation of foreign knowledge. Living in Japan, many thinkers learned ideas from foreign scholars in addition to Spencer, such as Max Müller and Pyotr Kropotkin, and used them to examine evolution theory. Some Japanese biologists studied abroad, for example, those who worked under August Weismann in Germany.

Previous scholarship has summed up what happened in Japan during the Meiji era, and it has examined Japanese experiences and students studying abroad.[36] The Meiji intellectuals learned the views of intellectuals from the United States, the United Kingdom, Germany, and Russia. Thus, the Japanese responses to evolutionary theory should not always be considered "Japanese." But the main analytical focus of earlier scholarship is not religious responses to evolutionary theory. This section analyzes the more complicated, multinational responses to Spencer by Japanese intellectuals who traveled abroad and by those who had access to foreign books and ideas in Japan.

EVOLUTIONARY THEORY IN JAPAN

Many foreign books were brought to Japan due to the rapid modernization promoted by the new Meiji government. The first translation was Spencer's *Social Statics* (1851) in 1877 by Ozaki Yukio (1859–1954), who would become one of the leading liberal politicians.[37] Most of the translations of Spencer's work appeared between 1877 and 1886. Because Spencer's *Social Statics* was often discussed in parallel with the works of John Stuart Mill, it gained attention from liberal activists, such as Ozaki and the leader of this liberal movement Itagaki Taisuke (1837–1919). They called Spencer's work a "textbook of social rights."[38]

At the beginning of the Meiji period, many skilled workers and scholars were invited to Japan to facilitate its rapid modernization. Edward Sylvester Morse was appointed as the first professor of zoology at Tokyo Imperial University and lectured on evolutionary biology between 1877 and 1879.[39] Listening to his lectures, biologist Ishikawa Chiyomatsu (1860–1935) translated the works of Darwin and Huxley into Japanese.[40] One of the most influential Spencerian books written and published in Japan was *The Future Japan* (*Shōrai no nihon*, 1886) by Tokutomi Sohō (1863–1957), writer, historian, and journalist. He read the works of Mill, Bentham, and others at the Dōshisha, a Christian university in Kyoto. Tokutomi embraced Spencer's vision of social evolution, as he combined Spencer's concept of "social evolution with an enthusiastic endorsement of liberalism, democracy (also relying on J.S. Mill), and the free market (also inspired by the Manchester school)."[41] In addition, Tokutomi saw Spencer's theories as relevant when explaining Japan's past. Writing about the history of Japan, Tokutomi adapted Spencer's concept of the "militant society," which helped to explain the Tokugawa period.[42] At the time, when the government aimed to develop a strong nation with increased military capacity (*fukoku kyōhei*, or "rich nation, strong army"), Spencer's idea of evolution of nations from military to industrial proved to be persuasive for Japanese intellectuals and provided a new type of argument for political activists who were critical of the power of the military.[43] Spencer's work was spread purposefully for the benefit of liberal activists.

The "acceptance," "adaptation," and "critique" of Spencer's work happened concurrently in Meiji Japan, however.[44] Many Japanese Christians accepted evolution, in contrast to Christian thinkers in the West who often reacted strongly against evolutionary theory. After examining the works of Japanese Christians, including Uchimura Kanzō (1861–1930) and Uemura Masahisa (1858–1925), several scholars, notably Ōuchi Saburō, Murakami Yōichirō, and Koyama Chūmaru, have commented that the "scientific" aspect of evolutionary theory was what enabled Christians in Japan to accept evolutionary theory.[45] Although Huxley considered Buddhism the most congenial religion to modern science, Christians in Japan examined and appropriated the evolutionary theory.

Yulia Burenina further points out that there was a common response to evolutionary theory from Japanese intellectuals. Christians and Buddhists agreed that any concept of evolution based on linear progression contained a critical flaw. Since Spencer's concept of evolution was neo-Lamarckian, teleological, and progressive, while both Darwin and Huxley were more ambivalent about a directional theory of evolution, this criticism applied to Spencer and those who embraced his synthetic philosophy. Observing the religious practices in Japan more broadly, G. Clinton Godart states that both Japanese Christian thinkers and Buddhists "helped to introduce evolutionary theory in Japan." However, he sees changes in their attitude in the 1930s. Japanese reactions to evolutionary theory "became more controversial and became a subject of debates in the first decades of the twentieth century and into the 1930s," he says, because the Marxists and anarchists used the evolutionary theory to support materialism and atheism.[46]

All of the above studies are based on the observation of Japanese thinkers in Japan and are quite valid. Previous scholarship has shown that both Christians and Buddhists in Japan demonstrated a flexible attitude when they were assessing evolutionary theory. They were selective about what to take in and what to reject, but they were in general agreement that the investigation of evolutionary theory was a critical step toward understanding science. What previous scholars have not examined is the case of Japanese intellectuals who traveled abroad and learned about Spencer and consequently reflected on their religious beliefs.

MEETING SPENCER OUTSIDE OF JAPAN

The initial contact with Spencer was a political one. It was not scholarly, religious, or philosophical. Japanese diplomats and scholars met Spencer in person before any translation works came out in Japan. The first Japanese who had contact with Spencer was Mori Arinori (1847–1889), a diplomat assigned to stay in the United States who made a short trip to the UK. Mori met Spencer in 1873 and asked him for advice on the reforms being considered by the Japanese government. Around the same time, Kaneko Kentarō (1853–1942) began his study at Harvard Law School and learned from John Fiske, a disciple of Spencer.[47] After becoming an official drafter of the Japanese Constitution, Kaneko made a trip to the UK in 1889, met Spencer, and asked for more advice. Their communication continued, as Spencer corresponded with Kaneko in 1892.[48]

The next Japanese intellectual to encounter Spencer's work was Minakata Kumagusu, who had access to the British evolutionist's writings while traveling in the United States. Literate in multiple languages, Minakata read the works of Spencer in English from 1887 to 1892 and while in the UK from 1892 to 1900. He never met Spencer, but he became a committed Spencerian. Minakata was originally a Shingon Buddhist, but his religious views did not seem to interfere with his reading Spencer in the United States. For example, Minakata read Spencer's description of totemism in the essay entitled "The Origin of Animal Worship" and agreed with Spencer's explanation for the "strange" worship of plants or animals. Spencer claims that certain tribes believed they were the descendants of these plants or animals. Minakata thought of his own name; it consists of two characters, a bear and a camphor tree, and he thought of traditions of his hometown where babies received the names of plants or animals that can protect them from evil spirits. Minakata concluded that the Japanese were also influenced by totemism, and he wholeheartedly accepted Spencer's observation.[49]

Minakata read more of Spencer's works during his years in England from 1892 to 1900.[50] At that time, Minakata started to write and publish articles in English in *Nature*, trying to include his observations on East Asian beliefs.[51] In one of these articles, "On Chinese Beliefs about the North" (1894), he took up the ideas of Spencer's *Principles of Sociology* and quoted him.[52] During his stay in London, Minakata read Spencer's *The Study of Sociology*, which contained a discussion of science and religion. Minakata was intrigued by Spencer's point about "confusing observations and conjectures that could happen from misunderstanding on the facts."[53] Minakata must have stopped reading for a moment, as he wrote a note on the side: "I now remember my time in the US; Captain Jones told me that Japanese modernization has been largely hindered by Confucianism. But I thought of how the modernization was detrimental in most developed countries in Europe—the damage that Christianity had done in Europe was no comparison to what Confucianism had done to Japan."[54] Minakata believed that Christianity had done extensive harm to modern Europe. Christianity resisted new scientific discoveries in Europe, but Confucianism had not done the same in East Asia.[55] Minakata started to be more conscious about the relationship between science and religion from this time because his interests in both Buddhism and foreign religions grew after he met Doki Hōryū (1854–1923) in 1893.

AGNOSTICISM, THE UNKNOWABLE, AND BUDDHIST DOCTRINES

Doki was a Buddhist scholar and ordained monk of Shingon who visited London as he was making his way to Paris. Doki became a great influence on Minakata for the rest of his life and helped deepen his thoughts on Buddhism and Spencer's evolutionary theory as they corresponded between London and

Paris. When they met in London, Doki was eager to learn the wisdom contained in the works of the great Western thinkers, but he did not speak any Western languages. Doki thus relied on Minakata to learn more about the texts written in those languages. The multilingual Minakata was interested in learning more about Buddhist sculptures because he was asked to organize them at the British Museum in London. Doki was an ordained Buddhist monk, so he was able to provide Minakata relevant information on Buddhist arts. They had mutual interests to explore in their correspondence.

In his letters to Doki, Minakata brought up the question of the compatibility of evolutionary theory and Shingon Buddhism.[56] A Spencerian who valued science, Minakata wrote to Doki that the non-scientific elements of Shingon Buddhism had to be eliminated as soon as possible; scientific knowledge, in general, was becoming important during the modernization of Japan, and the teachings of Shingon Buddhism could be strengthened if science was seen as backing up its doctrines. "Science was not an enemy of Buddhism," so Minakata claimed.[57]

Minakata was especially struck by how the concepts of both causality (*ingaritsu*) and reincarnation (*rinne*) existed in Spencer's writings. In "Progress: Its Law and Cause (1857)," *First Principles* (1862), and *Ecclesiastical Institutions* (1886), all of which Minakata bought in the United States and read in English, Spencer discussed the Unknowable and the Knowable. Spencer wrote, "By continually seeking to know and being continually thrown back with a deepened conviction of the impossibility of knowing, we may keep alive the consciousness that it is alike our highest wisdom and our highest duty to regard that through which all things exist as The Unknowable."[58]

The epistemological basis of Spencer's entire evolutionary system is the realization of the existence of both the Knowable and the Unknowable. When one cannot find a natural cause, the Unknowable emerges as the final cause. For Spencer, the Unknowable guaranteed that the evolutionary process was progressive and teleological. Minakata saw this model as an advantage for all religions. As long as humans encounter the Unknowable, religion would have a place to provide what science cannot offer. In other words, Minakata interpreted Spencer as claiming that both science and religion have value and are not in conflict. The purpose of religion is to acknowledge the Unknowable, while the Knowable is that part of our world that can be explained by science using the concept of causality.

Minakata also believed that Spencer's work contained an idea analogous to reincarnation. Minakata went back to *First Principles* and read a chapter on "The Rhythm of Motion." There, Minakata understood that Spencer saw that everything was interrelated and that it had a certain rhythm. Quoting such ideas discussed in Spencer, Minakata looked to reconcile both modern scientific theories, in particular that of evolutionary theory, and Shingon Buddhism.[59] Minakata wrote to Doki, "I am not the person who talks about science based on the European knowledge. Basically, Europe is only one of five continents, and science covers a wider region [than Europe]. If there is an Eastern science, as opposed to European science, I think we shall study it. Science itself seems to me a little part of what the doctrines of Shingon Buddhism can cover anyway."[60] Minakata praised Shingon Buddhism and gave it the highest status above all. However, Doki responded negatively to Minakata, furious with him. Doki thought that Minakata did not understand Buddhist principles; Buddhist doctrines could not be similar to Spencer's concepts, Doki believed. He referred to Spencer's theory as the "non-sense of a foreigner" and accused Minakata of offering an inaccurate comparison of it to Buddhist doctrines.[61]

TEACHING SPENCER AT TOKYO IMPERIAL UNIVERSITY

Buddhist thinkers who did not go abroad also had a chance to learn about evolutionary theory from the foreigners who came to Japan. In the 1890s, while Minakata was still in London, Edward Morse's successor, Ernest Francisco Fenollosa (1853–1908), taught at Tokyo Imperial University.[62] Prior to his stay in Japan, Fenollosa studied philosophy at Harvard University and became a devoted supporter of Spencer's work. Fenollosa's lectures and research were titled "Seitaigaku" (theory of social evolution), which later would be called sociology in Japan.[63]

Fenollosa, after quitting Tokyo Imperial University, became a Buddhist and a keen collector and curator of Japanese arts. One of Fenollosa's students, Kiyozawa Manshi (1863–1903), graduated from Tokyo Imperial University in 1887. While at the university he learned about Spencer's evolutionary theory from Fenollosa. On Spencer's theory of the origin of religion, Fenollosa quoted Spencer and wrote, "The early humans were similar to animals. They did not have a family, language, government, nor law, let alone a religion."[64] Kiyozawa learned this idea, without casting doubt on the "religion" at first, but he later decided to clarify the ideas of "philosophy," "religious principles," and "comparative philosophy" as follows: "Philosophy is a study of thoughts and principles. Thus, it needs logic. Science is based on experiments and observations. Thus, it relies on facts. . . . Religious philosophy looks at thoughts and principles and reveals the fundamentals of how religions came about. Religious studies examine the various beliefs and compare the actual practices and phenomena that came about from each religion."[65] Kiyozawa, in other words, thought that practicing religion could lead one to reveal more unseen principles.[66] But Kiyozawa was not the first one to come up with this idea of clarifying what religion was. He was influenced by Max Müller, who worked on

Sanskrit and religion at Oxford. Kiyozawa's senior fellows at his Buddhist sect, Ōtani, namely Kasahara Kenju (1852–1883) and Nanjō Bun'yū (1849–1927), had learned directly from Müller at Oxford University in 1876. It can be surmised that Kiyozawa had heard of Müller and some of his ideas from Kasahara and Nanjō. Kiyozawa did own Müller's book, *An Introduction to Religion* (1873). The critique of Spencer, therefore, did not come from a "Japanese" intellectual or "Japanese" religions alone. The Buddhist thinkers were influenced by Max Müller.[67]

It is important to recognize the impact of foreign ideas imported to Japan at the beginning of the Meiji era. Kiyozawa was a Buddhist in Japan, but that did not mean that his ideas were formed based solely on the existing religious beliefs in Japan. As Fenollosa taught Spencer and as Kiyozawa was inspired by Müller, Japanese thinkers took in the approaches of foreign thinkers when formulating their views on religion.

THE INFLUENCES FROM A RUSSIAN THINKER AND A PUBLICATION IN LONDON

Another large sect of Japanese Buddhism, Nichiren, rejected evolutionary theory based on the notion of competition. This sect originated in the fourteenth century and spread in Japan by the end of the nineteenth century. Among the Buddhists who advocated Nichirenism, Tanaka Chigaku (1861–1939) rejected the evolutionary theory that had been supported by the political activists, especially by Katō Hiroyuki (1836–1916). Though evolutionary theory was largely supported by the Japanese intellectuals by that time, Tanaka warned that evolutionary theory was basically wrong. The main reason was because it contradicted Nichiren's idea of *mappō*, which predicted the degeneration of the world and the subsequent end of the Dharma era. The evolutionary model based on science predicted that the world continuously evolved,[68] and for Tanaka and other Nichiren Buddhists, that could not be right.

Tanaka reexamined evolutionary theory, together with the idea of mutual aid developed by Pyotr Kropotkin (1842–1921); however, when Kropotkin's *Mutual Aid: A Factor of Evolution* (1902) was brought into Japan, Tanaka accepted Kropotkin and his re-interpretation of the evolutionary process. Where Darwin emphasized competition in nature, Kropotkin highlighted how mutual cooperation between species was the driving force behind the evolutionary process. Reading the translated work from Russian, Tanaka continued to criticize evolutionary theory but started to praise the principle of mutual aid, which was key to evolution itself.

Tanaka's son, Satomi Kishio (1897–1974), also found the idea of mutual aid interesting and also saw it in the teachings of Nichiren. Further, Satomi questioned the whole practice of examining evolution as a "modern" exercise and claimed that Nichiren, the founder of the Nichiren sect, had already had the "modern view" seven hundred years ago. Nichiren saw the importance of mutual aid long before Kropotkin. Satomi wished to demonstrate how visionary Nichirenism was to the scholars of religious studies in the West and decided to write about it in his book in English in 1923.[69]

THE PARTICIPATION IN A GLOBAL DISCUSSION ON EVOLUTIONARY THEORY

This section on Japanese intellectuals examining Spencer's works and evolutionary theory has highlighted the facts that the early discussion took place outside of Japan and that some of the key thinkers in Japan were influenced by the ideas brought to Japan during the Meiji era. Minakata learned about Spencer and developed his responses in London. While staying in Japan, other thinkers, such as Kiyozawa Manshi and Satomi Kishio, had a global vision from learning and speaking to intellectuals outside of Japan. The discussions on evolutionary theory in Japan became more and more outward-looking at the end of the Meiji, and the nationalist sentiments added momentum. It is important to recognize the foreign travels and books that had a major impact on Japanese intellectuals and their religious responses to evolutionary theory.

EVOLUTION AND RELIGION IN CHINA

It is often commented that when evolutionary theory was introduced to China in the late nineteenth century, it was quickly accepted.[70] This observation is a gross oversimplification. While it is true that there wasn't a widespread and persistent opposition to evolutionary theory, responses were complex and varied. Vocabulary such as acceptance, resistance, or reception fails to capture the dynamics. At a time when the modern categories of religion and science were still being constructed, the encounter of evolution and religion in China was much more than a process of translation and appropriation.[71] Chinese intellectuals interrogated the new knowledge and reconfigured their cultural traditions. This section offers a brief but comprehensive survey of the relationship between religion and evolutionary theory in China in the late nineteenth and early twentieth centuries. It situates the relationship in the transnational flow of ideas described in the previous sections but broadens to include China. It considers the ascendence and the criticisms of modern science as cultural authority, a phenomenon of both global and local features. And it emphasizes that both religion and evolution were being reimagined, often in relation to each other, in the discourse of modern Chinese intellectuals at the time.

To focus on the role of translation, either linguistically or conceptually, is inadequate because it was only a small, though important, part of the story. In the case of China, as well as Japan, the process of recreating evolutionism reflected a broad politico-cultural reconfiguration, in which cultural traditions were reimagined and new cultural domains, such as science and religion, were constructed. What happened was a reworking of old and new knowledge. It was not so much about translation as about *worldmaking*.[72] As will be explained below, for many Chinese intellectuals in the late nineteenth century, evolution provided an intellectual framework that helped them make sense of the world at a time of crisis and that appealed to their urgent desire for change. What mattered most to them was not the particulars of various authors and their evolutionary theories, with which they were largely unfamiliar. Instead, evolution became a worldview and a syncretic system that allowed them to navigate a rapidly changing world.

EVOLUTION AS *TIANYAN*

It should be noted that there wasn't one but several introductions of evolutionary theory to China. As early as the 1870s and 1880s, translated Western works already referred to the nebular theory, Lamarck, and Darwin. Protestant missionaries, notably John Fryer and William A.P. Martin, introduced these theories in translated works.[73] In a text written in 1877, Fryer, who had left his mission to join a translation bureau of the Qing government, mentioned that animals evolved, from lower to higher and from simpler to more complex forms, including the evolution of apes to humans. In an 1884 publication, Martin, an American Presbyterian missionary, brought up both Lamarck and Darwin. In the passage on Darwin, he touched upon the notion of struggle for survival. Neither Fryer nor Martin was critical of evolutionary theory. However, these references were scattered and fragmentary, and they largely went unnoticed. There wasn't anything in China comparable to the impact of Edward S. Morse's lectures on evolution in Japan.[74] Not until Yan Fu's writings on evolution appeared in 1896–1898 did the theory become widely known among Chinese intellectuals. Soon after, there were new venues for introducing evolution to China—first, overseas students in Japan and new-style schools in the 1900s–1910s and, later, the professional scientific community in the 1920s and beyond. Unlike in some other places, in early twentieth-century China, evolution did not cause any notable controversy that involved either public schools (as in the United States) or the scientific community (as in Victorian Britain). For our topic, it is most appropriate to start with the ramifications of Yan's work and the aftermath, roughly from the 1890s to the 1920s.

Yan Fu (1854–1921) was a product of the Self-Strengthening Movement, which consisted of a series of modernization programs launched by the Qing state in the 1860s–1890s. After graduating from the Fuzhou Naval Shipyard, Yan was later sent to study at the Royal Naval College in Britain from 1877 to 1879. For the next fifteen years or so, he taught at a naval school in Tianjin, feeling stranded and unhappy. China's defeat in the Sino-Japanese War (1894–1895) shocked many Chinese intellectuals, who were plunged into a dire sense of crisis. It was at this moment that Yan published a series of influential writings, including an interpretive translation of Huxley's "Evolution and Ethics."[75]

Titled *Tianyanlun* (On evolution, 1896), Yan's book had an enormous impact (see figure 9.1). A writer commented that he had read "more than 500 memoirs" by Chinese intellectuals of the early twentieth century, and few of authors he sampled were not influenced by *Tianyanlun*.[76] Still a schoolboy, Hu Shi, who would become one of the most iconic intellectuals in Republican China, changed his name to "Shizi" (meaning "one who fits") because of the powerful image of the survival of the fittest.[77] Scholars have written voluminously about Yan's rendition of and commentaries on the original text. As discussed, Huxley's essay was a critique of Spencer's theory of laissez-faire competition. Concerned about its ethical implications, Huxley asserted the ability of humans to practice an ethics in opposition to the "cosmic process" of survival of the fittest. Yan's interest in evolution reflected the political reality of China at the time. Both Yan and his readers recognized the harsh Spencerian world of competition—mainly in struggle among nations and races. Yet they were also drawn to Huxley's view of human capability. If the fate of China could be saved at all, it would have to come from the will and ability of the Chinese people to better themselves. After immersing himself in *Tianyanlun*, a reader reflected, "Darwin's evolutionary theory proposes the struggle for existence and survival of the fittest. The principles certainly apply to the biological world. But humans are different from other animals, in that they are able to cooperate." He further reasoned that the principle of the survival of the fittest describes the competition between nations, but the people of the same nation should cooperate among themselves.[78]

Yan's term for evolution, *tian yan*, opened the door for various interpretations and criticisms. *Tian* wasn't an exact counterpart of "nature," as in natural selection; nor was it the same as "heaven," or a realm beyond this world. For Yan's readers at the time, *tian* would likely have invoked the impression of neo-Confucian moral cosmology (which could well have been Yan's intention, with the goal to better reach the Chinese readership).[79] The second character, *yan*, indicated a process of gradual change or unfolding. Thus, the term presented not an image of random mutations but change in accordance with a certain cosmic order.

In the original text, Huxley had much to say about Buddhist ethics: "A system which knows no God in the western sense . . . which refuses any efficacy to prayer and sacrifice; which bids men look to nothing but their own efforts for salvation."[80] It is clear that Huxley considered Buddhism a much better form of ethics than those proselytized both by Christian institutions

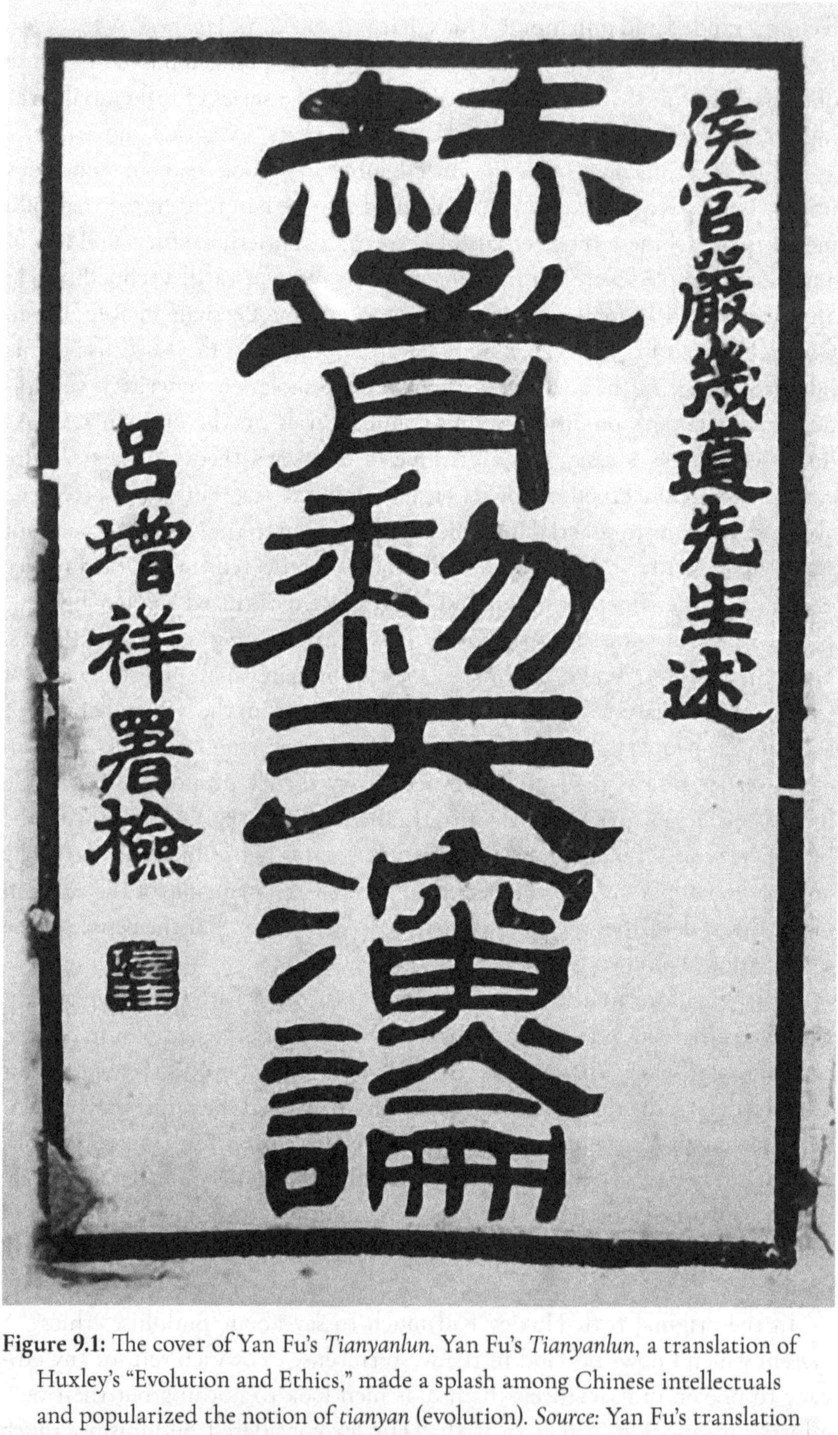

Figure 9.1: The cover of Yan Fu's *Tianyanlun*. Yan Fu's *Tianyanlun*, a translation of Huxley's "Evolution and Ethics," made a splash among Chinese intellectuals and popularized the notion of *tianyan* (evolution). *Source:* Yan Fu's translation of Evolution and Ethics, May 3, 2023, 8:47 a.m., https://en.wikipedia.org/wiki/Yan_Fu#/media/File:Tianyanlun.jpg.

and by Spencer. This feature might have also been a factor in Yan's decision to translate this particular text. Indeed, in *Tianyanlun*, Yan frequently drew on Buddhist, Daoist, and Confucian thought to expound evolution, as had Huxley in his original essay.[81]

The terms Yan adopted to translate evolutionary theory didn't last and were soon replaced by their Japanese counterparts. The influx of translated Japanese school textbooks and the large numbers of Chinese students in Japan in the 1900s brought about the change. *Tianyan*, or evolution, was now *jinhua*, a term that strongly suggested progress (or advancement). This image strongly affected how the Chinese understood evolution.[82]

INTRA-REGIONAL CONNECTIONS

Beginning in the 1890s, tens of thousands of Chinese intellectuals, including students and political exiles, flocked to Japan. These intellectuals played a vital role in introducing a new wave of Western and Japanese learning to China. The best known of them all was Liang Qichao (1873–1929), who fled China because the political reform he and others had engineered was crushed in 1898. Deft with the pen, Liang made the most of his time in Japan by churning out a torrent of essays, often drawn on unspecified Japanese sources, on far-ranging topics including politics, law, history, and science.[83] Liang was a nimble and eclectic thinker. He mixed and matched ideas from different traditions with little hesitation.

Many of his publications during this period further amplified the notion of evolution for an eager audience. Liang was one of the few readers to whom Yan Fu had sent his manuscript of *Tianyanlun* for comments. Liang praised it, though he would for some years insist that the theory of *tianyan* shared major similarities with his teacher Kang Youwei's Confucian theory of Three Eras, which charted the progress of human civilization from chaos to the Confucian utopia of the Great Unity (Kang's Confucianism is discussed at length in Wan Zhaoyuan's chapter in this volume).[84] Later, in one of his well-known essays, Liang compared the Enlightenment stages of civilization (advancing from savagery to civilization), probably taken from the Japanese reformer Yukichi Fukukawa's work, to the scheme of Kang's Three Eras.[85]

Liang used evolution, adopting the Japanese term *jinhua* (*shinka* in Japanese), to expound new social and political ideas, such as nationalism and society. Even his biographical sketch of Charles Darwin, published in 1902, reflected this tendency. Based on a Japanese source, the short essay might have been the place where many Chinese readers first learned about Darwin's life leading to his publication of *On the Origin of Species*. The title of the essay described Darwin as "the originator of tianyan studies," but throughout the paper, evolution was rendered as *jinhua*.[86] The piece contained a surprising amount of information about Darwin's research on natural history and evolution, including his investigation of orchids, his interest in pigeon breeding, and his comparison of artificial and natural selection. Liang's main concern was, however, evolution in human societies.

Comparing natural and artificial selection, Liang remarked, "Just look at dogs. There are hunting dogs, fighting dogs, sheepdogs. . . . This is not because their ancestors were born this way. It was the accumulation of many years of human effort to tame and train them."[87] The same principle applied to humans as well. Ancient Spartans regularly used this method to select their people. "When the babies were born, they tested their physical abilities. The weak and disabled were often abandoned or killed." This process of artificial selection produced the strong nation of Sparta. "Although this cruel method . . . should not be tolerated in today's civilized world, it is still possible to improve the spiritual and physical strengths of a people by proper training." Liang went on to advocate school education toward this end. This way, "not only the people of a nation, but of the whole world can progress together." Liang commented, "This is the contribution of Darwin's theory." Liang effortlessly squeezed a variety of ideas—Darwinian natural selection, artificial selection and domestic animals, competition and struggle for existence, eugenics, Lamarckian evolution, and education and human progress—all into a few pages. It might seem head-spinning to a modern reader, but it was precisely this take of evolution—all-encompassing and all-powerful, as much prescriptive as descriptive—that spoke to the needs and wishes of his Chinese audience. Because of this tendency, many Chinese thinkers (including Yan, Kang, and Liang) blended various religious or philosophical strands into their evolutionism.

Liang was only one of the many Chinese intellectuals in Japan who reworked and popularized studies on evolutionary thought (defined broadly, including sociology, history, political theory, and natural science). They eagerly translated works by, for example, the politician Katō Hiroyuki and the jurist Nagao Ariga on social evolution. Liang's brief sketch of Darwin was quickly followed by Ma Junwu's piece on a list of evolutionary thinkers, from Buffon and Lamarck to Darwin's contemporaries, published in the same popular periodical, *Xinmin congbao*, edited by none other than Liang himself.[88] The vibrant scene of the new-style periodicals, from newspapers to weeklies, in the Chinese community in Japan and in Shanghai and other major Chinese cities helped the wide circulation of these writings.[89]

RELIGION

Like evolution, religion erupted as a major topic in China in the context of modernization and national crisis. In 1898, when Yan published *Tianyanlun*, Zhang Zhidong, a prominent reformer and high-ranking official, brought out an influential text, *Quanxuepian*, or an exhortation to learning. The work staked a position regarding Chinese and Western learning, namely, "Chinese learning as essence and Western techniques for application," and called for *baojiao* or preserving Confucianism (as cultural essence).[90] In the same year, a more radical reformer, Kang Youwei, implored the emperor to establish Confucianism as a national religion, modeled largely after Christian institutions. He insisted that religion was a key factor in the success of Western civilization. Both to emulate and to resist the West, China must do the same. Yet many Chinese intellectuals recoiled from his proposal to erect Confucianism into a national religion. They believed that the move would be a departure from what Confucianism was about. It had functioned as a widely shared cultural and ethical framework, but it had never been a monolithic state institution. Not surprisingly, the proposal also faced opposition from leaders of other religions.[91]

In the meantime, there was a revival of Buddhism among Chinese intellectuals. The event that helped trigger this movement was a remarkable instance of a transregional intellectual encounter. Yang Wenhui, a pivotal figure of the movement, was hardly a traditional Buddhist. As a minor official, he visited Europe on diplomatic missions in the 1870s and 1880s. In Britain, he met the Japanese Buddhist scholar Nanjō Bunyū, who was there working with Max Müller on Indology. Through Nanjō, Yang acquired hundreds of Buddhist texts, which he published in China and spurred a surge of interest in Buddhism among young intellectuals.[92] The intellectuals who delved into Buddhism were often the same reformers who also tried to fathom Western learning; they were hungry for new knowledge and new ways of thinking. Yan Fu himself was an example. The boldest attempt at making sense of all this whirlpool of knowledge was Tan Sitong's *Renxue* (1896) or "On Benevolence."[93] A reformer who would soon die for his political cause, Tan tried to unify Buddhism, Confucianism, Christianity, and modern science. Tan argued that the universe is permeated with ether, which manifested itself in myriad forms, including *ren* (Confucian benevolence), mercy (Buddhism), soul (Christianity), and gravitation (science), to name only a few.

CRITICAL ENGAGEMENT WITH EVOLUTION

Because there wasn't a dominant religion or, indeed, a shared notion of what religion consisted of, the process and outcome of the encounter varied. For this reason, it would be hard and even counterproductive to pigeonhole different major "religions" in China, with the probable exception of Christianity, at least when it comes to examining the encounter of evolution and religion. To keep things simple, we'll first consider Christianity and then discuss other major religions.

The earliest and fiercest criticisms of evolution in China came from Jesuit priests at Zikawei (in Shanghai). The missionary station, comprising various religious, educational, and scientific departments, had a meteorological observatory and a natural history museum. In the 1870s and 1880s, Father Pierre Heude (1836–1902), a naturalist directing the museum, attacked Darwin's evolutionary theory in his taxonomical work on Chinese specimens (which remained broadly in the Cuvierian tradition).[94] Although the Vatican had not pronounced an official position on Darwin's theory, there was a vocal group of Jesuits campaigning against it.[95] Heude might have been influenced by this group. Heude didn't write for a Chinese audience, but his stance on this issue was clear. Later, alarmed by the popularity of Yan's work, a Chinese Jesuit at Zikawei, Li Wenyu, unleashed arrows at evolution. Li's opinions were straightforward. He defended creationism against the origin and diversification of life forms contained in Darwinian theory.[96] If in the beginning there were only one or two organisms, each possessed its own particular nature—just as "cattle have their nature and goats have their own nature." "How, then, could the one or two things have produced tens of thousands of descendants each with its own distinctive nature?" He insisted that there was no evidence supporting Darwin's theory of transmutation of species by citing examples such as grains found in ancient Egyptian tombs resembling those of today. Besides, there are no creatures in immediate forms of evolution. Li also rebutted the supposed claim that monkeys evolved into or shared ancestors with humans. Some superficial resemblances are no evidence, he opined. "There are many similar creatures in the world. Just because tigers look like cats, does it mean that tigers produced cats?"[97]

Protestant missionaries in China came from various denominations and countries and often held diverse opinions on many issues. However, by the end of the nineteenth century, there was an increase in criticisms of evolution among their fold. Many of them were troubled by the remarkable popularity of certain evolutionary views among the Chinese. Like the Jesuits in Zikawei, Protestant missionaries also had their own publishing houses and periodicals. In 1899, the missionary publisher, Guangxuehui in Shanghai, serialized a partial translation of Benjamin Kidd's *Social Evolution* (1894) in its periodical, *Wanguo gongbao* (*Globe Magazine*). Although Kidd was a social evolutionist, his book became wildly popular in Britain and the United States because it argued for the importance of Christianity, with its emphasis on altruism, to social evolution. For many, Kidd offered a welcome alternative to the sociology of Spencer and of Marx, both of which foregrounded competition and struggle. (It was also for this reason that Liang Qichao praised Kidd's theory, though he had little to say about Christianity. He reckoned that Kidd offered a scientific explanation and a hopeful direction for advancing the greater good for humanity than the other social evolutionists.[98]) Along similar lines, the missionary Young J. Allen published a translation of a little-known book by an American minister, James Wideman Lee, called *The Making of a Man* (1893), in *Wanguo gongbao* in 1903–1904. Lee's primary target was Spencer, particularly his worldview of survival of the fittest and his notion of the Unknowable, both of which Lee believed contradicted Christian faith and values.[99] Just like back home, there were relatively liberal or conservative groups of Protestant missionaries in China. This situation might help explain the fact that Nathaniel Gist Gee's textbook on biology, published in English for college students in China, never mentioned Darwin or used the word "evolution." Gee, a missionary biologist, was clearly sympathetic to evolutionary theory, but he was also aware of the conservative stance of some of his colleagues. He cautiously described biological evolution in such terms: "All of each of the kingdoms are supposed to be related, and it is commonly held that the highest forms that are so abundant today have, through a process of gradual development, come into existence from the lower and simpler forms that preceded them. If this is really true, the natural way to treat of these various forms is to begin with the simplest ones and then take up the succeeding groups in the regular order of their accepted relationships."[100] He organized the contents of his book accordingly.

In the early twentieth century, some Chinese Protestants also railed against evolution. One of the more thoughtful challenges was from Li Chunsheng (1838–1924). Li, a Presbyterian, was a successful tea merchant in Fujian and Taiwan. After retiring from business, he became a prolific Christian writer. Incensed by Yan's *Tianyanlun*, he tried to refute Yan's commentaries paragraph by paragraph. Evidently, Li had given serious thought to the topic. In disputing the theory of natural selection, for example, he analyzed a scenario of birds and bugs in a desert, an example briefly mentioned in Liang Qichao's essay on Darwin.[101]

At first, there were bugs of various colors in the desert. Over time, the ones that were easy for birds to spot and prey on died out, and only bugs of the color that blended into the sandy background have survived. A similar process happened to the birds. The ones that could not be easily seen by the prey (bugs) outcompeted the more colorful ones. Eventually, only bugs and birds of different shades of yellow and gray remained because they were most adapted to their surroundings. To this theory, Li advanced several objections. First, how do we know there were originally birds and bugs of different colors? It was little more than a speculation. Moreover, after the process of natural selection, why do the bugs still have the misfortune of being eaten by the birds? And why do the birds now still have the good fortunate of being able to find the bugs and eat them (otherwise wouldn't the birds have died out)? Is it because they have evolved the eyes to see bugs of the adapted color? If so, shouldn't have the bugs similarly evolved the eyes to see birds of the adapted color? He went on with this dialectic analysis until he reached the conclusion that it simply makes more sense to see the whole thing as designed by a creator. "The wisdom of the creator specially created the bugs for the birds . . . it has nothing to do with evolution."[102] Yet Li's biggest beef with Yan was his translation of evolution as *tianyan*. Li protested that *tian* was not nature, because *tian* had the connotations of a higher order. He accused Yan of playing a sleight of hand. Yan was trying to lure the Chinese into accepting evolution because they would find *tian* familiar and easily comprehensible.[103]

Overall, the readership of these Christian publications was limited to certain communities. Among the Chinese intellectuals, Confucianism, Buddhism, and Daoism, in varied forms, remained more widespread. It must be noted, however, that in this context, the Chinese intellectuals often used such terms as *foxue* (Buddhist learning) and *ruxue* or *rujia* (Confucian learning and Confucian school of thought, respectively). Mostly, their first concern was with the learning or thought rather than an institutionalized religion.

Compared with Christian writers, supporters of Confucianism, Buddhism, and Daoism were usually less hostile to evolution. Some nuance is needed, however. As we have seen, missionary writers accepted Kidd's social evolution, and many Chinese intellectuals favored certain interpretations of evolution over others. They also incorporated evolution into their preferred religious or philosophical thought, and vice versa. Yan injected *tian* into evolution, and Kang believed that his brand of Confucianism encompassed evolution. Some others found parallels. Liang pointed out the similarities between evolutionary theory and Buddhism. In evolution, the traits selected or acquired in this generation will be passed on to the next generation. There is a law that governs the accumulation of traits or modifications. This is similar to the Buddhist idea of *yinguo* or karmic action and result. The action in this life remains consequential in the future. Thus, heredity in evolution can be likened to karma in Buddhism. The evolution of humanity—including those of human groups such as nations—follow these cosmic rules.[104] Similar arguments had been made by Huxley and Japanese Buddhist thinkers (discussed above). It was certainly possible that Liang had been influenced by this discourse on Buddhism and evolution.

Yet Chinese intellectuals could certainly be critical of evolutionism. When evolutionary theory swept across young Chinese intellectuals at the turn of the twentieth century, Zhang Taiyan, a prominent scholar of the classics, questioned evolution in a famous essay, "Jufen jinhua lun" (1906). Not that he dismissed evolution. He supported Yan's *Tianyanlun*. And he accepted the theory of struggle for existence and the reality that the superior ones defeat the inferior ones. His main problem with evolution was its assertion of progress. (As mentioned, the Japanese and now Chinese term for evolution, *jinhua*, strongly suggested progress.) It isn't all progress, Zhang observed. "Good and evil, misery and happiness," and other opposite states "evolve together," like a double-edged sword. Humans might have become more intelligent over time, but have they become more moral? Influenced by Buddhism and German philosophy, Zhang proposed his dialectic theory of double-edged evolution as a corrective to the naive progressivism.[105] Like Liang, Zhang was a political exile, had Japanese friends, and read Japanese books on a range of subjects. It was probably no coincidence that Zhang's criticism of linear evolution appeared to resemble those of Japanese intellectuals mentioned in the previous section.

Like Kang's attempt at "modernizing" Confucianism (by making it relevant to a new world), some Chinese intellectuals committed to Buddhism and Daoism also tried to modernize them, often by scientizing them. One of the first Buddhist monks who invoked modern science in his interpretation of Buddhist texts was Taixu (1890–1947). Part of his approach to modernizing Buddhism was to find science in Buddhist texts, a common strategy adopted by many non-Western intellectuals to domesticate modern science. Nineteenth-century Chinese intellectuals discovered in their classics fragments reminding them of modern science, just as Indian intellectuals unearthed scientific elements from ancient Vedic texts.[106] Similarly, Taixu emphasized the common grounds between Buddhism and astronomy, modern physics, and microbiology. For example, he cited as proof a remark from a Buddhist text, "When Buddha looks at a bowl of water, he sees eighty-four thousand worms." Taixu recounted his own experience of observing microbes swimming in a droplet of water through a microscope. Taixu had no background in science.[107] However, in the 1920s–1930s, a respected scientist, Wang Jitong (also known as Wang Xiaoxu) compared certain elements of science to those of Buddhism. Wang studied in Britain and spent some time at Siemens in Germany, but he had also been interested in Buddhism since his youth. In his writings, Wang didn't directly discuss evolution. His focus was at the epistemological and ontological levels (with references to atoms, Einstein's relativity, the limits of materialism, etc.).[108] But it's evident that he would have had no problem with evolution as a natural phenomenon—though a phenomenon that should and could be transcended.

In the meantime, some other intellectuals tried to modernize and reimagine Daoism.[109] It is worth noting that in China at the turn of the twentieth century, the boundaries between science, religion, and superstition were not clearly demarcated. So many new ideas, knowledge traditions, and cultural practices intersected and converged in a rapidly changing world. Yan Fu referred to Confucian, Buddhist, and Daoist ideas in his translation of evolution. Later, he became fascinated with spiritualism, which had attracted certain prominent Chinese intellectuals aware of the kinds of activities pursued by the Spiritualist Association of Great Britain and the like. Not surprisingly, the Chinese incorporated a great deal of Daoism and Buddhism, which had rich lores of their own on the spirits of the dead, into their interpretation of spiritualism.[110]

A major modernizer of Daoism was Chen Yingning (1881–1969), a doctor of Chinese medicine. In the 1920s, Chen began championing what he called *xianxue* or the study of immortality. With his disciples, he experimented with different methods to reach immortality, including elixirs and life-nourishing techniques.[111] Like the efforts to modernize Buddhism and Confucianism, Chen's mission sprang in part from a sense of national crisis. Chen maintained that *xianxue* was a vital part of traditional Chinese culture and could aid national survival. In today's war, Chen argued, the new weapons—e.g., poison gas, airplanes, and death rays—were extremely destructive. "If you want to save our nation, what can be better than becoming an immortal" invulnerable to such horrors?[112] Like many Chinese intellectuals, Chen accepted a version of evolution that emphasized struggle for existence. In the ultimate contest between races and nations—war!—immortals obviously had an advantage. Chen's idea of evolution was ladder-like. "If apes could become humans, why can't humans become immortals?" It was also a process that required intent and effort. "If we just let nature take charge, not only might we not evolve, but we might actually degenerate. The kinds of apes that were creative have evolved into humans; the ones who lacked the creativity are still apes . . . and may even become extinct one day."[113] Just like those of Yan and many other Chinese intellectuals who were inspired by evolution, the take-home message here was twofold: the reality of struggle for existence and the necessity of improvement to become the fittest in order to survive.

Much of this genre of discourse can be understood in the context of the May Fourth Movement in the late 1910s and 1920s, a youth protest movement that attacked both foreign imperialism and traditional Chinese culture. The movement exalted modern science as one of the central ideologies that could save China from its demise, and many of its followers regarded religious practices as little better than superstition. This intellectual climate added new pressure on advocates of traditional religions to further appropriate scientific discourse and refashion themselves. Very soon, however, new questions arose from the fallout of World War I. The devastating war shook European intellectuals' confidence in their civilization, which, to generations of Chinese intellectuals, was best represented by modern science and technology. The carnage of modern warfare caused many intellectuals, in Europe as well as in China, to criticize materialism and naive scientism. In China, the famous "Science and the Philosophy of Life" debate in the 1920s was triggered by an argument about the limitations of science, which suggested that there were realms in humanity that science was ill suited for or incapable of reaching and fulfilling. Yet other intellectuals feared that this view was wrong-headed, particularly because it would undermine the urgency of China to pursue science and modernization.[114]

In any case, the debate over scientism yielded spaces for reimagined cultural traditions to claim modern relevance. Works by Chen Yingning and Wang Xiaoxu directly reflected this intellectual milieu. They both appropriated modern science and aimed to transcend it. This phenomenon captured the creative tensions among transnational intellectual currents, including evolution and religion, in late Qing and Republican China.

To sum up, other than certain Christian writers, the Chinese intellectuals of the early twentieth century weren't alarmed by the problems central to the controversy over Darwinism in Victorian Britain, such as creationism, natural selection or design, and the place of humans in evolution. They didn't fret over the seeming cruelty of the worldview (as did, say, Victorian followers of natural theology); in fact, they felt that they were already living it. In their intellectual traditions, the critical boundaries between humans and animals were ethical and behavioral. What makes humans truly human and not fowl and beasts is their moral conscience and conduct. Therefore, most Chinese intellectuals were not troubled by the idea that humans had evolved from monkeys (which did not contradict the central doctrines of Confucianism, Buddhism, or Daoism). Many Victorians, including scientists, found the randomness of natural selection meaningless and incomprehensible. In dominant Chinese cosmologies, however, the problems of creation, design, and origin were minor. In any case, they tended to construe evolution in linear, progressive terms. This doesn't mean that they simply accepted evolution without question. Rather, as we have seen, they engaged it critically in creative endeavors to reconfigure their intellectual and religious traditions.

REFLECTIONS ON THE GLOBAL SPACES OF EVOLUTION AND RELIGION

The British scientific naturalists who engaged extensively with Eastern religions were Huxley and Spencer. Both of them were also among the most important evolutionists for Japanese and Chinese intellectuals. Both Huxley and Spencer approached Eastern religions as sociologists and anthropologists attempting to place the world religions into a larger evolutionary framework. They made extensive use of the theory of ancestor or ghost worship put forward by figures who contributed to the new scientific study of religion. For Huxley and Spencer, discussing Eastern religions provided a way to subtly critique the shortcomings of Christianity in accommodating modern science, which was tied to their attempt to create a new scientific worldview as a replacement for the older Christian one. Japanese and Chinese intellectuals turned to Huxley and Spencer when discussing the relationship between evolution and religion, and some of these discussions took place within a global context. For Japanese intellectuals like Minakata, who read Spencer when traveling outside Japan, Spencer's studies were influential when they worked through the relationship between evolution and Buddhism. Not all Buddhists were as welcoming to evolutionary thought. The Chinese response to evolution was also complex. Since evolution was first introduced to China by Yan Fu through a translation of Huxley's "Evolution and Ethics," Huxley became a central evolutionary figure, although Spencer was no less important and ultimately better known. Yan Fu drew on Buddhist, Daoist, and Confucian thought to explain evolutionary theory. While Christians in China were increasingly worried about the impact of evolution, supporters of Confucianism, Buddhism, and Daoism were inspired by it and incorporated evolutionary ideas into their attempts to modernize Chinese culture and religion.

Some similarities appear in the way Japanese and Chinese intellectuals responded to evolution. In both cases, evolutionism, often construed in terms of competition, survival, and progress, had an enormous impact. That evolution became such a potent worldview in both countries had much to do with the might of Western imperialism and the status of modern science and technology that accompanied it. However, there were also criticisms. Some critics strove to find a place for cooperation and altruism. Others questioned the linear progress suggested by evolutionary theory as they understood it. Japanese and Chinese intellectuals mobilized their cultural resources to engage with both religion and evolution. They tried to modernize Buddhism, Daoism, and other religious traditions. Evolutionary theory inserted itself into the process, yet it was also transformed or reinterpreted in new intellectual or religious terms.

Differences were in part due to the dissimilar national contexts. Chinese intellectuals discovered evolution later than the Japanese. By that time, the geopolitical situation in East Asia had changed, and many Chinese intellectuals were still reeling from the shock of the defeat in the Sino-Japanese War. Indeed, many Chinese intellectuals now studied Western thought, including evolution, in Japan or through Japanese translations and interpretations. The complicated transmission and translation of evolution took place across political and cultural boundaries and across intellectual traditions. Historical actors moved in many directions, from Japan to Britain, from China to Japan, and so on, but were shaped by the different historical conditions in which they found themselves.

Evolution was both a destabilizing force and an overarching framework for making sense of a rapidly changing world. It was a major factor in how both science and religion came to be understood in different cultures in a period when a new global space for intellectual exchange was being created. This intellectual exchange was made possible, in part, due to the imperialism of Western powers in combination with the Industrial Revolution, which paved the way to increased international travel by curious intellectuals as well as the circulation of printed materials on science around the world. At the end of the nineteenth century, Chinese and Japanese intellectuals were included extensively in this new global intellectual space as never before. In the process, Chinese, Japanese, and British intellectuals participated in the global dialogue on science and religion in novel ways. Christianity, which had previously been the bearer of Western scientific and religious values as transmitted by European imperialism, was now subjected to a rigorous critique, while Eastern religions were seen in a more positive light, even by British evolutionists, thanks to how evolution became central to the emerging scientific worldview of global modernity. But there were many different varieties of evolutionism, and Japanese and Chinese intellectuals shaped their theories to fit their own religious, social, and political needs. Evolution in its many forms played a crucial role in how entire cultural traditions were reconfigured worldwide, especially in the second half of the nineteenth century and the first half of the twentieth century.

10

The Jōdo Shinshū Embrace of Science in Late Meiji and Taishō Japan

Science, Secularism, and Buddhism in the Thought of Ishikawa Seishō and Fujikawa Yū

Tomoko Yoshida and Stephen P. Weldon

On a late November afternoon in 1909, several dozen intellectuals filled a classroom in the law school of the University of Tokyo. On the platform was a portrait of Charles Darwin adorned by flowers and a collection of his books.[1] It was one of many similar events taking place around the world, celebrating the fiftieth anniversary of the publication of *On the Origin of Species* and the hundredth birthday of Darwin himself. By this time in Japan, two complete translations of the *Origin* had appeared, and the work was widely read and considered to be one of the most significant scientific publications of the nineteenth century.[2]

Prominent members of the academy from widely different disciplines attended—men who studied fields as wide-ranging as zoology and physiology, psychology and anthropology, ethics and law. Many of the men held prominent posts, being firsts of their kind in the nation: the first professor of psychology, the first physiologist, the first anthropologist, and a leading pioneer in Mendelian genetics.[3] Born during the early years of Japan's rapid modernization efforts, many of these men were educated in a brand-new, and strikingly international, educational system, one designed to launch them into the modern era and allow them to compete as equals on the world stage.

The Darwin celebration was not a huge event, but it was indicative of how

Japanese intellectuals conceived of their place in the world at this time. It also shows how science (and especially evolution) had come to play a central role in all areas of modern academic disciplines. The main instigator of this event was a forty-four-year-old medical doctor and polymath by the name of Fujikawa Yū (1865–1940). Fujikawa was already recognized as a pioneer in the field of the history of medicine in Japan. An institution builder, Fujikawa was also the prime mover behind a new topical publication that focused on the scientific understanding of humankind, the journal *Jinsei*. As the editor of this journal, Fujikawa became the hub of a large network of Japanese intellectuals. He established an academic society called Jinsei gakkai (Society for Studies of Humanity), and it was this society that organized the Darwin conference.

One aspect of Fujikawa's life that was not apparent at this conference was his commitment to Buddhism. He was an adherent of Jōdo Shinshū (the True School of Pure Land; hereafter Shinshū), a large, lay-oriented denomination. In later life, Fujikawa frequently spoke and wrote about Buddhism, especially about its relation to science. Fujikawa viewed both science and religion as indispensable for human life. Not surprisingly, he devoted his apologetic work to demonstrating the compatibility of science and Shinshū belief. At about the same time in the western region of Japan, the geologist Ishikawa Seishō (1872–1945) was pursuing a similar line of argument vis-à-vis the compatibility of science and Buddhism. Ishikawa had worked as a mining engineer before he assumed a teaching post at the University of Kyoto. Though he was not part of Fujikawa's circle, Ishikawa too was an ardent follower of Shinshū. Outside of his professional life as a scientist, he gave public talks and wrote books on the relationship between science and religion. In this chapter, we focus on these two Buddhist scientists, Fujikawa and Ishikawa, and examine how they viewed the relationship between science and religion in the early twentieth century.

The relation between science and religion in non-Western cultural contexts is a fascinating and yet still much understudied subject. G. Clinton Godart's recent book, *Darwin, Dharma, and the Divine*, surveys evolutionary thinking in Japan across a spectrum of intellectuals spanning over a century.[4] Our study comes on the heels of this book, turning attention to some of the nuances that a close focus can afford. By turning to some lesser-known intellectuals who were both scientists and religious apologists, we have illuminated a train of thinking that has not yet been discussed, namely how Jōdo Shinshū Buddhists found ways to harmonize science with Buddhist thinking.

Our study explores the faith of two devout scientists. There are many commentaries on the life and work of Fujikawa, but few of them focus on his ideas about science and religion. In fact, according to one biography of Fujikawa, he was quite unique in being actively engaged in both science and religion.[5]

In contrast to Fujikawa, Ishikawa is virtually forgotten these days, despite his relative prominence as a public intellectual in his lifetime. The activities and writings of these two men shine a light on a tradition of religious writing that has not received much attention, namely, a form of popularization based on the understanding of Buddhism and its relation to modern science.

Evolutionary thinking was instrumental in this relationship. In Japan, just as in the West, evolutionary ideas spread well beyond the field of biology, encouraging the reassessment of the nature and understanding of humanity as a whole. At the same time, the Western sciences spawned a deep faith among many late Meiji- and Taishō-era intellectuals in empiricism and scientific method. The new faith in scientific empiricism in turn led progressive intellectuals in Japan to brand traditional religions obsolete, unscientific, and superstitious. Fujikawa and Ishikawa stood out among their educated cohort in seeking to integrate modern science, including, of course, evolutionary theory, with their traditional religious values. They criticized the tendency of secular intellectuals to bash religion and presented a very different picture of the way that modern science could be harmonized with Buddhism, not unlike the way that Christians have used natural theology to defend their views.

THE DARWIN MEMORIAL CONFERENCE AND THE SECULAR ACADEMIC OUTLOOK

The Meiji period was a time of rapid transformation for Japan. The military government that had ruled the country for two and a half centuries was removed, and a constitutional monarchy took its place. The isolationist policy of the previous regime was also lifted. To cope with the pressure from the Western colonial powers, Japan strengthened its own power by quickly adopting Western science and technology. Foreign teachers were hired from Europe and North America. Japanese students traveled to the West to study in all branches of modern science, including the humanistic and social sciences. Beneath the material importation of modern Western culture, intellectual adaptation was much harder. Many Western concepts did not have exact Japanese equivalents, and it took much trial and error to find or create suitable translations. Terms such as "religion," "science," "evolution," "nature," and "society" all posed difficulties. Overall, however, Japan fared quite well in comparison to other Asian nations. By modernizing institutions across the society, they had reached a kind of parity with the West in the early twentieth century. Japanese scientists began making international-level contributions. By winning wars against China and Russia, Japan in turn was becoming a colonial power of its own.

In this modernization effort, the University of Tokyo played a leading role as a flagship institution for educating the new generation of elites. It was established in the spring of 1877 with legal scholar Katō Hiroyuki (1836–1916) as its president. Biology, including the theory of evolution, was taught right from the beginning. In the fall of that year and the next, a newly hired American zoologist Edward S. Morse (1838–1925) gave a series of public lectures on evolution. Although the theory of evolution had been trickling into Japan, this was its first systematic introduction. His lectures were well received, and a young student of Morse, Ishikawa Chiyomatsu (1861–1935), recorded those lectures and published the Japanese translation as *Evolution of Animals* in 1883, making it the first textbook in Japan outlining the theory of evolution.[6] Ishikawa (not to be confused with the geologist Ishikawa Seishō) went on to study with the German zoologist August Weismann (1834–1914) in the late 1880s and even coauthored several articles with him.[7] Back in Japan, Ishikawa then taught Toyama Kametarō (1867–1918). Toyama's work on silkworms brought him international recognition in 1906 as one of the first experiments to confirm Mendel's law.[8] Toyama himself was at the 1909 Darwin Conference and later joined the Jinsei gakkai. His pathbreaking genetics work was occasionally reported in the *Jinsei* journal.

Outside of the strictly scientific aspects of biological evolution, Katō Hiroyuki's experience was indicative of the way scientific ideas were embraced by many secular intellectuals. He had been a materialist since the 1880s. He read and embraced Darwin, Spencer, and Haeckel, all of whom became central to his thinking about humankind as well as to his political ideas.[9] Fast-forward three decades to 1909. At seventy-three, Katō was a respected and honored guest at the Darwin celebration. He gave a talk called "Mental and Social Evolution," encouraging more empirical approaches to subjects like human psychology and society.[10] Even at this mature age, Katō kept up with contemporary scholarship. Indeed, he had recently edited and written a preface for the first Japanese translation of *The Riddle of the Universe* by Ernst Haeckel (1834–1919), one of the most widely read popular science books of its age.[11]

The other featured speaker at the Darwin celebration was zoologist Oka Asajirō (1868–1944), a major popularizer of evolution within Japan. After studying at the Tokyo University, he had earned his doctorate in Germany and later taught at the Tokyo Normal College. His best-selling *Lectures on Evolutionary Theory* of 1904 was well known throughout Japan. At the conference, Oka spoke on his "Group Competition for Survival," sharing his observation that while competition occurred between groups, there was much cooperation among individuals within a group. In addition to these two papers, the philosopher Fujii Kenjirō (1872–1931) read his paper "Evolution and Ethics Studies," and psychiatrist Kure Shūzō (1865–1932) discussed his "Heredity and Psychiatric Problems."[12] As the titles alone indicate, the main thrust of the talks concerned the role of evolution as a framework for understanding humanity in the broad sense, our biological, moral, and social world. Another notable aspect of the event was that it points to the strong German intellectual influences within the group. Many of the attendees had studied in Germany, which became the choice destination for Japanese students especially in science.[13]

These tendencies also manifested in Fujikawa's journal *Jinsei*. He had launched the journal four years earlier in 1905. The German influence was obvious right from the cover, which showed *Jinsei*'s alternate title as *Der Mensch* (both words meaning *mankind*). (See figure 10.1.) It proclaimed that its purpose was to "find solutions to problems concerning social and mental lives of human beings based on the sciences." It printed both new and previously published academic articles from a wide variety of disciplines, including biology, anthropology, psychology, medicine, social and racial hygiene, law, sociology, education, philosophy, religion, and statistics.[14] The word *kagaku* used in his journal was a translation of the German word *Wissenschaft* and was broader in meaning than the English word *science*, which usually referred to natural science.[15] The journal included short summaries of recent books and articles by both Japanese and Western (mostly German) scholars. It was a kind of scientific *Reader's Digest* on the topic of humanity and its place in the cosmos. Katō subscribed to *Jinsei* and called it "my most favorite magazine these days." He liked the journal because its articles were science based and yet were written in an easily understandable manner for non-specialists.[16]

At the Darwin commemoration event, Katō emphasized that competition was the key factor in evolution and in social progress. He quickly linked this issue to his anti-Christian agenda. A staunch materialist, Katō disliked all religions save the nationalistic State Shinto. He attributed Christian reluctance to accept evolution to the idea of a loving God. In his opinion, the Darwinian survival-of-the-fittest worldview was incompatible with that notion: "In today's world of advanced science, these people cannot leave their [religious] heritage of thousands of years and still believe in a ghost-like God. Even if they do not literally believe in the personified God of the Bible, they still believe in a mysterious, supernatural, ideal deity and a force of benevolence and love. Therefore, they just cannot admit the fact that all things in the universe are subject to the struggle for survival, and evolve." More fundamentally, in Katō's view, the Christian idea of God and, indeed, all metaphysical claims of religion were undermined by scientific empiricism. "God's benevolence and God's love are nothing but dreamy ideas," he said, the reality of which "cannot by any means be verified experientially among the phenomenal facts of the universe."[17]

Oka, in his talk, did not mention religion at all, but in other places it is clear that he held a similarly dismissive view of it. His religiosity presents a confusing case for historians, because some of his ideas are resonant with

Figure 10.1: Cover of the journal *Jinsei/Der Mensch* ("Mankind") established by the medical doctor Fujikawa Yū in 1905. The journal highlighted the human sciences and reviewed pathbreaking work around the globe—with a heavy Germanic influence. It covered a wide variety of scientific and intellectual disciplines. Fujikawa's devout Jōdo Shinshū Buddhist beliefs were not evident in this professional journal of science, and it was not until later that he began writing tracts with a strong denominational focus. Front cover, *Jinsei* 1, no. 1 (1905).

Buddhist ideas. Masao Watanabe and G. Clinton Godart considered Oka's view of nature to be imbued with Buddhist elements, ideas such as impermanence and non-essentialism.[18] To be fair, both scholars were somewhat tentative about Oka's views on Buddhism. Godart called Oka's attitude ambivalent, and Watanabe asserted that Oka "probably unconsciously adopted" the idea about the transient nature of life from the famous medieval ballad *The Tale of Heike*.[19] In fact, however, though Oka was probably culturally informed by Buddhist ideas, a close inspection of his writing shows him to be disinterested and unsympathetic toward the religion. *The Tale of Heike* by itself has a strong Buddhist undertone, but Oka's reference to the tale in his essay indicates no particular appreciation of Buddhism. He simply used the decline of the Heike clan to illustrate the dinosaurs' demise. Indeed, in the same essay, Oka mentioned the futility of religion and predicted its inevitable decline.[20]

Oka's negative view of religion is even more apparent in the final sections of his book *Lectures on Evolutionary Theory*. He considered religion an institution for the uneducated. In this sense, he acknowledged, somewhat cynically, that religions could be useful in promoting social progress, but only if the right sort of religions were employed for the task. Since he believed that "all current religions are superstitions" and "superstitions are powerful things," the best path forward was to "eliminate the superstitions that contradict [progress] and protect the ones that agree with it." Oka was especially despairing of the ignorant masses: "No matter how true a scientific truth is, it is not suited for those who have below-average intellect. It is useless to teach it." He predicted, however, that as people became more educated, old-fashioned religious dogma would dissipate. Christian creationism was particularly vulnerable to education because people would simply cease to believe in it. That said, Buddhism was not an especially helpful religious outlook either. Because of its emphasis on impermanence, it encouraged people to abandon worldly life and seek escape.[21] He considered Buddhism, in essence, a pessimistic religion, a view that was quite common among young people in the country at this time. About 80 percent of the Japanese college students surveyed in the mid-1920s described Buddhism negatively, considering it pessimistic, depressing, and superstitious.[22]

Oka's secular, dismissive attitude toward religion is especially clear when it comes to his reaction to rituals, which tend to remain unchanged for centuries and lose meaning over time. Consider Oka's sardonic comment on a presumably Buddhist funeral: "People go to a funeral service of a friend and see a monk wearing a strange costume and chanting a song-like thing in a strange tone of voice. Although we keep a straight face and listen, few of us truly appreciate it."[23]

All in all, Oka and Katō represent a type of secular outlook, not uncommon in the Japanese academy, that had little concern in, or interest for, Buddhism as a practice or an institution. What we want to emphasize here is that Darwinism and evolution were not at all incidental to the way these secular intellectuals thought about science and religion in general. Over fifty years ago, historian of science Murakami Yōichirō argued that Darwinism provided an impetus for a dogmatic scientism in which evolution was regarded as an "almighty hypothesis."[24] Although his work is quite old now, and his generalization seems to be overly sweeping, in our estimation, it contains a kernel of truth, which we see emerge from the Darwin conference. The talks at that conference highlight connections between the scientific understanding of evolutionary theory and its philosophical/cultural implications, connections that sometimes surfaced explicitly. It was not an accident that Oka presented his dismissive attitude toward religion in his work on evolution. Nor was it an accident that Katō linked the theory of biological evolution directly to human social and psychological progress and the deficiencies of religion. Evolution played a central role in their beliefs. "Be it ethics, law, sociology, history, or any other science," Katō stated in his speech, "it is obvious that you cannot solve questions about them without depending upon the principle of evolution. Ignoring the logic of evolution, none of these disciplines becomes science."[25]

The two men whose ideas we explore below, Fujikawa and Ishikawa, noticed this scientism in the academy as well and staunchly opposed it. Evolution itself was not the problem. They were both undisturbed by the theory—indeed, it was clearly correct, in their view. What they objected to was its use as a means of attacking religion. By considering the thought of these religious men, we open up a window into an understudied area of thinking that complicates the overly simplistic outlook of historians like Murakami, whose picture of the scientistic use of Darwinism was based on the positivistic views of nationally prominent secular thinkers like Oka and Katō. By contrast, we turn to two scientists who were also devout religious men.

PURE LAND BUDDHISM

Ishikawa and Fujikawa embraced religious tradition, committed as they were to the Buddhist denomination of Jōdo Shinshū. Ishikawa Seishō was the son of a Shinshū priest in Aichi Prefecture in central Japan. After attending a local Shinshū school, he went to the University of Tokyo, where he studied geology, graduating at the top of his class in 1899. Professionally, he oscillated between academia and industry. For several years he lectured at various colleges in Tokyo, including the Tokyo Normal College, where Oka was teaching. In 1907, Ishikawa became an engineer at the Fukuoka Mining Bureau in far western Japan. He intended to retire after ten years but was persuaded by his friends to teach geology at the University of Kyoto. From 1917 until his retirement in 1932, he taught in both the engineering and the science departments of the university. While in Kyoto, he took on an advisory position for one of the Shinshū head temples, Higashi Honganji. He also briefly assumed the priesthood at his home temple between 1916 and 1918. Throughout his career, he wrote professionally on the subject of geology, from research articles to textbooks, but he also devoted a great deal of time to his second passion: exploring the intersection between science and religion for a popular audience. He was an eloquent speaker and was invited to give talks all over the country.[26]

Fujikawa Yū was a physician, historian of medicine, journalist, educator, and religious thinker. He was born in what is now Hiroshima City, one of the regions of Japan where Shinshū piety was strong. His family, though not the temple-going kind, was affiliated with Shinshū, and Fujikawa was familiar with its literature from his youth. The son of a doctor, Fujikawa pursued medicine at Hiroshima Medical School. Upon graduation in 1887, he went to Tokyo and began his career as a physician and medical journalist. At that point, he traveled to Germany and, from 1898 to 1900, studied medicine at the University of Jena, receiving his doctor of medicine degree there. After his return he expanded his activities to university teaching, writing on Japanese medical history, and organizing the society Jinsei gakkai mentioned earlier. Fujikawa's magnum opus, *History of Japanese Medicine* (1904), earned him a doctorate in literature from Tokyo University in 1914. In 1915, he was granted another doctorate from Kyoto University for his book *History of Illness in Japan* (1912). At around this time, however, Fujikawa shifted his focus from medicine to education and religion. In 1915, he began writing about his own Shinshū faith, and three years later his connection to the secular *Jinsei* project ended when the journal was forced to close its doors as a result of the rising cost of paper. During the last twenty-five years of his life—with the same enthusiasm he had had as a younger man—he created a religious society, published a Buddhist journal, headed a cultural institute for women, gave public talks, and wrote numerous articles and books on religion. The main topics of his interest were the relationship between science and religion and the religious lives of ordinary people.[27]

A form of Mahāyāna Buddhism, Shinshū doesn't conform to the typical austere philosophical image of Buddhism common in the West. The Western conception of Buddhism came about in the nineteenth century among intellectuals studying the philological traditions found in ancient South Asian texts. It was this version of Buddhism that Victorian intellectuals in Europe became attracted to, as discussed in the chapter by Fan and Lightman in this volume, because of its non-theistic and agnostic outlook. Contemporary Mahāyāna schools of Buddhism prominent in East Asia diverged significantly from this image. Unlike the abstract meditation-based, monastic practice discussed in the ancient texts, Mahāyāna schools invoke many deities (Buddhas and bodhisattvas), and they serve lay devotees. Jōdo Shinshū fits this description. It is a faith-based religion that teaches the potential of salvation for all by Amida Buddha, who resides in the Pure Land of Bliss, a mythical place similar to the Christian Heaven. Followers do not meditate but rather attend sermons at local temples to learn the implications of Amida's compassion for their daily lives. And yet, if examined more closely, it becomes apparent that this is not a simple otherworldly deity worship. Shinshū shares its goal—attainment of *nirvāṇa*, peaceful mental equanimity—with other less-theistic traditions such as Zen. It is only that their methods are different.

The Pure Land scriptures that originated in India tell a unique salvation story. Amida Buddha, who now resides in the Western Paradise of Pure Land, once vowed to welcome anyone to his land who sincerely wished to be born there after death. Unlike the defiled land of ours, conditions in the Pure Land are ideal, and, once there, people can eventually reach enlightenment. Note here, Amida is neither a creator nor a judge but simply a savior. His sole function is to aid believers to attain *nirvāṇa*. Pure Land is an example of many other imaginary Buddha Lands in Mahāyāna Buddhism. The cult of Amida and his Pure Land became extremely popular in East Asia and inspired many Buddhist arts.

In Japan, two medieval masters, Hōnen (1133–1212) and his student Shinran (1173–1262), reformulated Pure Land Buddhism into a distinct tradition that was widely accessible to the common people. Hōnen taught that salvation was available by invoking Amida's name, in the formular *Namu-amida-butsu* [homage to Amida Buddha]. No other practice was needed. Shinran distilled Hōnen's teaching further and asserted that as soon as one accepts Amida's compassion, that person is already saved. No need to wait until death. Furthermore, faith is spontaneous because it is a gift of Amida. One need not pray even for salvation because prayer is a selfish contrivance—Amida's compassion is already working. The only thing that believers can do is to simply say the name of Amida in gratitude. It is to Shinran that Shinshū followers trace the foundation of their tradition. Shinran himself never thought of establishing his own school of Buddhism, but his teaching gained wide popularity in the fifteenth and sixteenth centuries, making Shinshū a powerful, dominant sect.[28]

The hallmark of Shinran's teaching is his emphasis on the spontaneity of faith. This makes sense when we think about Buddhist ideals, such as *nirvāṇa*, suchness, and Buddha-nature. All point to the original, holistic oneness that defies description and that cannot be strived for but can only be realized by intuition. As Buddhist studies scholar Paul Williams puts it, "Faith is not a volitional belief in something, but an articulation of our Buddha-nature."[29]

Shinran used the word *jinen* (naturally so) or *jinen hōni* (natural flow of dharma) to express the spontaneous nature of this faith. In his pastoral letter of 1258, Shinran said, "*Jinen* signifies being made so from the very beginning. Amida's Vow is, from the very beginning, designed to bring each of us to entrust ourselves to it . . . and to receive us into the Pure Land; none of this is through our calculation." In the same passage, Shinran went on to explain that the specific name of Amida is used as a devise for us to see the Supreme Buddha, who is formless. But then Shinran warned people not to fall into a trap of doctrinal debates: "After we have realized this, we should not be forever talking about *jinen*," for to do so would rob the spontaneity from the spontaneous working of Buddha's vow.[30]

Both Ishikawa and Fujikawa read Shinran avidly and found meaningful links between Shinshū discourse and modern science. Metaphorical references to Amida Buddha and the Pure Land, as well as admonishment against dogmatic debates, gave the apologists flexibility to accept science without it contradicting their faith. Furthermore, to say we are already embraced by Amida's compassion collapses the dualism between Buddha and sentient beings—it was this aspect of Shinshū thinking that made Fujikawa so interested in monism. Finally, both men highlighted a linguistic connection that further encouraged a bridge between Shinshū and natural science, namely, Shinran's use of the word *jinen*. As we discuss more in detail later, this word became the translation of the English word "nature" in the late nineteenth century; this time though, the pronunciation shifted to *shizen*.

Although secularism reigned in the academy in the early twentieth century, Japan remained a country with a historically eclectic religious tradition. For hundreds of years, Buddhism, Shintoism, and Confucianism had existed side by side, more often than not overlapping with one another. To this mix, once-banned Christianity was allowed back in 1873, adding more complexity to the religious dynamics. In the early years of Meiji, Buddhism came under fire by nationalistic Shintoists, and the number of Buddhist temples was drastically curtailed. The persecution did not last long; by the turn of the twentieth century, Buddhism had made peace with the political attacks, accepting imperial rule and promoting various types of reforms to reframe the religion for modern conditions.

Statistics from the late nineteenth and early twentieth centuries show the dominance of Jōdo Shinshū among the Buddhist denominations. For example, according to a 1922 government survey of Buddhist denominations, Shinshū was the largest in the country, in terms of both the number of temples (making up 28 percent of all Buddhist temples) and the number of parishioners served (37 percent of all households).[31] It is not surprising, then, to find that many of the Meiji Buddhist leaders had Shinshū roots. We name just a few individuals relevant to our discussion: Inoue Enryō (1858–1919) split institutionally with Shinshū and established a philosophy university of his own. He used Western philosophy to reinterpret and revitalize Buddhism and became one of the primary instigators of the Buddhist modernization movement. Kiyozawa Manshi (1863–1903), Inoue's friend and fellow philosopher, took a more inner-looking, spiritual path, calling for an austere "Spiritualism." Unlike Inoue, Kiyozawa maintained his ties with the traditional temple organizations. In the 1890s, both Inoue and Kiyozawa wrote critiques on evolution from the Buddhist philosophical points of view.[32] Buddhist studies scholar Takakusu Junjirō (1866–1945) studied with the famed Orientalist Max Müller at Oxford and later taught Sanskrit at the University of Tokyo. Takakusu is remembered for his hundred-volume *Taishō Tripiṭaka* (1924–1934), a compendium of Buddhist scriptures still used as the standard source in Buddhist studies. Takakusu was a friend of Fujikawa's, and they collaborated on a number of projects. Although Shinshū was a dominant Buddhist denomination, and the achievement of its leaders was impressive, it was still a minority religion. Their distinct religious view was not well understood by secular scholars. This situation prompted Ishikawa and Fujikawa to write apologetic works.

ISHIKAWA SEISHŌ

Ishikawa wrote a number of books on science and religion over the years, all with similar titles: *Revelations of the Universe* (1906), *Wonders of Nature* (1907), *Natural Science and Buddhism* (1907), *Nature and Buddhism* (1916), *Natural Science and Religion* (1919).[33] These books were written when Ishikawa was in his thirties and forties, about the time he worked as a mining engineer, and some appear to be based on his talks. The advertisement for one of his books describes him as follows: "Professor Ishikawa is a notable scientist of our country, but also an ardent believer of religion. He observes nature with religious eyes and critiques religion with scientific eyes."[34] Even when the book title said "religion," Ishikawa mostly spoke about Buddhism, especially Shinshū. Below, we summarize Ishikawa's thoughts on science and Buddhism. Since his books on science and religion appeared in a relatively short period of time and there is a great deal of overlap in content, we treat them together as a single group.

The way that Ishikawa discusses science and religion shows how he understood the similarities and differences between them. On a fundamental level, the two practices were alike insofar as they each begin from our observation of nature and our desire to know more about it. Zoologists, physicists, and chemists alike, when they study nature, come to appreciate its intricacy and realize the existence of universal natural law behind it. In this way, he argues, they "enter the gate of religion without even knowing."[35]

Ishikawa criticizes scholars who boast of their knowledge and claim that religion is no longer necessary. As one reads Ishikawa's critique here, it is hard to imagine that he wasn't referring implicitly to Oka's extremely popular *Lectures*, published only a few years earlier. These intellectuals, according to Ishikawa, hold totally dismissive ideas about religion, considering it merely a tool for edifying the uneducated masses, or, alternatively, believe that it is entirely useless and will simply disappear as science advances. Such views seriously underestimate the importance of religion, he argues; these scholars just do not understand religion.[36]

To contradict these misperceptions, Ishikawa points to the case of Buddhism. There is something fundamentally scientific about Buddhism, he says, insofar as this religion arose out of Buddha's observation of the natural world—namely, witnessing the natural course of human life: birth, aging, sickness, and death.[37] It was that observation of humanity's existential nature that launched the historical Buddha on his quest for enlightenment. This may be a challenging idea for readers in the Western Judeo-Christian tradition, for whom the human world and the natural world are entirely separate, but those whose outlook is Buddhist find nothing strange at all about situating the human world within the purview of nature.

Although religion and science have similar beginnings, Ishikawa sees crucial differences in their approaches and their end results. The scientific approach is analytical, depending upon our intellect; by contrast, the religious approach is intuitive and holistic, derived from emotion. Ishikawa likens the two to organ systems in the body. Science is like the nervous system, religion the circulatory system. The former is cool, the latter warm. Notably, he asks his readers to consider the warm religious circulatory system as the place where religious sentiments flow: namely Buddha's compassion and—in a striking effort to be more universalistic—God's love! The upshot of these differences, he explains, is that science and religion, though they have similar origins, cover different areas of human life, intellectual and emotional, and, in the final analysis, cannot conflict with one another.[38]

One aspect of life only religion can successfully address, Ishikawa maintains, is the emotional or psychological needs of humans. The primary goal of religion (which, again, in his writings refers primarily to Buddhism) is to provide emotional satisfaction or peace of mind. And in this respect, science fails spectacularly. Because scientific theories are never final and because they are always subject to revision, science can never provide complete satisfaction. For example, when we look at the science of botany, despite all of its great advances, we still do not know why a plant only produces seeds of the same plant: "Melon vines do not produce eggplants," he notes.[39] And the same problem exists throughout the scientific disciplines. It is a function of the scientific method that our knowledge about the world is always tentative and incomplete, and that is deeply unsatisfying psychologically for human beings.

In the early years of the twentieth century, this concern about the failure of modern rationalistic epistemology took center stage in the popular mind as a result of a tragic suicide at the site of one of the nation's most prominent landmarks, the majestic Kegon Falls. In the spring of 1903, a promising high school student threw himself off the waterfall to his death, leaving a note at the site, which read in part "truth about all things is incomprehensible." This highly publicized suicide became a source of intense contemplation for Japanese intellectuals for years following the incident.

The secular zoologist Oka Asajirō, mentioned earlier, commented on this event in his *Lectures*, defending science against the sweeping concerns of philosophy. From a scientific point of view, he explained, human brains have evolved only to the current limited capacity, and it is folly to think that we can know everything about the universe. Philosophers, however, create unreasonable expectations these days by making overly ambitious and speculative claims. The proper attitude to take is to adopt a humble and realistic outlook on our knowledge and stick with scientific facts.[40]

Ishikawa, too, addressed the psychological predicament of modern knowledge: "No wonder skeptics have to shout 'heaven and earth are incomprehensible!'"[41] Yet, in contrast to Oka, he looked to religion for the ultimate psychological solution.[42] How important is scientific knowledge for a person's life?, he asked. Not essential. Consider an old woman who knows nothing about physical sciences but is content with her faith. "The old woman who is at peace with the question of life and death, . . . has satisfied her fundamental desire for knowledge. We can call her a very wise person."[43]

As if to respond to Oka's claim that religion is obsolete and unscientific, Ishikawa highlights the ways in which Buddhism can be harmonized with natural science. He agrees that some beliefs do contradict scientific facts, which makes them difficult to reconcile with modern knowledge and therefore problematic for their adherents. He points to the biblical creation story and the ancient Buddhist cosmology of Mount Sumeru as examples of religious views that need to be revised in light of modern knowledge.[44] True faith, he says, must satisfy all aspects of the human mind—emotion, intention, and intellect. If a religious claim contradicts empirical science, then dogmatic adherence to it will only result in an unhealthy faith, a delusion.[45] In this respect, Buddhism comes out ahead of other faith systems. Buddhism, in comparison to some forms of Christianity, offers a distinctly healthier outlook: "Unlike Catholicism, Buddhism does not object to natural science or try to stop its development. Therefore, believing in Buddhism does not at all interfere with the study of natural science; nor does scientific advancement become an obstacle to Buddhism."[46] Indeed, "natural science and the religion of the future should be harmonized and made closer to each other."[47]

Ishikawa also countered a common, negative view of Buddhism that was widely held among Japanese people of his day. The Buddhist idea of impermanence was often considered pessimistic, but Ishikawa didn't agree. As a geologist, he was fascinated by nature and its changing faces. Impermanence was not a problem for him. Changes and transformations are observed everywhere in nature, from rocks and stars to animals and plants. Inorganic matter also changes: igneous rocks are broken down by water to become sedimentary rocks, and they over time transform into metamorphic rocks by pressure and heat.[48] Similarly, scientific theories also change. Throughout his writings, Ishikawa explores those transformations in the natural world as well as in the scientific understanding of it. Diamonds are not forever, and you are mistaken if you think the earth is a perfect sphere, and so on.

At this point, several very interesting conclusions emerge. First of all, he believed that all things have a purpose or meaning. Second, he believed that the purpose is change itself. In other words, to transform, to become different, is the purpose of all things in the world. And third, he assumed that change and transformation was progressive. This is especially significant in light of the understanding of evolution. Ishikawa was unbothered by impermanence and change because he thought, without any clear explanation, that the result of change would be progress: "The significance of all things clearly lies in their changes in the direction of progress. The purpose of all things is their own betterment and development."[49]

Another reason for his equanimity regarding the Buddhist concept of impermanence is that this notion is not an endpoint in Buddhism. It is only the first step. Buddhist doctrine explains this in terms of the three seals of Buddhism (*sanbōin*): *shogyō mujō* (all phenomena are not constant), *shohō muga* (all existence lacks permanent self), and *nehan jakujō* (*nirvāṇa* is peace). One must first realize the impermanence of things and the contingency of the self before moving on to achieve the ultimate peace of *nirvāṇa*. By highlighting the third seal, with its focus on achieving peace, he refuted the common image of Buddhism as a pessimistic resignation. Putting it together, he explains that "the true meaning of all things is evolution and progress, while each individual's goal lies in the peaceful realm of absolute *nirvāṇa*."[50]

Perhaps most important to the consideration of science and Buddhism is Ishikawa's use of the term "nature." The word is not easy to deal with because it was translated in multiple ways in the post-Meiji period, only belatedly becoming fixed. The translation of the English word "nature" took decades to settle. Until around the turn of the century, the word *banyū* (all things) was commonly used, which was gradually taken over by the current *shizen*.[51] But in the early twentieth century both were still used, sometimes occurring side by side in the same text, as we see happen in Ishikawa's essays below. More important is the semantic shift in the word *shizen*. In premodern times, the word was more commonly pronounced as *jinen* and was usually used adverbially to mean "naturally" or "spontaneously." In Chinese characters, both *jinen* and *shizen* are written in the same way, as 自然. By the late nineteenth century, this medieval word, now pronounced *shizen*, was used as a noun, referring to "nature" as the object of study in natural science. Differences in the nuance and implications of the two words are quite significant. While *jinen* stresses spontaneity (natural occurrence), *shizen* is more objectified and deals with a natural world separate from humans. Shinran, as we saw in his letter, used the word *jinen* to describe how salvation happened *naturally* by the faith given by Buddha. By conflating the two uses of this word, the scientific and the religious, Ishikawa manages to bring Shinshū doctrine into alignment with modern scientific notions.

In Ishikawa's writings, we see the blurring of the distinction between the two words, the medieval *jinen* and the modern *shizen*. "What is called *shizen*," he says, "is the state of the natural world (*tennenkai*) as it is. . . . The 'caw caw' of a crow and the 'tweet tweet' of a sparrow are all *shizen*. These are *shizen* as they are without any decoration or deception. In other words, these are the conditions of nature (*banyū*) revealing their original, and true face."[52] From here, it is easy to slide into the concept of *jinen* in its adverbial sense, meaning "naturally." Ishikawa points to a number of uses of *jinen* in the Larger Pure Land sūtra: in the Land of Bliss, music sounds *naturally*, hunger is relieved *naturally*, and so on. Ishikawa then cites Shinran's letter on *jinen* and comments on how important it is to entrust ourselves to Amida's compassion and let go of our own calculating minds. There is an overlap here between the natural world as it is and the blissful mental state void of contrivances. Then, it is not so strange for Ishikawa to declare that Mahāyāna Buddhism and, in particular, Pure Land Buddhism is a natural religion (*shizen shūkyō*). All natural phenomena are workings of Buddha: "Blooming of flowers and singing of birds . . . these are all manifestations of Amida's great Vow."[53] To see nature as animated and sacred, however, is not uncommon in Buddhism. For example, Shinran's contemporary, Zen master Dōgen (1200–1253), also identified Buddha-nature with natural phenomena.

In addition to the traditional Shinshū teachings, Ishikawa absorbed modernizing trends from a variety of contemporary religious and intellectual movements. For example, Ishikawa's ideas about the future of Buddhism expressed in his 1907 book were almost analogous to the manifesto of the New Buddhist (*shinbukkyō*) movement, which proposed reforms like making Buddhism scientific, simplifying rituals, and promoting dialogue with

Christians.[54] In his eagerness to adopt the new ideas, Ishikawa's arguments sometimes diverged from the traditional Shinshū piety, but his basic arguments were consistent. As a scientist, he observed and appreciated nature in its constant transformation. He recognized the insufficiency of the scientific method for satisfying human psychological needs. For that reason, he turned to his Shinshū faith, all the while acknowledging the necessity to reform the traditional religion to meet the intellectual standards set by modern science.

FUJIKAWA YŪ

Fujikawa's views on science and religion are similar to Ishikawa's. It is possible that the two were aware of each other's works, but they were geographically separate and probably not personally acquainted. Unlike Ishikawa, Fujikawa did not start writing about religion until he was around fifty years of age and, as a result, his works show more maturity and are more carefully and deliberately constructed than those of Ishikawa. Fujikawa was also more directly influenced by German literature. He read widely in the German language on subjects ranging from medicine to philosophy, and he kept up with the most recent publications from that country.

His earliest writings on religion developed upon the similarities between Shinshū teachings and Ernst Haeckel's monism. Although Fujikawa studied in Jena during Haeckel's tenure, there is little indication that he had any close interaction with the controversial German biologist. Haeckel's works were well known in Japan, but some Japanese scientists—including Oka—cautioned against his larger philosophical agenda, considering it to be scientifically dubious.[55] Fujikawa was not among those critics. In his early encounter with Haeckel's works, he saw some ideas that he appreciated, in particular the German biologist's strong religious outlook, a worldview that many Western intellectuals interpreted as unorthodox and, in fact, irreligious. (We note the vastly different way that American biologists encountered him, as recounted in Halverson's chapter in this volume.) For Fujikawa, Haeckel's monism provided a bridge between science and religion, and in *Jinsei*, he printed a number of articles discussing it. Most notably, in 1906, the journal carried a serial translation of Haeckel's famous 1892 Altenburg address, "Monism as a Bond between Religion and Science."[56]

In 1914, Fujikawa himself wrote an article titled "Monistic Religion." In it, he outlines current monistic philosophy in Germany and expresses his support for that view: "Professor Haeckel renders monism as a chain to connect religion and science. As a faith for natural scientists, it does not require a personified deity; it finds divine nature in all things; it defines God as an expression of God-nature (Truth, Goodness, and Beauty); it is a type of pantheism."[57] He mentioned that Buddhism, and especially Shinshū, did not conflict with natural science and could be considered a monistic religion, and he indicated his desire to write about Shinshū from this perspective.[58] Fujikawa kept his promise and wrote an article, "Holy Man Shinran," in a popular magazine, *Chūōkōron*, in 1915,[59] which he revised and published as a book by the same title the following year.[60]

Holy Man Shinran grew out of, and was written as a response to, his experience with the secular academy. In this sense, Fujikawa was not unlike the German Protestant theologian Friedrich Schleiermacher more than a century before, who explained religion to "its cultured despisers."[61] Similarly, Fujikawa explained that his "intention for writing this little essay was to recommend Shinran's teaching to the new generation of educated people, especially those engaged in natural sciences like myself."[62] Without his naming names, it is clear that fellow intellectuals like Katō and Oka were on his mind, men who were quick to dismiss religion as superstitious and obsolete without even knowing what religion was all about. Fujikawa also said, "I have been listening to Shinran's teaching for thirty years. I regret, however, that his teachings seem to be misunderstood by today's educated class."[63] He probably thought that by demonstrating that Shinshū was a type of monism, its ideas would appeal to secular intellectuals.

How, then, did he relate Shinran's ideas to monism? Fujikawa begins by quoting a long passage from Shinran, which partially reads as follows: "*Nirvāṇa* is called extinction of passions, the uncreated, peaceful happiness, eternal bliss, true reality, dharma-body, dharma-nature, suchness, oneness, and Buddha-nature. Buddha-nature is none other than Tathagata [Buddha]. This Tathagata pervades the countless worlds; it fills the hearts and minds of the ocean of all beings. Thus, plants, trees, and land all attain Buddhahood."[64] Fujikawa claims that all these concepts Shinran enlists—dharma-nature, suchness,[65] oneness, etc.—are, in fact, none other than what the monists call *Weltwesen* (universal essence, world-nature).[66] Fujikawa agrees with Haeckel and his followers that the modern scientific understanding of the world is incompatible with a dualistic view that separates humans and nature. Darwinism destroyed the human-centered worldview, he acknowledges, and it is now understood that animals and plants, inorganic and organic matter, all originated from single unified substance and are governed by the same law of nature.[67] When Fujikawa talks about monism as a type of pantheism, he means that it recognizes a universal spirit existing in all things, and he likens this to the idea of a ubiquitous Buddha-nature, as explained by Shinran.[68]

Fujikawa also finds a similarity between the monists' understanding of the soul and the Buddhist view of the self. Both reject the idea of a personal and eternal soul that is separate from the body and that survives the body's death. As we've seen, one of the crucial aspects of Pure Land Buddhism is the notion of a person's rebirth in the Pure Land. Fujikawa accepts this idea but does not believe that it is a literal rebirth in a physical place. Rather, "being born [in the Pure Land] is to become Buddha, [or in other words] to go back to the universal energy."[69] The Pure Land, in Fujikawa's hands, is a metaphor for a state of being, not a location: "There is no need to debate where the Pure Land of Bliss is. . . . What [has previously been] said about the afterlife and human rebirth in the Pure Land of Bliss means that parts of the universal energy that constituted my body dissolve and go back to the universal energy."[70] Fujikawa nonetheless prefers to retain the religious terminology because he sees it as necessary: "To [carry on] religious life, it is useful to add religious color and talk about the Pure Land."[71] The geographical metaphor of Pure Land provides a concrete image for believers that helps them grasp the idea of world-nature.

The conclusion of *Holy Man Shinran* deals directly with the relationship between science and religion. Similar to Ishikawa, Fujikawa acknowledges the differences in the purposes of the two practices, the former based on intellect and the latter on emotion. We need science to give us understanding. We need religion because it provides us with an ideal that shows us how to live: religion "can turn a selfish, materialistic life to an altruistic spiritual life"; "supreme art and noblest action come from religion." Though different, religion and science are both necessary for a coherent life. They both must work together to develop a complete person: "Emotion and intellect are both wondrous functions of our spirit"; "to study [science] and understand the real intention of the Buddha has a great effect in satisfying one's religious sentiment."[72]

Later in 1916, Fujikawa wrote another book on Buddhism called *Diamond Mind*, which was in a much more colloquial style. He admitted in the preface that he wrote his previous book *Holy Man Shinran* in an "objective and superficial" manner in an attempt to reach out to secular intellectuals. And it was only when some readers suggested he publish a book expressing his personal understanding of Shinran's teaching that he put together *Diamond Mind*, the title of which refers to unwavering faith. This book is a collection of his talks, probably given to lay Shinshū audiences. His speeches are down to earth, and the chapters read like sermons delivered at a village temple. There is no philosophical discussion comparing monism and Shinshū here. Instead, he gives accessible lessons on current scientific theories accompanied by passages explaining their meaning in a Buddhist framework. The point was to show that the study of science could provide an understanding of "the real intention of the Buddha." This work, more than *Holy Man Shinran*, provides insight into how Fujikawa made sense of the world through Buddhist teachings.

The recurrent topic in *Diamond Mind* is the interconnectedness of all things. The concept of dependent origination and its concomitant idea of no-self are fundamental to Buddhist ontology. Dependent origination refers to how things arise by cause and effect, while the notion of no-self describes the idea of self as a temporary construct existing only in the interrelationships with other things in the world. We suffer because of our clinging to our own existence, wrongly believing it to be independent, free, and unchanging. The traditional Shinshū remedy for this problem is to let go of our selfish attachment by accepting Amida's compassion. Fujikawa elaborates this Buddhist path by turning to examples in contemporary biology. For example, we cannot control the natural aging process of our own body because of the physiology we inherited from past generations: "Although we consider our body and mind as 'mine,' they are products of karmic ties [*sanze inga*] and we cannot control them at will." At another point, he explains that humans are by no means special or separate from the rest of nature. "Our body and mind are a part of the universe, and a part of the universe is our body and mind." Humans, animals, and plants all share the atoms making up the universe and are all "under the control of the wonderous laws of the universe."[73]

Genetics, in particular, helps us realize that we are only a small part of a long chain of beings. Our germ cells come from our parents, and theirs from their parents, and so on. The process works in the same way with plants: As if to answer Ishikawa's quandary of why melon vines do not produce eggplants, Fujikawa says, "It is not that a rice plant produces rice seeds, but it is a rice seed that produces a rice plant." So the true parents are not the plants but the seeds. These seeds, from the "beginningless beginning," have been transmitted for generations and are still producing rice plants. Likewise, for human beings, in addition to our mother and father, he talks about our true parents—referring to our germ cells. Fujikawa compares these invisible primordial parents to Buddha. Just as we know the compassion of our mother and father, we can reflect on how our germ cells, our true parents, have given us being, and we can feel the compassionate embrace of the Buddha.[74]

From the recognition of the interconnected nature of our existence flow emotions such as indebtedness, humility, gratitude, and the appreciation of the Buddha's compassion. These are the type of emotions Fujikawa considers religious and essential to humanity. Once we realize that we are dependent on others and that "we cannot accomplish anything on our own, we will inevitably turn to the absolute, unlimited, great power."[75] And yet, many intellectuals of his time were too proud of their own abilities and did not care to acknowledge and rely on a power greater than themselves: "There are many people who say religion is for the ignorant and illiterate, and it has no use for the learned. Many of them are arrogant people. They are attached to themselves, think on their own, and believe they can control the affairs of the world. Many of them hold a crooked view, thinking about themselves and not paying attention to others around them."[76] Such selfish pride needs to be tamed: "Once we realize that we are surrounded by the compassion of all things around us, we come to realize that we cannot behave as we please."[77]

Fujikawa thought religious emotions were essential and universal to humanity. In contrast, elements of religion, such as dogmas and rituals, vary and may become outdated over time. Like Ishikawa, Fujikawa noted that Buddhism arose out of the insights gained when the historical Buddha observed nature (i.e., the nature of human life). In this respect, Buddhism was not fundamentally different from natural science in the way that it sought to understand the world. Buddhism, like science, admitted change. Any elements of Buddhism that no longer are in accord with modern scientific knowledge can be revised.[78]

Fujikawa, like so many other Japanese Buddhists, believed that his religious tradition was superior to Christianity in its capacity to accept scientific truths. During a talk in August 1925 on science and religion at a local Buddhist association, Fujikawa highlighted this by turning to the news from across the Pacific. The notorious Scopes trial had just ended, making headlines around the world. "According to the newspaper," Fujikawa noted, "this kind of view of nature [Christian creationism] is causing a problem in America. In our country, those who finished elementary school and middle school were not taught such a view, and probably no one believes in it." Fujikawa explained that Buddhism was much more easily aligned with science because it originated through straightforward observation of nature.[79]

What he didn't mention, however, was the elephant in the room, namely the country's increasingly repressive enforcement of State Shinto ideology, which demanded complete loyalty to the divine emperor and the ancient Shinto mythology that justified it. Although a crackdown against the teaching of evolution due to its contradiction of imperial mythology did not take place until the mid-1930s, by 1910, the writing was already on the wall. The High Treason Incident that year, which led to the execution of twelve activists on the grounds that they plotted the assassination of the emperor, was a chilling reminder that certain topics were too dangerous to discuss publicly. Around this time, even Oka's bestseller, *Lectures on Evolutionary Theory*, was temporarily banned.[80] Despite this, Fujikawa, even as late as 1925, seemed still oblivious to and unconcerned about the political challenges evolution created under the pressure of State Shinto.

Fujikawa's thoughts on religious tradition were complex. When he heard intellectuals dismiss religion, he recognized that many of them were conflating the antiquated rituals and dogmas with religion itself. Fujikawa believed that the rituals and dogmas were merely the outer forms of religion, not anything fundamental to human spirituality. Take, for example, the positivist Auguste Comte (1798–1857), or Haeckel, or Haeckel's monist protégé

Wilhelm Ostwald (1853–1932). All were "simply saying that beliefs or rituals of traditional religions no longer have meaning in today's society." At the same time, none of them denied the essence of religion, Fujikawa thought. They all came up with substitutes; Comte created his Religion of Humanity, Haeckel and Ostwald pushed monism. These religious surrogates helped "fill in the vacuum" that came to exist in the spiritual life of secular modernity.[81] Fujikawa's thinking in these works seems to parallel similar sentiments regarding monism and positivism in other countries at this time, such as the way that progressive thinkers in both Ottoman Türkiye and Argentina drew inspiration from these two radically non-Christian philosophies. In all three countries, Haeckel and Comte were embraced for the ways that their worldviews bridged the apparent divide between science and religion, even though the ways in which they used these science-based worldviews differed.

Fujikawa's ideas developed over time, however, and he began to back away from Haeckel. After his initial endorsement of monism in 1914, his opinion on Haeckel shifted. He began to read the man critically, putting Fujikawa more in line with Haeckel's religious critics in America and elsewhere. In his later writings, Fujikawa remarked that Haeckel's understanding of religion was superficial and his criticism of Christianity overly harsh. Given Fujikawa's strong attachment to his own religious tradition, one can imagine that he must have winced when he read Haeckel's insulting description of God as a "gaseous vertebrate."[82] For his part, Fujikawa always adopted a pious attitude toward traditional religious forms, a point that was clear in his lay sermons, where he always used honorifics to refer to the Buddha.

Fujikawa returned to the concept of religious feeling once again in his book *Medicine and Religion*, written in 1937, three years before his death. The book opens with the words of Christoph Wilhelm Hufeland (1762–1836), a great German physician whom Fujikawa admired: "If medicine does not become religion, it is a most difficult, hopeless, and unappreciated craft." Hufeland viewed the physician's job as a sacred duty to God, and Fujikawa appreciated the sincerity and humbleness of the German physician.[83] Such humility could only come from a religious mind. Indeed, religious feeling arises from the depth of the human mind and not from a particular belief, ritual, or knowledge. It is an awareness and appreciation of something greater than oneself. Fujikawa used a variety of words for this "something"—*Weltwesen*, Dharma-nature, Buddha, and so on. This "something," he explained, is not separate from human beings but rather directly connected to us insofar as it is the mind that creates Buddha, not the other way around.[84] There is no supernatural creator deity, nothing supernatural at all. Indeed, in the conclusion of this book, Fujikawa suggests that this something is nature itself. In science, philosophy, and art, we treat nature outside of ourselves as the material for study.

If we want to know "the purpose of nature," however, our power of reason fails, and we need to rely on our intuition. We can do this because we are also a part of nature. "Through the function of feeling, we sense something great in nature, and treat all things with sincere mind, and then we can accomplish that goal [of nature]. What is called religion is none other than this kind of subjective attitude."[85]

CONCLUSION

This chapter has shown how religiously devout scientists in early twentieth-century Japan integrated a modern scientific worldview with their specific Buddhist tradition. Doctrinally speaking, Buddhism had a number of advantages over Christianity in accommodating modern science. Its rejection of hardened dogmas, its naturalistic view of the world, and the absence of creationism, which afforded humans a special place apart from nature, all helped harmonize Buddhism and natural science. In addition, the Buddhist perception of the world as transformational and interconnected was particularly pertinent in accepting evolutionary biology.

The differences between Christianity and Buddhism meant that the overall cultural climate was quite different in the West and the East. In countries where Christianity was dominant, there were frequently strong reactionary forces that pushed back against the reception of scientific ideas. Both Catholic and Protestant movements fought against the authority of science, and evolution was often especially singled out. The popular image of a war between Christianity and science in the West was known to Japanese scholars and was often a cause for their boasting about the superiority of Japanese traditions.[86] And they were in this respect not too wrong in their analysis. With a few minor exceptions, Buddhism in Japan never formed any aggressively anti-scientific movements.

Reflecting on Hugh Slotten's ideas as discussed in the introduction by the editors of this volume, we too have been impressed by ways in which the circulation of knowledge across national borders plays out in local contexts. Despite vastly different conditions and histories, similar patterns have shown up in lands quite distant from each other. In Japan, just as elsewhere in the world, evolutionary worldviews were often central to larger debates over the relationships between science, religion, and society. The reception of Ernst Haeckel's monism is a case in point. As we have seen, the Japanese reception was mixed. Some irreligious scientists like Oka agreed with the American biologists that Haeckel's thought was fuzzy and speculative—not scientific at all. By contrast, Haeckel was read enthusiastically by others. On the one hand, there was Katō, who found Haeckel's irreligious remarks engaging and promoted his ideas for their aggressive anti-clericalism. On the other hand, the Shinshū Buddhist Fujikawa initially ignored Haeckel's irreligion, focusing instead on his pantheism, which seemed to resonate so well with medieval Buddhist ideas. In this case, there is a parallel with how the Ottoman Turks also responded to Haeckel, linking his modern science-based worldview to that of a thirteenth-century Islamic Gnostic. In ways that would have confounded both Haeckel's religious critics *and* his ardent atheist devotees, these religious men found his ideas helpful in harmonizing science and their own religious traditions.

Just as striking, we found a strong parallel between Japanese and Western responses to an encroaching secularism in different national contexts. The fears among religious people over the dangers of a science-based secularization have been widely explored by historians looking at Western countries with a Christian heritage. Those same fears manifest elsewhere as well. Even in Japan, where modern science was continually embraced by Buddhist thinkers and where Buddhism was frequently elevated above the more "superstitious" Christianity, there were concerns that materialism and secularism posed a serious threat to the overall well-being of society. Indeed, it was that central divide that we have tried to explain here, a divide that pitted ardent secularists like Katō and Oka against believers like Ishikawa and Fujikawa. These latter two Buddhists scientists worried deeply about the adverse effect of encroaching secularism. Fujikawa's continual effort to rebut anti-religious secular intellectuals in his midst is not very different from the way that Protestant and Catholic devotees in the West—as seen in this volume, from Argentina and Spain to England and Germany—have tried to defend their traditions against what they feared to be a growing unbelieving scientific establishment.

What makes Ishikawa and Fujikawa so interesting is that they were both religious apologists *and* scientific modernizers. On the one hand, they showed that their Buddhist faith did not—and indeed could not—contradict scientific truths, and on the other, they pointed out the insufficiency of science in addressing human spiritual needs. They, like so many religious progressives of the modern period, contrasted a stark and ultimately unsatisfying scientific materialism with a faith that was open to new knowledge and that provided psychological comfort in a changing world.

11

Cosmic, Theistic Evolution and Kang Youwei

"The Martin Luther of Confucianism"

Wan Zhaoyuan

On October 12, 1904, the eminent Chinese reformer and political exile Kang Youwei (1858–1927), who was then primarily based in Canada, visited the Natural History Museum in London at the close of his year-long tour of Europe. When his eyes fell on the statues of Charles Darwin and Thomas Henry Huxley upon entering into the main hall, he felt great pleasure, as if he had met "old friends"—as he subsequently wrote in his travelogue. He commented that the evolutionary theories of the two great men were regarded as "new inventions" in the countries of Europe, but he had already developed similar, though less elaborate, ideas even before he "had ever read a word of text in any Western book." Asserting that the principles of evolution as espoused by Huxley, and Darwin's doctrine of humanity being evolved from "things," had already been touched upon in ancient Chinese classics, he stressed that he had explored similar principles with his own students in China and had done so "to a considerable degree" in the early 1890s.[1]

This claim was substantiated by Liang Qichao (1873–1929), Kang's renowned disciple and right-hand man, an influential journalist and social and political activist around the turn of the century. Liang was among the earliest readers of Yan Fu's commentary translation—known as the *Tianyanlun* (On natural evolution)—of Huxley's "Evolution and Ethics," through which the Darwinian theory of biological evolution, along with its social implications, was effectively introduced to China in the mid-1890s. As a result, an "evolutionary sensation" was created among Chinese intellectuals.[2] After perusing Yan's translation, Liang was deeply impressed, but he nevertheless noted that he and his fellow students had already learned about similar ideas from their teacher, Kang Youwei, though not to the same depth as examined in the *Tianyanlun*. "Don't think," Kang was reported to have cautioned his students, "that these [notions discussed by me] are original principles, for there might be many Westerners who have studied them for many years."[3]

So what are the ideas pertaining to the concept of evolution that Kang Youwei claimed to have developed in his youth independently of Western learning and explored with his students? How do they fit into his overall thought as a religious as well as political reformer? And did his views on evolution and religion evolve as a result of his closer contact with Western civilization? These are interesting and significant questions, for they lead to answers that will enrich our understanding of the mind of such a prominent figure in modern Chinese intellectual history as Kang, who came to be widely considered one of the most representative thinkers of the period of transition from Imperial to Republican China.[4] Furthermore, as this discussion is being carried out in the context of China's unique intellectual tradition, it can also contribute to the study of the global history of evolution and religion by presenting a Confucian-inspired perspective. It should be emphasized that this perspective, although part of an earlier Chinese intellectual tradition, was in dialogue with Western evolutionary and religious notions beginning at the end of the nineteenth century. That was a time when imperialist European powers, and Japan as well, vied to encroach upon China, and as a result, the Sino-centric view of the world long held by many Chinese intellectuals was shifted to a view of competing nations.

A firm believer in the religious origin of morality and in the moral basis of social polity, Kang concurrently embarked on two interrelated realms of reformation: political and religious. Galvanized by China's debacle in the Sino-Japanese War (1894–1895), the same catalyst behind Yan Fu's work on Huxley and evolution, Kang's political campaign aimed to transform the imperial government into a constitutional monarchy through gradual reforms by drawing on the European (especially the British) and Japanese precedents, so as to better prepare China for "survival" among the competing and often militarily aggressive nations. His efforts culminated in the Reform of 1898, which was abruptly ended by a palace coup within one hundred days of its establishment (hence the name "Hundred Days' Reform"). Since a high reward was placed on his head, Kang was forced into a subsequent fifteen-year exile. He was not able to return to China until 1913, a year after the Qing dynasty had been replaced

Figure 11.1: Portrait of Kang Youwei, ca. 1905. Image available from the US Library of Congress (http://hdl.loc.gov/loc.pnp/cph.3a45114). Public domain.

by a republic, during which he resided in foreign lands, chiefly Canada and Southeastern Asia. Throughout this entire period, he remained committed to his political vision, continuing to advocate for reform through a constitutional monarchy among overseas Chinese.[5]

His religious quest was to "restore" Confucianism to its "pristine" form (*Kongjiao*) as originally created by Confucius and to institutionalize it as a state religion. Referred to by his disciples as "the Martin Luther of Confucianism,"[6]

Kang sought to recast Confucianism through a radical reinterpretation of the Confucian canon, revitalizing the ancient Chinese belief in *Shangdi* as a universal concept, equated with "God." At the institutional level, he proposed the creation of a national Confucian Church modeled on Christian practices, with branches set up across the country. His proposal was partially realized through the establishment, with the assistance of his young disciple, Columbia graduate Chen Huanzhang, of the Confucian Association during the early Republican period; but, within half a dozen years, the association slid into decline and eventually lost its appeal to the public. A remnant of the association, though, continues to exist even today in Hong Kong as the Confucian Academy.[7]

Kang (see figure 11.1) penned his earliest manuscript writings in the 1880s, most of which were published posthumously and not until nearly a century later in the late 1970s. These manuscripts have recently been included in the *Complete Works of Kang Youwei* (*KYWQJ*).[8] A close examination of those writings reveals that Kang's early thinking on evolution was closely related to his Confucian training, his religious leanings, and his pursuit of Western knowledge in his youth and that his fascination with "cosmic evolution" was embedded in his vision of a future utopian world of the Great Unity (*Datong*). In this chapter, I start with a brief account of his early intellectual growth, in which he attempted to integrate the Confucian, Buddhist, and Western sources of his knowledge, before I delve deeper into his conception of evolution—a cosmic, theistic evolution that is informed with a strong sense of teleology.

EARLY SYNTHESIS

Born into a family with a scholarly tradition, Kang was well trained in orthodox Confucianism. In his early youth he embraced the belief in *Shangdi* and Heaven taught in the Confucian classics and was exposed to the local folk beliefs, made up of an admixture of Buddhist and Daoist concepts. Despite being a diligent and talented scholar, the young Kang failed repeatedly in the civil service examinations, and eventually, he developed a revulsion to orthodox book learning. He began to explore other avenues of knowledge outside the Neo-Confucian tradition. He resumed his early interest in Buddhism and started to practice Chan (Zen) meditation, through which practice he "suddenly perceived heaven and earth and whatever is between them come into oneness with me, radiating bright light in great profusion."[9] From this mystical experience Kang became convinced that he was born to be a sage with a bodhisattva-like mission to deliver humankind from suffering, a conviction that would from then on serve as a relentless driving force in his life.

He then went into a short period of semi-seclusion on the mountain near his home, so as to nourish his spirit through immersing himself in reading

Buddhist texts. In this Kang was not without predecessors and comrades. Quite a few prominent Confucian scholars before him and some of his contemporaries were deeply attracted to Buddhism, and some were even converted. It has been argued that there was a "Buddhist revival" among Chinese intellectuals in the second half of the nineteenth century, a revival that could, in a sense, be considered a (perhaps subconsciously nationalistic) response to the growing presence of Christianity following China's defeat in the Opium Wars (1840–1842).[10] Kang did not convert to Buddhism, however, for he thought that Confucianism would better prepare him for his self-appointed mission to deliver his fellow human beings from suffering.[11] Furthermore, Kang made efforts to learn about Christianity, unlike many of his educated contemporaries, who customarily resented or ignored the Western religion. Through analytical comparisons of Christianity with Buddhism, he concluded that many doctrines of the former had in fact "stemmed" from the latter.[12] In this Kang seems to have followed, perhaps in an unconscious way, the comparative science of religion proposed by Max Müller (1823–1900), as well as the practices of Christian missionaries in China who tended to compare the "superstitious" Buddhist and Chinese folk beliefs with the "merits" of their own "true" religion.

Besides the avenue of religious exploration, Western science became a new source of knowledge for Kang. Initially, like the generality of the Chinese literati of his time, Kang did not take Western learning seriously. However, after his 1882 visit to Shanghai, a treaty port with a large international settlement, he completely changed his mind. As Kang did not read in any foreign language, the only viable way for him to learn from the West was through reading translations of Western books, and he could not find a better place than Shanghai in which to purchase them. Since the 1860s, the city had become the center for the translation, publication, and distribution of Western books. The driving force of the scientific translation enterprise was an Englishman, John Fryer (1839–1928). As the chief translator of the Translation Department at the Jiangnan Arsenal, Fryer, along with his staff, had translated over 140 titles by the year 1880, covering a wide range of subjects.[13] Kang purchased a large number of those translations before embarking on a study of such subjects as acoustics, optics, chemistry, electricity, and mechanics. As a result, he turned himself into a well-informed self-taught amateur in Western science.

By the mid-1880s, Kang seemed to have achieved an intellectual breakthrough—an epiphany of sorts, or what might be called an experience of a Buddhistic enlightenment, consequent upon which he attempted to integrate all the sources of knowledge he had been pursuing. He later reported that he speculated about the "mystery and infinity of time and space," examined the "profound meanings of Confucianism and Buddhism," and compared the teachings of various religions. This process of reading and reflection, coupled with some sort of mystic experiences, led Kang to the conception of himself as having been called to the mission to "follow the will of Heaven" in bringing about "the unity of nations, the unity of races, and the unity of religions" on the earth.[14] Kang lost no time in noting down his insights from those experiences, and he produced a number of works in manuscript, which expressed, besides his general philosophical and religious thought, his ideas on cosmic evolution, including the history of natural and human evolution.

COSMIC EVOLUTION

By the time Kang penned his reflections on the "evolution of Heaven and Earth" and the "origin of living things" in the mid-1880s, he had two sources of knowledge to draw on: Chinese tradition and Western science. Chinese cosmological theories of an evolutionary nature are often found in classical Confucian and Daoist texts. Although accounts in different texts vary in details, it is generally assumed that out of a prior state of oneness and non-differentiation a primal duality evolved, consisting of the yin and the yang, through the interactions of which myriads of things were brought into being. The *qi* (breath, air, ether) is taken as the primordial substance of the cosmos, with Heaven being formed through the accumulation of the light and clear qi, while Earth was formed by the congealing of the heavy and turbid qi. Although all things between Heaven and Earth became differentiated by virtue of their natures and the variety of their properties, they all coexisted in an organic cosmos undergoing constant change.[15]

As for biological evolution, there is a well-known passage included in both the *Zhuangzi* and the *Liezi*, two key Daoist classics secondary only to the *Laozi* (commonly rendered as *Tao Te Ching* in English). The young Kang was very familiar with the passage and would refer to it several times in his later writings. It presents a process of metamorphoses from tiny organisms to various plants and animals, including legendary abnormal births involving humans. Subject to the vagaries of translation, metaphorical expression, and linguistic imprecision, the Daoist discourse goes as follows:

> Among the various species, there is a microorganism which propagates in water. It becomes moss on the water margin and it becomes plantain on the highlands, the root of which becomes the larva of the dung-beetle and the leaf of which becomes the butterfly. Soon afterwards, the butterfly becomes an insect. . . . The *Yangxi* grass lives with the bamboo that no longer sprouts, which gives birth to an insect by the name of *Qingning*, which in turn gives birth to the leopard, and which again in turn gives birth to the horse, which again in turn gives birth to the man. The man, in turn, reverts to the microorganism.[16]

This narrative offers an interrelated organic worldview in which the boundaries between plants, insects, animals, and humanity are dissolved in the process of transmutations between species, with every thing in the world coming out of microorganisms and going back to them upon death. This fits in well with the general themes of ancient Chinese philosophy, such as all things being in constant eternal change or flux, natural creation without a creator, and the harmony of humanity and the cosmos (*tianren heyi*).[17] Although none of the species is considered definitely more advanced than the others in the biological sense, in this scheme of things, humanity's moral superiority is nevertheless widely acknowledged and emphasized in the general Chinese tradition.

In addition to Chinese evolutionary ideas, Kang had also learned about European evolutionary notions. As is also discussed in chapter 9 by Fa-ti Fan and Bernard Lightman, there were "several introductions" of evolutionary theory to China. By the early 1880s, fragmentary information on the Lamarckian and Darwinian theories on evolution had already been introduced into China, primarily through the popular *Dixue qianshi*, the 1871 translation of Lyell's *Elements of Geology*, which was read as a geological textbook in China, and Fryer's popular *Chinese Scientific and Industrial Magazine* (*Gezhi huibian*),[18] both of which were among Kang's favorite reading titles. In the former, where the original author mentioned in passing Darwin and Lamarck, the two translators, the American Baptist medical missionary Daniel MacGowan and the Chinese mathematician Hua Hengfang, expanded on the theme: "The old doctrine holds that when species were created, their forms and characters were fixed. . . . Later Lamarck maintained that all species were capable of gradual transmutation. . . . Recently Darwin maintained that each of the living things was capable of survival by adapting to the environment." Nevertheless, the translators commented that neither of the doctrines had then been firmly established.[19]

In an 1877 article entitled "Hundun shuo" (On the primordial chaos) published in his magazine, Fryer gave a brief introduction to the "recent" European investigations into humanity's origin from various approaches. "It has been maintained," he wrote, "that animals had started in rather simple forms and gradually evolved in complexity. Initially, there were only infusoria, and then fish, birds and beasts emerged. Among the beasts were large apes, which evolved into humans. That was a process from the base to the noble, from the simple to the complex." Although the author promptly switched to a lengthy discussion on humanity's final destiny in the rest of the article, these introductory remarks were very informative, not least because of its explicit statement that humans evolved from apes, a notion rarely found in Chinese literature.[20] As we will see below, this notion finds echoes in Kang's early writings.

It is worth noting that a systematic, organic worldview is also presented in the serialized introduction to modern science in Fryer's magazine. The introduction consists of 301 items treating such branches of scientific knowledge as astronomy, natural philosophy, geology and physical geography, meteorology, electricity and magnetism, chemistry, botany, zoology, human physiology, and mental philosophy. They were translations by Fryer of the *Introduction to the Sciences*, a popular science primer widely used in mid-nineteenth-century Britain. Its editor and author, Robert Chambers,[21] was a Scottish geologist and evolutionary thinker, whose anonymous work, *Vestiges of the Natural History of Creation* (1844), presents the view of progressive transmutation of species and helped prepare the British public for the Darwinian theory of evolution by natural selection. Chambers's *Introduction* aims to present the reader with a "systematic view of nature" instead of "a superficial view of a few unconnected phenomena."[22] This is, for example, evident in the section on animals, which reveals an evolutionary perspective on the relationships between plants, animals, and humans. Humans are placed within the class of animated beings, but at the head of it. The Chinese translation expands here by stressing man's intellectual and spiritual distinction: "Within animals, humans are the most superior, for they possess not only the animating principle but also intelligence and a soul."[23]

Thus, equipped with the Chinese traditional cosmology and his fragmentary knowledge of Western geology and evolutionary notions, Kang developed a comprehensive conception of evolution that consists of the progressive development of the material universe, the biological kingdoms, and then the human species—this is exactly what Chambers did in his *Vestiges* by starting with the nebular hypothesis and then moving to geology/paleontology, biology, and then human evolution.[24] While acknowledging that the origin of the cosmos was beyond his knowledge, Kang assumed that it had originated in the accumulation of qi, as was conventionally believed in the Chinese tradition. He considered the invisible, formless qi to be of two kinds: the humid and hot, and the dry and cold. The former was the source of "vigor and birth," while the latter, "decay and death." These two kinds of qi, he asserted, were known to ancient Chinese as the yin and the yang, through whose eternal interaction a myriad of things in the universe were formed. While reiterating that the firmament or Heaven was formed by the rising humid and hot qi at the beginning of the ultimate reaches of primordial time (*wuji*), Kang deviated from the conventional track and attempted to explain the early stages of cosmic creation by appropriating Western scientific concepts such as gravity and the elements. "From the persistent friction and interaction of *qi*," he wrote, "heat, gravity, light and electricity were generated respectively, and the elementary substance (*yuanzhi*) came into being through a process of transmutations.

Thereafter the sun was brought into existence, which in turn gave birth to the earth."[25]

The same creative qi was responsible, according to Kang, for the progressive generation of life on the earth, by virtue of which minerals, plants, animals, and even humans came into existence, progressively and respectively, each of the latter being the result of the accumulation of the former. In his own words: "The accumulation of *qi* results in the generation of heat, the accumulation of heat in the creation of minerals, the accumulation of minerals in the formation of soil and rocks, the accumulation of soil and rocks in the growth of grasses and trees, the accumulation of grasses and trees in the birth of worms, the accumulation of worms in the birth of birds and beasts, the accumulation of birds and beasts in the coming into being of man."[26] The underlying principle of those transformations, according to Kang's later summary, is that "the accumulation of what is coarse eventually results in the birth of what is fine, and the accumulation of what is mean in the birth of what is noble, and the accumulation of what is stupid in the birth of what is wise."[27]

Kang was very vague on the mechanism of those transformations, however. Perhaps his ideas were merely preliminary conjectural theories. After he had read more about the Darwinian theory of natural selection, he agreed with the idea that nature plays a decisive role in the development of things on the Earth according to their specific qualities. He argued that this principle had been anticipated and, actually, expounded in ancient Chinese classics. In making this assertion, he quoted Zisi (ca. 483–402 BCE), the grandson of Confucius, as stating in the *Doctrine of the Mean*, one of the Four Books of Confucianism: "Thus it is that Heaven, in the production of things, is sure to be bountiful to them, according to their qualities. Hence the tree that is flourishing, it nourishes, while that which is ready to fall, it overthrows."[28] Later, in his work on a utopian future world, he explicitly acknowledged that "survival of the fittest" was the rule in the biological realm of the jungle, but it may not necessarily hold true in the human society, where cooperation and collaboration should prevail.[29]

HUMAN EVOLUTION

According to Kang's principle of cosmic evolution, humans, being the most intelligent, could only appear after other creatures. He stressed humanity's intellectual superiority in a way that echoes the simplicity-to-complexity tone of Chambers: Plants and trees are able to grow but have, unlike animals, no senses or movement; beasts and birds are similar to humans in possessing a vertebral structure, but they are lacking in intelligence which distinguished humans. That distinction, in Kang's view, had a biological basis. Because animals "move in a horizontal way," more blood would gather in their brain and therefore "hinder the development of their intelligence." In contrast, humans "live in an upright manner," and as a result, their intelligence is not hindered, which has enabled them to dominate other creatures.[30]

In his early writings, Kang was silent on the specific types of animals from which humans initially evolved. When he started a private academy in Guangzhou in the early 1890s, Kang told his students that "humans could produce the horse and vice versa" and that "man had evolved from apes."[31] As indicated above, the first idea can be found in several pieces of ancient Chinese literature, and the second was most probably picked up from his reading of Fryer's work. By juxtaposing the two sources, but with the Chinese followed by the European, Kang may have echoed the prevailing view of "Oriental/Chinese origins of Occidental sciences" as upheld among some of the missionaries in China, including William A.P. Martin, as well as many of the Chinese literati of that time. Kang claimed that all of the branches of learning had once been highly developed in ancient China, but they had not been passed down to posterity, and yet their remnants were somehow spread to the foreign lands and were further developed by the Westerners.[32]

Compared with the issue of human origin, however, Kang was much more concerned with the fate of his own race during the closing decade of the century. Believing that China was in imminent danger of partition by foreign powers, Kang and his fellow reformers, including Liang Qichao and Yan Fu, called for the preservation of the Chinese race (*baozhong*). As Frank Dikötter has neatly summarized, they developed the traditional lineage discourse into a racial identity. All inhabitants of China were considered to be descendants of the Yellow Emperor and members of the yellow race. In face of the threat of racial extinction, "yellows" had to compete with "whites" in the racial hierarchy that included the degenerate breeds of "browns" and "blacks."[33] This view of racial struggle, as we shall see, would soon give way to a vision of racial unity as Kang went into exile on foreign lands following the failure of his reform program.

By the time of Kang's 1904 visit to the Natural History Museum in London, he was convinced that humans had evolved from none other than the orangutans, through "countless changes and adaptations." He made an interesting observation that humans became distinguished from the orangutans by getting rid of the tail, the possession of which would make impossible upright walking, a condition for the development of intelligence. He further concluded that different human races today demonstrated different degrees of perfection and intelligence. This, according to Kang, could be gauged by their facial resemblance to orangutans. A closer resemblance indicated a shorter history of evolution from the orangutans and, as a result, a lower degree of intelligence, and vice versa. In his view, the Aboriginal inhabitants in Australia, the black people in Africa, and the indigenes of America, Tibet, and Southeastern Asia all bore closer appearance to the orangutans, with flat noses and broad faces, prominent brow ridges, and long philtra, and therefore were considered barbarians. In contrast, the people of the more civilized regions, such as the Europeans, looked different, with full foreheads, aquiline noses, and shorter philtra. In terms of intelligence and physical beauty, according to Kang, Europeans ranked the highest, followed by Arabians and Persians, Indians, Chinese, Japanese, and so on.[34]

By today's criteria, Kang's views are full of perhaps unwitting racial prejudice and appear to be statements asserted without being grounded in scientific evidence. However, without in any way justifying Kang's conjectures, the historical record shows that similar views were popular at his time in both Europe and China. As a matter of fact, it was highly likely that Kang had developed his initial ideas on race theory from Chambers's *Introductions*, in which people in the world were classified into four races. The white-skinned Caucasian variety was "the highest in intelligence, and the most elegant in form." The olive-colored Mongolian variety was "of less intelligence and vigor of character." The third race was "of black skin, coarse features and less intelligence." The last was the copper-colored indigenous race in America, "of no great intelligence" and "in a generally barbarous condition."[35]

Kang, however, rose above the simplistic, highly racist classification of races by Chambers. His intention was not to discriminate some races from others (he was not lenient in judging his own race); rather, he aimed at a higher vision, the unification of races on the Earth, a vision that was further bolstered by his personal observations of peoples of different races during his extensive travels to American and European countries. The racial inequalities in human evolution, he claimed, could be remedied. Marriage between a more evolved race and a less evolved one, say, between the white and the black in the United States, was conducive to the improvement of the less advanced race. More important, interracial marriage could help solve racial enmity and contention. The conflict and contention between the white (Westerners) and the yellow (Chinese), he stressed, would become more intense, with dire consequences in the future, but if white-yellow marriages could be encouraged, the two races would become more united in one or two centuries, and both would enter the world of Great Unity (*Datong*) and Peace. "Here lies," he asserted, "the means of racial evolution" and "the greatest of all matters in the future."[36]

Kang pointed out that although humans were distinguished from other creatures by their superiority in intelligence, they were themselves unequal in intelligence and many other aspects. These inequalities would naturally, Kang argued, give rise to struggles for survival among human beings, just like as happens in the natural world. However, unlike other species, only favored varieties of which could survive through natural selection, humans were able to moderate their natural inequalities so as to survive together, thanks to the instruction and education of their sages.[37]

RELIGIOUS EVOLUTION

Who were those sages, and how did they first appear among the people? Kang did not resort to a supernatural account, as is sometimes found in the context of a revealed religion, as in the immaculate conception of Jesus, or even in the mythic account of the Buddha's birth. He explained that sages arose as a result of "the accumulation of the intelligence of the populace." With a relatively short history of several thousand years, he argued, humanity came into existence only after a universal deluge. As different races of people begin to propagate, the most intelligent among them arose and expounded on a set of instructions to teach their fellow beings. In the Chinese context, he reasoned, the earliest sage would probably be the legendary Fuxi, who proposed the doctrine of the Eight Diagrams (*bagua*) and set forth the principles for the regulation of people's lives. The counterpart of Fuxi in Western history, he speculated, was Adam described in the Old Testament. What the sages taught was then practiced and passed down to posterity through a community of followers, and that was how *jiao*, or religion, came into existence; for that reason, those sages could be described as founders of religions (*jiaozhu*).[38]

Kang considered the Chinese term *jiao* and the English word "religion," if properly understood, to be different linguistic references to the same universal concept. During his 1904 interview with Herbert A. Giles, the professor of Chinese at Cambridge, he defined religion in a broad sense. "The Chinese word *jiao* means emulation (*xiao*)," he observed. "In other words, one who is enlightened ahead of others (*xianjuezhe*) sets forth a doctrine (way), so that others can emulate the enlightened one by following the doctrine (way)."[39] He held the view that Christianity had, by virtue of its influence, narrowed the scope of the prevailing concept of religion, restricting it to systems of belief in doctrines pertaining to spirits. This does not mean, however, that Kang attempted to secularize religion or deprive it of spiritual connections. On the contrary, he stressed that a proper religion as described in his broad definition could well be said to have a heavenly origin, in the sense that it was based on an intimate grasp, by its sage founder, of the Heaven-endowed human nature.[40]

In this way Kang developed an evolutionary view of religion as a primary instrument for the education and advancement of humanity. In the earlier periods of human existence, Kang reasoned, there naturally appeared multiple religions in different regions of the inhabited world. As peoples in those regions became more connected, their religions would naturally come into contact. All of the sage founders of religions, according to Kang, aimed to achieve a moral life for their followers, but they necessarily differed in their means of instruction, in much the same way that great physicians choose different prescriptions for their patients according to their mental and physical state. If the people had been ignorant, their sage may have resorted to *shendao*, or the ways of instruction making use of such beliefs as those in spirits and ghosts, so as to make people avoid evil and do good. To an audience whose intellect had developed to a higher degree, however, the founder of their religion may have primarily employed *rendao*, or the ways of regulating human relations without relying on supernatural beliefs. In a comparative analysis of the major religions of his times, he argued that Buddhism, Christianity, and Islam were primarily confined to *shendao*, while Confucianism encompassed both of the ways, based on *rendao* yet retaining some aspects of *shendao*. Not only was Confucianism a religion, Kang asserted, but it also represented a more advanced form of religion, better suited to the modern world. That was his unequivocal position on the emerging debate over whether Confucianism was a religion or *zongjiao*, the Japanese neologism that was newly introduced into China and that had come to be generally accepted, but not completely by Kang himself, as the standard Chinese term for religion. Initiated by his erstwhile disciple Liang Qichao at the beginning of the twentieth century, the debate gained momentum in the 1910s, and its repercussions are still felt today among Chinese intellectuals.[41]

That was the rationale behind Kang's Confucian Religion Movement. He was convinced that orthodox Confucianism had been fundamentally corrupted, and he therefore endeavored to restore it to its "pristine" state. One of the profound doctrines of Confucius that had been misrepresented and largely forgotten, in Kang's view, was the theory of *sanshi* or the Three Ages. While the concept was conventionally understood as a description of the past, Kang considered it to be an evolutionary, prophetic view of human history, which could be used to furnish the basis of both his political and religious causes. According to his reinterpretation of the theory, a society would develop from the Age of Disorder (*juluan*) through the Age of Approaching Peace (*shengping*) to the Age of Universal Peace (*taiping*), and accordingly, the system of government would evolve historically from monarchy through constitutional monarchy to democracy. All societies would follow this path, but they may necessarily differ in their pace of advancement. The late Qing dynasty, for example, was still in the first age in his judgment, so its autocratic government ought to develop toward a constitutional monarchy instead of being precipitously changed into a republic form, which did not then suit China yet.[42]

Applying the progressive theory of the Three Ages to his analysis of religion, Kang asserted that the development of religion would go through three stages as well. In the case of the history of Chinese religion, pre-Confucius extravagant polytheism, which featured as the first stage, was turned to measured polytheism (the second stage) by Confucius, who dramatically reduced the number of spirits deemed worthy of receiving worship and sacrifices, which in turn prepared the general populace for the third stage, the eventual practice of monotheism.[43] In another place and in a somewhat different vein of thought, Kang also speculated on the future development of religion in the process of the unification of the nations into one world. Initially, a new religion would be created through the combination of the meritorious teachings in the various religions adopted by individual states, with Heaven as the primary object of worship. At a later stage, when the Great Unity was completely achieved, even Heaven would not be worshipped; instead, human souls would be revered, as the people would be recognized as being "Heaven's people, all endowed with absolute equality."[44] By that time, Kang further predicted, most of the existing religions would no longer be needed. Confucianism, Christianity, and Islam (which had a long presence in Chinese society since its introduction into China in the seventh century) would have served their purposes in the joyful society of the Great Unity, with universal love and equality naturally practiced in a society characterized by a general state of peace and harmony. "The shore having been reached," Kang explained, "the raft may be discarded."[45] There would, however, remain certain Daoist and Buddhist practices to cater to the human desire for longevity and the life beyond.[46]

That would not be the end of it, as the studies of immortality and Buddhahood would be succeeded by "the science of celestial peregrination," a doctrine that Kang expounded in his later years by synthesizing his astronomical knowledge with his metaphysical experiences and intellectual speculations. The realization of the vastness of the universe, coupled with the awareness of their being actually "denizens of the heavens" if observed from other planets, Kang argued, would enable human beings to transcend their trifling sufferings and vexations and eventually achieve utmost happiness—the ultimate goal of all religions, according to his own understanding.[47] With this summation, Kang had also become, at least within his own private mental universe and in his own terms, "a founder of a *jiao*/teaching/religion." He saw himself as the new sage for this age, carrying forward the mission of Confucius, who had, in the Age of Disorder, set up the progressive doctrine in order to guide humanity to advance toward the Age of Approaching Peace and finally enter into the Age of Universal Peace and the World of the Great Unity.

EVOLUTION, SCIENCE, AND RELIGION

In his early writings, Kang did not seem to be aware of the tension between Darwinian theory and Christian theology. This is understandable. For one thing, as observed by Elman and Jin, in late nineteenth-century China, Protestant missionaries and their Chinese aides tended to critique Darwin vaguely and were reluctant to highlight the potential threat of his theory to certain Christian beliefs.[48] For another, many scientific books translated or written by missionaries were often informed by a natural theology, which used scientific knowledge of the "design" in nature to glorify the Christian God. There are, for example, a number of places in Chambers's *Introduction* where the belief is voiced that the world was created and designed by an Intelligent Being.[49] Another example is Alexander Williamson's Chinese work, *Gezhi tanyuan*, compiled by drawing on William Paley's *Natural Theology* and James McCosh's *Method of the Divine Government*, in which the author tried to demonstrate the design of the Creator through science and stressed the harmony between science and religion.[50]

By the time he visited the Natural History Museum in London in 1904, Kang had clearly become informed about the tensions between the Darwinian theory of evolution by natural selection and the Christian doctrine of creation. He commented that "as natural history advances, the Christian doctrine of man's creation by God has fallen." "How extraordinary," he exclaimed, "that a mere biological principle was able to cause the collapse of a mighty religion with a long history!"[51] Kang was not troubled at all by this outcome. For him, the Darwinian theory did not challenge the validity of religion in general. In fact, it served to distinguish a "superior" religion—that is, the Confucian religion, which he believed had enshrined as well as anticipated the doctrine of evolution—from an "inferior" one like the Christian religion, which in his view opposed evolutionary theory because it was tied to an obsolete and unscientific cosmology. For him, religious doctrines stood in need of adjustment when scientific evidence made their correction imperative.[52] But there was a limit to scientific power. No matter how far science advanced, Kang declared, it could never disprove the existence of *Shangdi* or "God," whose infinite nature he believed lay beyond the grasp of finite—even if brilliant—scientific minds.[53]

This is a concept that had caught Kang's fascination throughout his life. From his childhood, he had developed a reverence for *Shangdi* (Supreme Ruler) or *Tian* (Heaven) through his solid training in Confucianism. Compared to *Shangdi* with a sense of a personal God, *Tian* carries naturalistic, fatalistic, and ethical connotations in addition to its anthropomorphic meaning. As *Tian* gradually and eventually replaced *Shangdi* and became the customary term, *Shangdi* was still retained as part of the name of the Supreme Ruler in the Confucian classics and the imperial rites of worship. As was the case in the Qing dynasty, the Supreme Ruler was represented by a wooden inscribed tablet placed at the imperial Temple of Heaven in the capital (see figure 11.2), to which annual prayers were offered by the emperor.

Figure 11.2: Tablet of August Heaven the Supreme Ruler (*Huangtian Shangdi*, inscribed in both Manchu and Mandarin), placed in the Hall of Prayer for Good Harvests (*Qiniandian*), Temple of Heaven, Beijing, where prayers and ritual sacrifices were offered by the emperors of the previous Ming and Qing dynasties at the winter and summer solstices. The tablet is still preserved in its original place in the Temple of Heaven, now a tourist attraction. Photograph by Mei Li Photographic Studio (1932). Image courtesy of Special Collections, University of Bristol Library (https://www.hpcbristol.net/visual/fh01-174).

In his youth, Kang became informed of and inspired by the debate among foreign missionaries and native Chinese Christians over the Term Question concerning the proper Chinese name for God, a protracted debate that was widely reported and discussed in *Globe Magazine* (*Wanguo gongbao*) in the late 1870s. Kang attempted to find a more suitable common term for naming the ultimate reality that he believed had been given different designations in different cultures. In addition to the Chinese terms *Shangdi* and *Tian*, he listed a dozen existing names from other religions, including Christian terms like Creator, God, Jehovah, Deus, and Elohim, Ahura Mazda in Zoroastrianism, and Brahma and Uma in Hinduism. None of them, however, were appropriate, in his view.

He then proposed three interchangeable terms as proper designations: *Qihua* (Qi-Based Transformation), *Yuanzhi* (Primordial Element), and *Dazhuzai* (Great Sovereign). Kang did not elaborate on them, each of which seems to have highlighted one aspect of ultimate reality as he conceived it. The first was a Neo-Confucian concept, something akin to the concept of spontaneous generation. The second was originally a Chinese neologism created to translate the chemical elements and a term that appeared frequently in Fryer's magazine. The third was one of the terms often employed by Christian missionaries in China to refer to their God. "In the world there are many names which fail to reflect realities," Kang wrote in one of his early manuscripts, "but because *Shangdi* is the origin of all things, his designations must be carefully chosen."[54] In his later years Kang had explicitly sought to prove the existence of God or *Shangdi* against the spread of strong anti-religion sentiments and the rise of scientific materialism in the 1920s, a period that witnessed the formation of a cadre of Chinese professional scientists. He devoted to the discourse on God a whole chapter of his last book, *Lectures on the Heavens*.[55] Having given an account of European utterances on God, Kang asserted, "That there is God high in heaven is a belief common to all nations and all religions."[56] It was his belief that there was but one God, who did not have a visible substance, but his notion of God was not completely congruent with Western monotheism, for he favored the conception of a pantheist God who existed with and within a cosmos in constant change.

Kang was aware of the attacks on the belief in God from mechanistic scientists of modern times. Citing Confucius's remarks on the limits of his own knowledge, Kang argued that the same humility should also be maintained by scientists in their position on the existence of God.[57] Therefore, it was inappropriate for a scientist to assert that God does not exist, a conclusion that in his view lay beyond the limits of scientific knowledge. In this connection, he set out to rebut Newton, Laplace, Darwin, and the like, whom he praised in glowing terms in other places in his writings, for their "arrogant denial" of

God: "Their knowledge being so limited, how could [they] hope to comprehend all under heaven? How could they assert there is no God? In so doing they only showed their ignorance of their own limitations."[58] It should be pointed out, however, that Kang was mistaken about Newton's disbelief in God. A firm Christian, not only did Newton regard his scientific work to be a way of knowing the mind of God, but he also shared Kang's position that "concerning things unknown nothing is rashly to be affirmed or denied."[59] Similarly, Kang was misinformed in his judgment of Darwin, who himself rejected the label of atheist and saw himself as an agnostic.[60]

Coupled with his construction of a universal concept of "God," Kang's conceptualization of science and religion relations extended beyond a simple conflict or harmony thesis. While maintaining that there was a fundamentally complementary relationship between religion and science, he also treated that relationship by taking into consideration the particulars of individual religions; and he strove intellectually to establish an integrated view of knowledge, not one in which the two bodies of knowledge were fragmented by misconceived tensions and false dichotomies. His acknowledgment of tension between scientific facts and certain religious doctrines did not make him reject religion in its entirety, nor did his recognition of the power and value of science convert him to scientism. This intellectually moderate and philosophically nuanced position was in stark contrast to the overly enthusiastic and oversimplified conflict thesis of the iconoclastic New Culture intellectuals of the late 1910s, who opted for a radical form of scientism and launched their vociferous attacks on religion in general and Confucianism in particular. In his holistic, associative thinking, however, Kang maintained that scientific principles and religious verities were in fact complementary aspects of one body of knowledge. Kang's intellectual legacy is important because he is one of the early thinkers to conceptualize the modern science-religion relationship in the Chinese context and he is representative of the tendency of Chinese intellectuals before the 1910s to view the relationship as one of harmony, not an easily assumed contradiction. This—together with his analysis of, and his position on, the religiosity of Confucianism in particular, and of Chinese culture in general—has attracted renewed scholarly attention at the turn of the new millennium.[61]

Kang Youwei is a syncretic thinker. Through his early intellectual breakthrough, he envisions the unity of all nations, religions, and cultures through the perspective of cosmic evolution. His notion of evolution is rooted in the Chinese traditional philosophy of an interrelated organic worldview. It also draws heavily on his knowledge of Western learning introduced into late nineteenth-century China through translations. In particular, Fryer's *Chinese Scientific and Industrial Magazine* supplied Kang with fragments of evolutionary concepts, especially a translation of Chambers's *Introduction to the Sciences*. In a way that echoes the simplicity-to-complexity tone of Chambers (and also of Spencer), Kang explains that the evolutionary process involves the progression from the coarse to the fine and from the mean to the noble. For Kang, evolution does not stop with the emergence of humans, the most intelligent of all beings. Not only do humans continue to evolve biologically, but they are also, culturally and collectively, on a path of a historical process in which all humanity is progressing toward a common future of unity and prosperity. To look for the blueprint for that future, Kang exhorts his fellow beings to turn to the teachings of the sages, the founders of religions, who in his view have a better grasp of the will of Heaven or God. In relation to the existence of God, he cautions that disbelieving scientific minds should take the humble or, in the Huxleian term, "agnostic," position as to the things beyond the limit of their comprehension. Kang attempts to find both the roots of modern evolutionary theory and the blueprint for human future in his own religious tradition. This allows him to conclude that Confucianism, or, more specifically, his reformed version of it, is a more highly evolved religion and better suited to both the Chinese people and the rest of humanity in their journey to the world of the Great Unity. A cosmic evolution, for Kang, is never complete without a religious teleology.

Kang's cosmic, theistic evolution—which involves the progressive and successive development of the universe, the biological organisms, and even the human world—is exactly the form of evolution posited by Chambers, who presents the reader with a systematic view of nature in which all phenomena are connected by the evolutionary process. It is also the form of evolution embraced by Herbert Spencer in his *System of Synthetic Philosophy*, and by his American disciple, John Fiske, in his summary of the synthetic philosophy titled *Outlines of Cosmic Philosophy* in 1874. There is no evidence, however, of the young Kang's knowledge of the evolutionary ideas of Spencer or Fiske.[62] Like those European and American evolution-oriented syncretic thinkers, though, Kang is interested not just in a narrow biological theory but in the possibility for turning evolution into a worldview that synthesizes various religious, cultural, political, and even scientific traditions into a vision for unity. To add to the attempts among historians to compare Kang with Western utopian writers and non-Chinese religious figures,[63] this study shows that modern China also had cosmic evolutionists (of whom Kang was a representative) like Chambers, Spencer, and Fiske. What distinguishes Kang is that he drew more from Confucianism than from Christianity and that he developed his cosmic evolutionism by grafting Western notions, scientific or evolutionary, onto Chinese religious traditions against the background of late nineteenth-century interactions between China and the West.

Kang's syncretic views on evolution and religion reflect a universalizing position on world cultures. As Kung-ch'uan Hsiao has insightfully observed, Kang rose above the common ways in which Chinese intellectuals in general reacted to the impact of the West in the late nineteenth century.[64] Unlike the pure conservatives who saw nothing wrong with the Chinese tradition and despised everything Western, or the radical Westernizers who rejected the Chinese tradition and called for unconditional Westernization, or the advocates of a version of *zhongti xiyong*, which regarded Chinese learning as the spiritual foundation and Western learning as the material force for China's development, Kang believed that reforms carried out in reference to the West in China's political, religious, and economic systems were not Westernization but in essence universalization. In other words, both Chinese culture and Western culture should be brought up to a universal civilization that all mankind should attain.

12

Evolution and Constructions of Islam in the Ottoman World, 1870–1920

M. Alper Yalçınkaya

Late nineteenth- to early twentieth-century Ottoman Muslim intellectuals wrote much on evolution. What the idea of evolution itself signified and the moral and religious implications of adopting an evolutionary perspective became topics of debate in many periodicals and attracted the attention of many of the prominent intellectuals of the period. Via a brief exploration of these debates, this chapter illustrates a point to which new approaches in the transnational and global historiography of science have also been paying increasing attention: the appropriation and use of scientific ideas such as evolution are shaped by local exigencies and power struggles. Existing cultural resources and institutional constellations inform processes of translation and appropriation, influencing the circumstances in which texts are encountered, read, and used. Responses that emerge during these processes are seldom strictly local, however. Texts and ideas from myriad transnational sources can, in the meantime, be transformed into local cultural resources and employed in the construction of these responses. In these respects, what appear to be local commentaries can also be seen as contributions to a broad transnational debate.

More specifically, this chapter argues that the Muslim debate on evolution can be better understood as an aspect of the much broader controversy on materialism that led to the rekindling of deep-rooted debates within Islam and the revival of particular esoteric traditions in Islamic thought. Within the debate, contributors deployed insights from a wide variety of resources, ranging from thirteenth-century mysticism and orthodox Islamic theology to the writings of German materialists and obscure French authors, and developed a debate that linked evolution to the political and social concerns of the late Ottoman era.

I will refer to cases from different parts of the Ottoman world and beyond to highlight parallel developments, but my examples will come predominantly from the Ottoman Turkish periodicals of the era, further illustrating the important role periodicals played in the region—as elsewhere—in the dissemination, interpretation, contestation, and appropriation of scientific ideas and in the production of discourses on science and religion.[1] Serving as venues for the Ottoman literati to encounter various interpretations of the views of figures like Lamarck or Darwin, as well as the opinions of one another, Ottoman periodicals played a key role in setting the terms of the debate and helped shape the ways in which Ottoman intellectuals of different generations and backgrounds would perceive the relations between evolutionary theories and religion.

To identify the themes that would come to dominate the Ottoman debate on evolution, it is helpful to dive into a series of texts on peculiar topics that the reading public in Istanbul came across in the early 1870s. The author of these pieces was the young journalist Ahmed Midhat (1844–1912). The son of a small merchant, Midhat later worked as a civil servant and edited the official journals of the Ottoman state in several provinces. Midhat, an autodidact with an insatiable curiosity, started what would become a thriving career as a writer and publisher upon his return to Istanbul in 1871, with works that include some of the first and most daring engagements with new scientific ideas about the origins of the universe and humans. In these works and in the polemics that they incited, we can find in a nutshell some basic characteristics of the nineteenth- and early twentieth-century Muslim engagements with evolutionary thought, both within and beyond the Ottoman Empire.

The introductory chapter to Midhat's fifteen-volume world history that he started to publish in 1871 is a head-on confrontation with the idea of a conflict between science and scripture. Midhat noted in this text that Muslims could not deny the account of creation as presented in scripture, yet they should also not shy away from acknowledging the apparent conflict between the new scientific ideas and scripture concerning key issues, such as the age of the Earth, the emergence of man, and the six-day-creation story.[2] The conflict regarding the six-day account was easy to resolve, since the Arabic word for "day" could be interpreted as an unspecified period of time, making it easily harmonizable with the concept of geological ages.[3] As for the emergence of life on Earth, Midhat stressed that the scientific idea of spontaneous generation was very convincing. Scientists perceived different layers of the Earth as having the capability to generate living organisms, which could suggest that the outermost layer had generated man, the most complex of organisms. This, Midhat contended, was in perfect harmony with the holy books, which asserted that Adam was created out of clay.[4] And this was the point where science could have no further comment, since the origin of the creative potential of nonliving things could be explained only by the holy books.[5]

While this chapter is significant in its acknowledgment that in appearance scientific and religious accounts contradicted one another, and in its openness to a materialistic understanding of the emergence of humans, Ahmed Midhat's more direct engagement with the idea of evolution occurred in the journal *Dağarcık* (The knapsack), which he started publishing in 1872 as an outlet in which to share with readers his views on the interesting ideas and debates that he came across in French books and periodicals.

The second issue of the journal contained an article entitled "Birth" that offered a strikingly materialist account, portraying humans as organisms entirely composed of chemical substances brought together in chance-driven ways. Substances found in nature generated the sperm and the egg, and by chance, particular sperms and particular eggs would unite to bring about new humans. In a humorous tone, Midhat wrote, "The various and diverse predispositions found in certain substances . . . that would generate a human could, by chance alone, end up in the stomach of a donkey, and go through their final development there, and a donkey would be born instead of a human. . . . There is no limit to the things chance does without motive." Giving so much agency to chance is an unusual approach in itself, but the account went even further, noting that the emergence and development of the substances that produced humans had started with the emergence of the universe, so every human body was ultimately a combination of matter that had gone through "hundreds of thousands of transformations."[6]

This piece was followed by an article discussing how a brick in the wall could in fact be but a re-composition of material that had once been part of a human being.[7] At this point, however, Midhat felt compelled to discuss the religious implications of this idea: "I almost burst into tears," Midhat wrote, "[when] I realized that the rapid progress of my thought seemed to invite me to believe in reincarnation . . . [even though] I find the creed in the Glorious Qur'an a thousand times superior to any other faith." Yet thinking further on the topic relieved him of this discomfort, as there was nothing in the Qur'an that contradicted scientific views, since it stated that man was created out of clay. It also noted that everything would return to its original state, which supported the view of matter in circulation.[8] The only issue that could appear difficult to resolve was that of corporeal resurrection—an issue that maintained its significance in Islamic thought after having been a major topic of controversy in the eleventh and twelfth centuries.[9] Midhat stated on this issue that while he could not answer this question, he had no doubt that esteemed scholars of Islam would be able to resolve it. Ostensibly an expression of deference, this comment was also an invitation to Muslim scholars for the development of a theology that took into account scientific views, an approach that became increasingly popular in Muslim societies from the late nineteenth century onward.[10]

But this wasn't all. In yet another piece, Midhat tackled the question of the emergence of humans on Earth. That humans were but animals was now the scientific consensus, Midhat wrote. "Since the human is an animal that is able to improve itself," he added, "it is possible to think backwards and reach a moment in time at which he had not yet improved." This earliest man most likely looked like an orangutan, Midhat wrote, offering *nesnas* as the Ottoman Turkish word for this creature. Here Midhat's reference was to Auguste Debay's work on natural history, discussing Lamarck's views on human evolution.[11] Interestingly, while Debay noted that such an understanding of the origin of man would outrage those who have faith in the Genesis account, Midhat proclaimed that no one who knew what *nesnas* look like could be offended by the idea that they could be "our original fathers." Refuting with several examples the idea of the inheritance of acquired characteristics, Midhat argued that a better way of linking the *nesnas* to man would be simply to assert that man is but a type of *nesnas*—one that, too, had gone through transformations but not as significant as gradually becoming bipedal, for instance. Here Midhat even claimed that humans with tails had been seen in Africa, showing the variety within the *nesnas* family; hence, man was but one type of tailless, bipedal *nesnas*. The key transformation this species experienced was not physical but social and cultural. It is worth quoting Midhat at length here, as he put forward his theory explicitly, in a way that would not be seen in the Ottoman Turkish press for almost three decades. "I believe that at the origin of life, humans were not four-legged creatures like the *nesnas* we see today. They were a type of *nesnas* that walked on two feet. As in all animals, some of them were black, some white, and some had other various complexions. The created nature of these had an extraordinary inclination toward curiosity and exploration, along with competition. . . . After much exploration and experience, they gradually abandoned the world of the quadrupeds and thousands of years later, reached the level of what history refers to as the 'first humans.'"[12] Midhat ended the discussion here, but he also promised a future article in which he would discuss how the "first humans" may have originated, which suggests that he did not expect a backlash to his arguments. But a backlash did come. Harputlu Ishak, a renowned Islamic scholar, noted in a letter published in a popular newspaper on March 4, 1873, that Midhat's publications could only confuse Muslims and threaten the unity of the community. To Ishak, Midhat appeared to deny corporeal resurrection; moreover, arguing that humans had descended from monkeys meant nothing but the denial of Adam as described in the holy books and, as such, constituted blasphemy. Midhat had reduced Adam, "the most glorious of the created," to whom all angels had been ordered to bow, to a product of the Earth's surface and a descendant of monkeys. It would be justified, Ishak contended, to suspect that the real aim of those disseminating such a baseless theory could be to lead the people to doubt Islamic principles. European nations had barely defeated the Communards, and what protected the Ottoman state at a time of grave threat was the faith of its people—hence, confused views like Midhat's could not be tolerated.[13]

Midhat published a much lengthier response in his journal due to the gravity of the blasphemy accusation. Stressing first that the Ottomans could regain their power neither by seeing Islam as the cause of their decline nor by condemning the new sciences behind European progress, Midhat complained that his critic knew less of the new sciences than schoolchildren did.[14] He had only mentioned, not endorsed, the scientific theories positing the *nesnas* as the origin of man, Midhat argued, adding that chastising people for discussing scientific ideas would give people the impression that Islam relied on ignorance.[15] After noting, with references to the commentaries of Ibn Arabi (1165–1240) and Qadi Baydawi (1286–1319), that the interpretation of Qur'anic verses as referring to human bodies made up of earthly substances had a long history, Midhat dwelled particularly on Ibn Arabi's commentary.[16] This work had showed that the Qur'an referred to the creation of Adam in Heaven, not on Earth, which meant that it was possible for the Earth to have existed with different species on it for long periods (that is, longer than the estimated age of the Earth based on scripture) before Adam the prophet emerged on Earth. The age of man-like apes had taken place long before the emergence of Adam, Midhat contended, implicitly building an interpretation that could allow the idea of pre-Adamites.[17]

The controversy between Ishak and Midhat illustrates the meaning and implications of engaging with evolutionary ideas for Muslim intellectuals in several ways. First, and most notably, Midhat was able to publish only two more issues of his journal after these essays, as he was exiled, along with several intellectuals famous for their political polemics and calls for a more representative government.[18] While Midhat's curiosity and interest in novel ideas would continue after his return from exile, he would assume a more cautious tone, advocating the harmony between science and Islam in more conventional ways and publishing many works on the vices of materialism. Indeed, references to these ideas would not appear so explicitly in the Ottoman Turkish press until the revolution of 1908 and the establishment of a constitutional monarchy. Second, Midhat's efforts to find in Ibn Arabi's views ways to build a science-friendly Islam are an early indication of what would become a more common trend, particularly in the early twentieth century, as I will discuss later. Additionally, while his references to earth or clay as the material out of which man was made by God are expressions of basic Islamic notions, their appearance in relation to evolution would also become much more common in subsequent periods. Most crucial in this context is the connection observed in Midhat's writings between evolutionary ideas and materialistic perspectives: as also seen in Ishak's response, perhaps even more concerning than the idea that the ancestor of humans was the monkey was the inclination to see humans as but material beings. Engaging with evolutionary ideas required one to address the issue of materialism, and seeking insights in a variety of Islamic traditions to handle this challenge gradually became common practice, particularly in the early twentieth century.

EVOLUTION AND MATERIALISM

The significance of the idea of evolution for Muslim intellectuals of the late nineteenth and early twentieth centuries, be they in Delhi, Tehran, or Istanbul, is anything but surprising. After all, no concept was more popular in this period than change—social, cultural, economic, and political change that would put an end to Muslim societies' "decline," domination by colonial powers, and moral and political "corruption." Granted the spectrum of views about the form and extent of the needed changes, one would be hard-pressed to find any Muslim intellectual who opposed change tout court in this period. But this diversity of conceptualizations of change is precisely what allowed for the various uses to which the concept of evolution was put. In the writings of some, *l'evolution* meant gradual, planned, and teleological change, while for others it was a force that operated outside of human will. To some the concept was not alien to Islam, whereas others believed it was a revolutionary idea that implied the necessity of radical social and cultural change, which could also involve the reinterpretation of parts of scripture.

In *Reading Darwin in Arabic, 1860–1950*, Marwa Elshakry has shown that the idea of evolution and the views of Darwin started to appear in Arabic texts in the 1870s, particularly in the pioneering journal *Al-Muqtataf*, a science monthly that originated at the Syrian Protestant College in Beirut.[19] A product of the representatives of the American mission in Syria and their students, the journal brought into discourse much material from American theological responses to Darwin, usually advocating a conciliatory tone that characterized science and religion as separate spheres.[20] But Elshakry also demonstrates the growing popularity of materialist views for young Arab intellectuals who wrote about evolution in the late nineteenth and especially the early twentieth centuries, while Şükrü Hanioğlu makes a similar observation concerning the Turkish case.[21]

Referring commonly to German materialists like Vogt, Moleschott, and in particular Büchner, many young intellectuals strove to build a broad worldview that could replace what they saw as the backward mentalities that had led to Ottoman decline. Rarely assuming the form of an overt antagonism to religion, their ideas primarily represented matter and force as the sole causes behind observable phenomena and science as the only way to explore and understand them. Holistic conceptualizations of matter and force, living and non-living beings, and, ultimately, the natural world and the social world constituted the core of their ways of thinking, and the significance of references to evolution was based more on their role within this worldview with specific social and political implications. A case in point is Shibli Shumayyil, a doctor trained at the Syrian Protestant College and the leading Arab advocate of scientism and materialism. Shumayyil translated Büchner's writings on Darwin into Arabic in 1884, with an introduction in which he argued that religions also needed to evolve and governments should not impose any faith on people.[22]

In turn, a vast literature arose to combat materialism and, as it was commonly argued, to protect Muslim societies from disintegration. The response to evolutionary ideas was frequently buried in texts on materialism and its political significance, as in the writings of the political activist and writer Jamal-al Din al-Afghani, whose scathing piece on Darwin and materialism was ultimately a condemnation of the Indian Muslim reformer Sayyid Ahmad Khan and a call for pan-Islamism.[23] Published in Calcutta in 1881, this work illustrates that local and particular political contexts generated a number of perspectives on evolution within the Muslim world, but materialism and its alleged political causes and consequences remained a central theme. Al-Afghani's writing also likely formed the basis of his connection with the Ottoman sultan Abdulhamid II, who both adopted a more pan-Islamic political discourse in this period and embarked on an onslaught on materialism himself.[24] Meanwhile, in Istanbul, an author well-known for his writings on modern science died by suicide in 1887, and none other than Ahmed Midhat, who had by that time become a publicist for the sultan in the Ottoman press, portrayed this act as a consequence of materialist views.[25] The palace ordered Ottoman newspapers not to publish news of suicides to prevent this allegedly materialist current.[26] The importation of numerous books was also prohibited, including French translations of Büchner's *Die Stellung des Menschen in der Natur*[27] and many works by Ernst Haeckel.[28]

Authors of anti-materialist works received much support from Abdulhamid II, as in the case of the Lebanese scholar Hussein al-Jisr. In his *Treatise on the Truth of Islam and the Shari'a*, which he published in 1888 and dedicated to the sultan, al-Jisr argued that evolutionary views, in their current state, were based on insufficient evidence. Bringing together natural theological ideas about the wondrous complexity of nature with an emphasis on the necessity of first and final causes, al-Jisr's account was primarily anti-materialist but in principle harmonious with theistic evolution.[29] In another text he wrote upon the encouragement by the sultan, he focused more on eschatology, making a case for the existence of the soul and corporeal resurrection—issues Ahmed Midhat had tackled in the 1870s.[30] In these respects, al-Jisr highlights the concerns of Muslim scholars in the age of materialism and evolutionary theory, alongside their tendency to respond to these currents in terms of established arguments in Islamic theology on issues like corporeal resurrection, as well as natural theological approaches.

But it was within the relatively free intellectual context of the post-1908 period that references to evolution and materialism became much more common in the Ottoman press, along with new efforts to combine traditional Islamic arguments with insights gained by a familiarity with scientific theories to make a case for Islam in the age of science.[31] In this freer context, the significance of evolutionary views for the broader projects of young intellectuals became clearer as well, as illustrated by the first work published in Turkish directly on Darwin: Subhi Edhem's *Darwinism*, published in 1910.[32] While it included a relatively detailed presentation of Darwin's perspective, this text, too, was about much more than Darwin: Subhi discussed evolutionary theory within a narrative that started with a discussion on Comte and positivism and ended with a discussion on materialism. Subhi had no qualms about representing evolution as an outcome and essential component of a positivist-materialist worldview. The age of theology and metaphysics had ended, Subhi argued, characterizing Darwin's views as a product of the positive age. Following the discussion of Darwin's views, Subhi noted lamentingly that "Moses's myths left deep marks on the contemporary civilization of Europe . . . as well as on the rational faculties of all the simple-minded races of the Orient." When one understood the theory of evolution, according to Subhi, one would see that all matter moved from the simple toward the more complex, and this was due to "a beneficial consequence of the laws of nature."[33] Subhi's account was unequivocal and explicit in its rejection of the story of creation as described in Genesis—a strategy that enabled him to avoid mentioning the Qur'an altogether. Indeed, while the logical consequence of his comments was that religious views on the emergence of the universe, if not religions themselves, were false, Subhi also added that his response to potential bigoted critics would simply be that "the Holy Scripture composed of the Old and the New Testaments is not a book of natural history."[34]

Even when they were not made with an explicit reference to materialism, many arguments that appeared in the Ottoman press commonly discussed issues pertaining to the origins of life in purely materialistic ways. A series of articles by the bacteriologist and chemist Osman Nuri (later Eralp) is striking in the way they fuse the standard Islamic narrative with a materialistic and teleological idea of evolution. "We humans are children of earth," Nuri wrote. "We were created out of a handful of earth. But many centuries passed until earth became human. . . . It was kneaded and molded. Nature did not like it. Kneaded and molded again. Ultimately the face of the human came into existence. Hence, all the organisms, plants, animals, and mankind that nature kneaded, molded, and brought about over millions of years are all made up of the same concoction. . . . [All] are products of the evolution of one cell, and its splitting into classes, orders, and species."[35] This account was primarily based on Lamarck and Darwin, Nuri noted. The two scientists' ideas did differ, but Lamarck's "use and disuse" and Darwin's "survival of the fittest" were both laws of nature.[36] Importantly, Nuri suggested that governments needed to operate according to the law of evolution, and in the hands of people who understood this law, illustrating the social and political implications of the views young men of science and intellectuals advocated.[37]

Most notable, however, is Ottoman intellectuals' interest in Ernst Haeckel's writings, in which they found a connection between materialism and evolution and an attempt to build a broad worldview out of these ideas. Ahmed Nebil's translation of Haeckel's *The Last Link* is striking in this respect, since the first three pages of the Turkish translation included the biographies and images not only of Haeckel but also of Büchner and Darwin.[38] The foreword by Baha Tevfik also noted that Haeckel's work elucidated the three foundations of modern science: the law of evolution, the eternality of matter, and cell theory. This was nothing but the essence of Darwinism, Tevfik wrote, thus presenting Darwin's work within a framework inspired by German materialism.[39]

I will return to the importance of Haeckel in late Ottoman representations, but here it is worth noting that some Ottoman anti-materialists themselves objected to the representation of Darwin as a materialist. Francis Darwin's publications on his father's ideas appear to have inspired these comments. For instance, Mehmed Şemseddin (later Şemseddin Günaltay), a religious scholar, not only stressed that the *Origin of Species* was a book that proved Darwin's faith in God but added that Darwin's letters also showed the same thing. Flaws in Darwin's views on evolution had already been identified, Şemseddin added, but the real culprits in his account were figures like Haeckel, Huxley, and Félix Le Dantec, who had abandoned science and entered the realm of speculation.[40] Similarly, in a series of articles published in the popular journal *Resimli Kitab*, the physiologist Said Cemil argued that the main problem of evolutionary theory was that of finality—a question life sciences could never neglect.

> Darwin's meticulous theory was like a lifesaver to materialists, since it appeared to explain the order in life without referring to finality. It would save them from imminent bankruptcy. Moreover, since it appeared to explain the entire living nature in terms of material laws with no reference to the idea of the divine, which is the sole purpose of materialists, they saw it like a savior. What they do not know is that Darwin himself states that he never went so far as to adopt atheism. Indeed, had he known that his theory would be interpreted so wrongly in order to deny finality, Darwin would not have expressed his theory; or, had he wanted to reach that conclusion himself, he would not have stated in one of his books [that he is a theist].[41]

Of course it should be added here that these authors, who based their comments on Darwin's autobiography included in Francis Darwin's *The Life and Letters of Charles Darwin* without reference, omitted the part where Darwin suggests that while he deserved to be called a theist when he wrote the *Origin of Species*, he had gradually become an agnostic.[42]

FROM MATERIALISM TO MYSTICISM

Ottoman Turkish representations of the theory of evolution that linked it to materialist thought justifiably attracted more interest in academic studies. But it was not the only strand, since a variety of efforts to express evolutionary views with an Islamic idiom were by no means rare. As the case of Ahmed Midhat illustrates, it was possible for commentators to find views put forward by past Islamic thinkers that were arguably in harmony with the idea of evolution, in particular exegetical approaches. Of these, I will discuss one that Midhat also drew on: that of Ibn Arabi.

Perhaps the most influential figure in Sufi Gnosticism, Ibn Arabi (1165–1240) developed a cosmological model later associated with the notion of the *wahdat al-vujud* (*vahdet-i vücud* in Ottoman Turkish), or the unity of existence. From a broadly Neoplatonic emanationist perspective, Ibn Arabi conceptualized all existence as manifestations of the divine and on a path commonly referred to as the "circle of existence." The corporeal world came into existence through manifestations on a "descending arc," at the bottom of which were minerals, and an "ascending arc," which followed the sequence of plants, animals, angels, the jinn, and ultimately, man. For man, further ascent was also possible, in the form of the *insan al-kamil* (*insan-ı kâmil* in Ottoman Turkish), or the perfect man. Adam was the perfect man in this model, yet Ibn Arabi also stated that there had been many Adams, thus suggesting that "Adam" stands for human existence rather than "historical" Adam in his thought.[43]

Crucially in this context, most Ottoman Turkish authors used the words *tekâmül* or *tekemmül* in their writings when they referred to evolution, and the former became the dominant term in the twentieth century.[44] Yet these words imply a path toward perfection and are conceptually linked to the notion of the *insan-ı kâmil*. This enabled a particular—and, from a Darwinian perspective, inaccurate—interpretation to arise, conflating biological evolution with the idea of spiritual growth and oneness with the divine.

Ibn Arabi and his followers also built on even earlier versions of such styles of thought centered on the notion of the chain of being. These models not only classified beings in terms of their complexity but proposed what can be called boundary, or transitional, forms between the different categories: corals or some types of fungi between minerals and plants, the date tree between plants and animals, and the *nasnas* (*nesnas* in Ottoman Turkish) between animals and man.[45] Possibly rooted in folkloric mythology, the *nasnas* was a monstrous humanoid creature commonly referred to in Arabic bestiaries.[46] In Islamic accounts like the history of al-Tabari (839–923), *nasnas* were described as humans who had turned their backs against the true faith and suffered God's vengeance.[47] The *nesnas* is referred to in the works of later Ottoman scholars like Kınalızade Ali (1510–1572) as well and appeared in the "circle of existence" model of the Sufi mystic Ibrahim Hakkı of Erzurum (1703–1780), which mentions the *nesnas* as the stage before man.[48] And as the example this chapter started with shows, it had become possible to use the word haphazardly to refer to apes, primates, or the ancestors of man in the hands of authors like Ahmed Midhat in the 1870s.

Thus, Gnostic traditions in Islam provided ample material for interpretation, making possible numerous ways of conceptualizing Adam, man, and the other species and beings that existed on Earth, as well as the potentials for transformation within and between kinds of beings. Nineteenth-century European and North American scholars noticed this: the German Orientalist Friedrich Dieterici wrote on the topic of ancient Islamic thinkers' views on evolution in 1878, and the *nasnas* attracted the attention of late nineteenth-century scholars interested in the precursors of the idea of human evolution.[49] George Sarton, too, would reproduce a model of the circle of existence in a section on evolution in his *Introduction to the History of Science*.[50] Note, however, that as is the case with Gnostic traditions, such Sufi interpretations and the views of Ibn Arabi and his followers were not always fully endorsed by established Muslim scholars, and at times they were viewed as too vulnerable to heretical interpretations.[51] While more research is needed to analyze the popularity and interpretations of these views in the early modern Ottoman world, it is clear that interest in them grew in Turkish thought in the late nineteenth century and especially the early twentieth century.[52] As Elshakry notes, ideas such as the chain of being and progress toward perfection, and esoteric understandings about the natural world, were popular among Arabic-speaking Muslim scholars as well.[53] The use of concepts from Sufi thought to make arguments about biological evolution and efforts to find in Sufism ideas that would be in harmony with modern science can be seen as the predecessors of the works of later spiritualists who produced versions of Islam for secularizing Muslim societies, not unlike the representation of Buddhism as a spiritual system most appropriate for adherents of science.[54]

Interestingly, the cover of Subhi Edhem's book *Darwinism* itself includes an epigraph that the author attributes to Ibn Arabi: "Just as it would be awkward to ask a geometrician if a triangle is red or black, it is equally inappropriate and absurd to ask a philosopher if a philosophical work is written in a religious or irreligious manner." In the 1910 re-edition of his translation of Büchner's writings on Darwin, Shibli Shumayyil included a new foreword in which he cited a statement with a pantheistic tone by Ibn Arabi.[55] And ultimately, the appeal of Ibn Arabi to advocates of broadly materialistic takes on evolution is inseparable from the appeal of Haeckel's monism. It is striking in this respect that Ahmed Nebil and Baha Tevfik translated Ernst Haeckel's *Der Monismus als band zwischen Religion und Wissenschaft*—a speech in which Haeckel made a case for monist pantheism, portraying matter and spirit as always coincident.[56] Nebil and Tevfik stated in their introduction that monism was to some extent similar to the Sufi concept *vahdet-i vücud* (unity of existence), but it was "more material, more scientific." While the Sufi concept involved seeing all that exist as vanishing in God, monism meant "seeing the creator as existing in the creature. And this is a fact identified through the study of the universe. It is scientific."[57] Similarly, in a section of Haeckel's speech where he proclaimed, "God is everywhere," the translators added a footnote with quotations both from the Qur'an and from Ibn Arabi making the same point.[58] Note, however, that Nebil and Tevfik also added an epilogue where they noted, in a way commonly seen in apologetic works, that some of Haeckel's comments could be in contradiction with Islamic faith, but this was entirely due to his ignorance of Islam. They had been assured by Ismail Hakkı, a religious scholar, that it was appropriate to publish this book so that the mistaken parts could be refuted.[59]

This hasty conclusion illustrates the cautiousness of Nebil and Tevfik but, more importantly, their interest not in Islam but in philosophical approaches within the Islamic tradition. This became clear when Ahmed Rıfkı, a journalist and writer who regularly attended Sufi lodges, published an essay on pantheism and Ibn Arabi, where he noted that progress and evolution were essential to Ibn Arabi's views. Copying an image representing the "circle of existence" from Ibrahim Hakkı of Erzurum, the eighteenth-century Sufi, Rıfkı concluded his essay with the phrase "here is an old example for those who can't stop screaming the word Darwin."[60] To this, Tevfik responded with a lengthy essay arguing that Ibn Arabi's views, while very significant, were ultimately based on those of Roman philosopher Plotinus; Tevfik concluded, "Here is an example that is ten centuries older but superior, for those striving to show an old example to those who scream about Darwinism."[61] The need Tevfik felt to underline that the merits of Sufi approaches lay in their *philosophical* underpinnings may be seen as an effort to treat "religion and evolution" as a purely philosophical topic about which dilettantes should not comment, likely in order to avoid any misunderstandings or accusations.

Another Sufi thinker who touched upon Darwin in his commentary on Ibn Arabi's works was Ahmed Avni. In the lengthy commentary he started to write in the 1910s, Ahmed Avni presented the story of creation in its literal form, then dwelled on mystical interpretations of the same sections from the Qur'an, finally discussing the contemporary scientific views on evolution. Avni argued that mystical interpretations regarded creation as a gradual process and could be in harmony with theistic evolution, mentioning "the monkey" as the boundary form between the animal world and man, in a similar fashion to the way the *nesnas* had been characterized in Sufi writings.[62] Additionally, Avni noted that the word *salsal*, commonly translated as dried clay—the material out of which Adam had been made by God—could also mean "fermented milk," which would be very similar to the protoplasm that "the school of Darwin" saw as the origin of man.[63] Ultimately, the problem was to jump to materialistic conclusions when scientific evidence was inconclusive: "Even if the case of the creation of Adam is as [described] in materialists' studies, there is no reason to deny God, the prophet, or religion. This denial is a judgment by [materialist] minds that are confused with delusions."[64] It was only due to pride that materialists saw evolution as a reason for denying God.

Similar engagements with evolutionary ideas can be found in the work of Izmirli Ismail Hakkı, a leading Islamic scholar who also had Sufi leanings and an interest in Ibn Arabi.[65] Ismail Hakkı regularly published on philosophy and Islam in the leading Islamist periodical of the post-1908 era, the *Sebilürreşad*, and taught at the Darülfünun (Istanbul University) after 1909. His lecture notes from this early period became the basis of his 1913 book on metaphysics, where he dwelled on the question of evolution. In this section, Ismail Hakkı noted that evolutionists like Darwin, Spencer, and Haeckel explained evolution through the struggle for life and natural selection, yet this theory could not explain the origin of the adaptations that made some members of a species more fit to survive. They had to either resort to making chance the main basis of their theory, which would never be a satisfactory explanation, or develop some theory of final causes.

Note, though, that Ismail Hakkı's analysis of evolutionary views was based on Émile Boirac's *Cours élémentaire de philosophie* (1888).[66] While Ismail Hakkı borrowed the philosophical commentary from Boirac, a philosopher much better known as a psychic, he added his own comments at the end of the section:

> The grace of God is nothing but the law that establishes the existence and dominion of the Earth. It is the divine reign over the universe, it is the exquisite order of the universe. . . . Let us look at nature: In nature we observe an undeniable progress. Plants come after minerals, and animals after plants. At the end of the sequence in plants and animals are the more perfect ones. In the order of animals, what comes last is man, since man is the most perfect being on Earth. *Laqad khalaqna alinsana fee ahsani taqweemin.* This is because naturalists say that perfection in beings has to do with the evolution of their organs, the unity of their characters. The more unified the character, the more evolved the organs, the more the perfection. . . . Today evolutionists acknowledge this progress in nature.[67]

Inserting a verse from the Qur'an (95:4, "We created humans in the best form.") in a discussion on evolution, Izmirli Ismail Hakkı presented scientific ideas as in harmony with the Qur'an, but he also alluded to the Sufi ideas about the chain of being. Evolution in this model remains teleological, with "perfection" as the purpose, which would also make relevant the Sufi notion of the *insan-ı kâmil*, the perfect man. In these respects, Ismail Hakkı's account builds a specific idea of evolution that is in harmony with a particular understanding of the Islamic conception of man and creation.[68]

Mehmed Ali Ayni was among the scholars who attempted to build an idea of evolution in harmony with Sufi Islam. In an essay devoted to Ibrahim Hakkı of Erzurum, the mystic whose model of the circle of existence was mentioned above, Ayni noted that Ibrahim Hakkı's views were nothing but the core beliefs of "evolutionnisme" (used in the original), since evolution meant "gradual and steady transformation from the simple to the complex, from homogeneity to heterogeneity, from a disorganized state to an organic and centralized state."[69] Yet to Ayni, evolution should never mean that apes were the ancestors of humans, which was an idea based on a misunderstanding of Darwin's views.[70] In his book devoted to Ibn Arabi, Ayni made similar points, this time citing Elie de Cyon, who, to Ayni, had demonstrated that religion had existed since the emergence of mankind, thus positing a categorical distinction between animals and humans.[71]

CONCLUSION

The thoughts of Darwin and other evolutionary theorists of the nineteenth century were shaped by the cultural resources of their time.[72] In turn, they themselves became resources with which to develop views about not only biology but also society and politics in different parts of the world. Despite their major differences, blaming Darwin, Haeckel, and evolutionary thought in general for racist and imperialist violence and making efforts to dispel such accusations are both types of boundary work that highlight the immense significance of these cultural resources.[73] The study of Ottoman readers' perceptions and representations of evolutionary thought reveals the same fact, since these readers consistently connected evolution to issues transcending biology during the period in which their first engagements with the theories concerning evolution took place.

This applies to Ottomans of both non-Muslim and Muslim backgrounds. As an example of the former, Shibli Shumayyil wrote that his "philosophy of growth and progress" would not be simply about the evolution of life but would be concerned "with all of nature and society."[74] Meanwhile, in a newspaper column about the state of Muslims in Morocco, Abdullah Cevdet, a leader of the Young Turk movement, stressed the need to emulate European societies and proclaimed, "The law of evolution is a law with no compassion or mercy. It condemns the weak, the ignorant, the incompetent to perish."[75] The nebulous system of ideas that such theorists constructed connected materialism, evolutionary theory, and scientism and served as the foundation of political projects concerning the future of the Ottoman Empire and the distribution of cultural authority within Ottoman society. Making Haeckel, even Büchner, rather than Darwin the main representative of evolutionary thought in the Ottoman world, such efforts to create a broad philosophical framework for a new political culture also strove to find applicable ideas within the Islamic tradition. This, in turn, transformed the pantheistic interpretations of Sufism, particularly those based on Ibn Arabi, into another set of cultural resources to be tapped. While scholars from the religious elite objected to some aspects of these characterizations, they, too, appropriated the "progressivist" understandings of evolutionary thought. Exemplifying further the importance of the transnational movement and appropriation of ideas, such efforts also made use of non-Ottoman or non-Islamic sources, citing figures like Émile Boirac and Elie de Cyon. Hence, what came about in the early twentieth century was a convergence around a teleological evolutionary approach that justified social and political reform; the question of man's ancestors could wait.

13

"Science Is Justified by Works, Not by Faith"

American Biologists and Ernst Haeckel's Evolutionary Science and Religion

Daniel Halverson

Ernst Haeckel (1834–1919) was one of the best-known public intellectuals of his time. He was a marine invertebrate zoologist, a best-selling popular science writer, and one of Darwin's earliest and most zealous allies in the struggle to establish evolutionary thought. His books were translated into two dozen languages, sold by the hundred thousand, and quoted as authoritative by intellectuals as diverse as Lenin, Gandhi, and Freud. The historian Henry Adams (1838–1918) called Haeckel the "clearest and steadiest spokesman of nineteenth-century mechanical convictions."[1] "There are not many personalities," the biologist-historian Erik Nordenskiöld has written, "who have so powerfully influenced the development of human culture."[2] According to a more recent biographer, the historian of science Robert Richards, he was "the foremost champion of Darwinism not only in Germany, but throughout the world."[3]

Haeckel conceived popular science writing as a vehicle for the dissemination of a new religion. "The true knowledge of the most general laws of nature," he wrote, "ought not to remain the private possession of a privileged class of savants, but ought to become the common property of all mankind."[4] In his most popular book, *The Riddle of the Universe* (German 1899, English translation 1900), he explained "the new monistic religion of the twentieth century."[5] God and Nature were one; there were not two substances (thought and matter) or two orders of reality (ideal and material) but only one eternal and indestructible substance, forever obedient to natural law. Humanity was in no sense separate or unique but subject to the same uniform principles of law-governed causality as all physical phenomena, most especially the law of evolution as revealed by Darwin. Belief in a transcendent deity, personal immortality, and free will were all ruled firmly out of bounds, and the Neo-Kantian philosophers and Christian theologians who defended them routinely mocked. "Every intelligent man with normal brain and senses finds his true revelation in nature on impartial study," Haeckel wrote, "and thus frees himself from the superstition with which the 'revelations' of religion had burdened him."[6]

Haeckel was in the front rank of nineteenth- and early twentieth-century anti-clerical polemicists. In his first successful popular science book, *The History of Creation* (German 1868, English translation 1876), he blasted Christians for their "low dualistic conception of God," which "corresponds with a low animal stage of development of the human organism."[7] In *The Riddle of the Universe* he assured his readers that Jesus was the illegitimate son of a Roman soldier and an illiterate "far below the level of classical culture";[8] that the biblical canon was determined by resort to a ridiculous miracle;[9] that Paul's epistles were "generally forged";[10] that Christians degraded the Creator by conceiving him as a "gaseous vertebrate";[11] and that whatever was worthwhile in their faith was, in any case, borrowed from others.[12] He especially loathed the Catholic Church, for "the deliberate and indiscriminate attacks of the ultramontane Church on science . . . are, on account of its powerful organization, much more severe and dangerous than those of other religions."[13] But he was not much better disposed toward Protestants, whose churches "do not fall far short of [the Roman Church] in ignorance and in the crass superstition of their dogmas."[14]

For Haeckel the central point was that the "unavoidable conflict between the discredited, dominant doctrines of Christianity and the illuminating, rational revelation of modern science" must end in victory for the latter, and that meant victory for atheism.[15] "*Pantheism is the world-system of the modern scientist*," he insisted, explaining a few pages later that "pantheism is only a polite form of atheism"—elsewhere, "this monistic pantheism is also Atheism."[16] Yet this did not make Haeckel irreligious per se. This comment appears at the end of a section titled "Monistic Religion," and in other places he described his views as "the simple religion of nature," "our monistic religion," the means by which to effect "the truly beatific union of religion and science," and so forth.[17] "We do not seek a mighty *revolution*," he explained, invoking Martin Luther, "but a rational *reformation*, of our religious life."[18] For Haeckel, true religion was veneration of Nature, expressed rationally through science and passionately through art. No divine interference in Nature's self-sufficiency could be contemplated, and he often stressed that Darwin had definitively excluded this possibility. "This final triumph of the monistic conception of nature," he wrote, "constitutes the highest and most general merit of the Theory of Descent, as reformed by Darwin."[19] Haeckel's idea was that since Darwin had shown that speciation could occur without positing separate acts of creation, and according to wholly materialistic principles, in order to be a logically consistent Darwinian, one had to exclude a transcendent Creator and any act of supernatural creation—just as speciation occurred via mechanical causation exclusively, so the first living things must have arisen in the same way; thus the unity of Nature was established by removing any supposed supernatural order or distinctive, non-material quality of living things (i.e., a soul, telos, or "life force"). All of this, according to Haeckel, followed logically and necessarily from Darwinism—hence his conviction that science and orthodox religion must clash and that a new and scientific religion (his evolutionary monism) must replace the latter.

Recent biographies of Haeckel by Mario Di Gregorio (*From Here to Eternity*, 2005) and Robert Richards (*The Tragic Sense of Life*, 2008) have substantially altered our view of his life and influence. While he had been defined almost entirely in relation to National Socialism (especially by Daniel Gasman in *Scientific Origins of National Socialism*, 1971), historians have recently placed him in his own nineteenth-century context.[20] This has allowed research on his influence in his own time to begin. Nick Hopwood's *Haeckel's Embryos* (2015) has described the history of his controversial drawings, especially in Germany and the United States, and a recent issue of *Theory in Biosciences* explores that influence in Germany, Russia, and Sweden.[21] This chapter is the first thorough scholarly examination of his influence on American biologists during his own lifetime.

Haeckel saw himself as a champion of science and reason, fighting bravely against the benighted "pseudo-Christianity of the nineteenth century."[22] But there is little evidence that American biologists took him at his own valuation. Within that community, support for Haeckel and his monistic religion was scattered, tepid, or submerged, while opposition was routine and strident. As we will see, leading American biologists thought Haeckel's ideas were dogmatic and speculative and in that sense opposed to science, rather than, as he thought, a logical extension of it. "Science is justified by works, not by faith," the Johns Hopkins zoologist William Keith Brooks warned his colleagues in the pages of *Science*. "When Haeckel says 'Credo' and not 'Scio' we need not discuss the value of his belief."[23] It was a characteristic repudiation.

Thus, American biologists of the Gilded Age and Progressive Era declined

Haeckel's union of Darwinian biology and a polemical, religiously infused atheism. Their religious opinions do not seem to have made any difference. Their rejection was uniform and often unequivocal. This shows that for more than sixty years after the publication of *Origin*, American biologists had little desire for conflict with the well-attended churches in the United States. If they had, they would have found an attractive ally in Ernst Haeckel. But far from following him in attempting a second, naturalistic Reformation in the name of science and progress, they consistently distanced themselves from him and his ideas. Louis Agassiz (1807–1873) rejected evolution, hence also Haeckel; Asa Gray (1810–1888) and Joseph LeConte (1818–1891) advocated for evolution within a Christian theological framework, which was wholly incompatible with Haeckel's atheism; Charles Sedgwick Minot (1852–1914), Alexander Agassiz (1835–1910), William Keith Brooks (1848–1908), David Starr Jordan (1851–1931), and Vernon Kellogg (1867–1937) kept their distance from religious questions but took exception to Haeckel's polemical and speculative tendencies. The only American biologist of whom I am aware who definitely supported Haeckel in these years was Carroll Lane Fenton (1900–1969), at the time a graduate student at the University of Chicago. This absence of support for Haeckel's union of religion, polemical atheism, and Darwinism shows that if we want to understand how conflict became endemic in this area of the science-religion dialogue in the United States, we will have to look outside the biological community of the Gilded Age and Progressive Era.

AGASSIZ AND HIS GENERATION

Few people did more to professionalize biology in the United States than the Harvard naturalist Louis Agassiz. In an article for the *Atlantic Monthly*, published the year after his death, he attacked Darwinism as unduly speculative. "The more I look at the great complex of the animal world," he wrote, "the more do I regret that the young and ardent spirits of our day give themselves to speculation rather than to close and accurate investigation."[24] These "young and ardent spirits" certainly included Haeckel, and Agassiz criticized him by name. "The other so-called discovery," he wrote, "is that of Haeckel, that starfishes are compound animals, made up, as it were, of worm-like beings united like rays in one organism. . . . There is not the slightest foundation for this assumption in the structure of the star-fish."[25] Agassiz's private comments on Haeckel were in the same vein. "To assert is not to prove," he wrote in the margins of his copy of Haeckel's *History of Creation*; later, "Haeckel is the Apostle . . . given in year one of the New World Order, E. Haeckel."[26] For Agassiz, Haeckel's science suffered from the same fatal flaw as Darwin's—it was based on speculation rather than careful observation.

Haeckel lacked the quiet modesty that served Darwin so well in the contentious debates over evolution. In his first successful popular science book, *The History of Creation*, he devoted an entire chapter to criticizing Agassiz's science as "absurd nonsense [leading to] utterly absurd inferences [and] many intrinsic contradictions and perversities," while Agassiz's *Essay on Classification* (1857) was, according to Haeckel, "undoubtedly a complete failure."[27] In his *Evolution of Man* (1874, trans. 1876), Haeckel claimed that Agassiz had been "gifted with too much genius actually to believe in the truth of the mystic nonsense which he preached. Crafty calculation, and well-judged reliance on the want of understanding of his credulous followers, can alone have given him courage to pass the juggler's pieces of his anthropomorphic Creator as true coin."[28] Elsewhere he described Agassiz as "the most ingenious and active confidence man in the entire field of natural history."[29] As late as 1904, Haeckel could not resist a swipe at Darwin's old adversary. "These modern creative theories," he wrote in *The Wonders of Life*, "have no more scientific value than that of Agassiz; they are equally based on pure superstition."[30] By then, Agassiz had been in his grave for more than thirty years.

If he had set out to deliberately antagonize American naturalists, insulting Agassiz in this way would have been a good strategy. Opposition to Darwin may have harmed Agassiz's reputation toward the end of his life, but he had nevertheless been a very eminent and well-loved naturalist.[31] He had worked at Harvard for more than three decades, founded the Harvard Museum of Comparative Zoology (America's "greatest science museum at the time," according to his biographer), helped found the National Academy of Sciences, published more than four hundred papers (including important papers on the geological effects of the ice ages), and taught some of the most prominent biologists of the next generation.[32]

In the 1860s and 1870s, the Harvard botanist Asa Gray wrote a series of anonymous articles combating Darwin's critics, such as Louis Agassiz and the American theologian Charles Hodge (1797–1878). The Protestant minister George Frederick Wright discerned Gray's authorship and helped him collect his essays into an anthology titled *Darwiniana*.[33] When we play pool, Gray argued, we cause one billiard ball to strike another and that one to fall into a pocket. Do we say that the billiard ball fell into the pocket by design or according to natural law? The answer, of course, is both: the balls move according to natural law, but only because the pool player strikes them. In the same way, according to Gray, species evolved according to natural law, but it did not follow that they did so apart from the Creator's design—for who designed the law? And who determined the initial state of the cosmos, such that life would evolve in one direction rather than another? An omnipotent and omniscient Creator surely could have arranged circumstances such that all subsequent history would be nothing other than the gradual unfolding of

His will and purposes.[34] Thus, Gray fit evolution via natural selection within a scheme that was already central to Presbyterian orthodoxy, the omnipotence of the Creator as expressed through the providential governance of creation.

In that case, Haeckel's demand that one choose between creation and law, God and Nature, was a false dichotomy. Gray conceded that Darwinism could be given an atheistic interpretation but not that it was inherently atheistic. After all, Newtonian physics could be given an atheistic interpretation, too, and on much the same grounds that Darwinism could—but it was not generally thought that Newtonianism implied atheism. Neither, according to Gray, should anyone think that Darwinism did.[35]

Darwiniana was published in the same year (1876) that Haeckel's first popular science book appeared in English, and most of its essays had been written beforehand, so it did not confront Haeckel's ideas directly. But as a member of Darwin's inner circle, Gray must have been aware of Haeckel's monistic views and their utter incompatibility with his own Presbyterian orthodoxy. According to Gray, Darwinism strengthened rather than (as per Haeckel) demolished the argument from design.[36]

Joseph LeConte, a naturalist at the University of California, likewise rejected the notion that natural law excludes the Creator. "If God is an honest worker, according to reason—i.e., according to law," he wrote in *Evolution* (1891), "ought not science rather to change gaping wonder into intelligent delight?—superstition into rational worship?"[37] Like Gray, he drew a favorable analogy with Newton: "If the law of gravitation be regarded as the Divine mode of sustentation," LeConte wrote, "there is no reason why we should not regard the law of evolution as the Divine process of creation."[38] True, some theologians objected to Darwinism, but he was not greatly disturbed. "A few more years, and Christians will not only accept, but love and cherish this also for the noble conceptions it gives of Nature and of God."[39] Haeckel rarely addressed the arguments of scientifically competent, theistic evolutionists such as Gray and LeConte.

JORDAN AND HIS GENERATION

The embryologist Charles Sedgwick Minot was one of Louis Agassiz's former students. By the time Haeckel's *Riddle of the Universe* was published, he had become an authority in embryology, a member of the National Academy of Sciences (1897), and president of the American Societies of Naturalists, of Anatomists, and for the Advancement of Science. In a short piece titled "Criticisms of Haeckel," which appeared in the popular science journal the *American Naturalist* (1877), Minot was unequivocal in his repudiation: "All of the published criticisms [i.e., of Haeckel's recent works, *The History of Creation* and *Evolution of Man*] by zoölogists of acknowledged high standing with which I am acquainted are unfavorable, while the praises, which the writer personally has heard, were bestowed for the most part by young persons."[40] He went on to quote half a dozen contemporary authorities against Haeckel on various points, concluding, "I think that the inaccuracy of Haeckel's pictures, of his statements of facts, and finally of his theories, has been sufficiently indicated."[41] Minot finished by quoting the eminent Swiss embryologist Wilhelm His, who had previously clashed with Haeckel over some of the latter's embryo drawings: "I myself have grown up in the faith that among all the qualifications of a naturalist, the only one which cannot be spared is accuracy and an unconditional respect for the truth. . . . Others may, therefore, admire Mr. Haeckel as an active and relentless party leader; in my judgment he has, by his very manner of attack, resigned his right to be reckoned an equal in the circle of serious investigators."[42] Minot's comments were so hostile that Alpheus Packard (1839–1905), editor of the *American Naturalist*, felt compelled to add a postscript reminding readers that "we should not lose sight of the fact that Haeckel stands as an original investigator far above some of his critics," and praised "his masterly drawings and elegant literary style."[43]

In another issue of the *American Naturalist*, Minot wrote that "Haeckel's work is inaccurate to a startling extent," for "he figures in detail things he cannot have seen, because they do not exist, and he describes phenomena that do not occur. His fault is to make very positive statements and give very diagrammatic figures after a hasty examination[;] consequently his writings contain so numerous errors, sometimes about fundamental points, that even a positive statement of his, until confirmed by other investigators, has no authoritative value. . . . It is unsafe to quote his writings as authorities in matters of fact."[44] The severity of Minot's criticism has to be seen against the backdrop of one of Haeckel's earliest setbacks as a naturalist. In the first edition of *The History of Creation*, he placed woodcuts of three embryos side by side and assured his readers that they would not be able to discover the difference between them, no matter how carefully they looked. Neither could professional embryologists, although one was an embryo of a dog, another of a chicken, the third of a turtle. Haeckel's goal was to break down the distinction between humans and other animals by showing how similar they were in the earliest stages of embryological development. If differences between them tended to disappear the further back one traced their development, and if the development of the individual recapitulated the evolutionary history of the species (as Haeckel believed he had established via his "biogenetic law"), then a common origin for humans and other animals (i.e., in an extremely simple single-celled organism) was established. It was an important point in debates over evolution and religion in the late nineteenth century, in which evolutionists stressed similarities between humans and animals (especially apes) in order to make a shared evolutionary history more plausible, and anti-evolutionists stressed differences in order to defend the necessity of a distinct creative act and the unique status of human beings within the created order. In the Darwinian scheme, "man and beast and potatoes are put on the same level," one clergyman complained.[45] The science of the 1870s supported Haeckel's contention, but there was a simpler explanation for the similarity of his embryo drawings—he had reproduced the same image three times and given them different labels.[46] The Swiss zoologist Ludwig Rütimeyer (1825–1895) pointed this out in a review of *History of Creation*, and Haeckel was greatly embarrassed.[47]

As later events would show, this incident fit within a larger pattern of taking shortcuts in pursuit of more expansive scientific and religious goals. Haeckel's clerical opponents seized on this and similar missteps to discredit Haeckel personally and undermine evolution more generally—a strategy that has been kept alive right up to the present. It is not the case, however, that criticisms of Haeckel's drawings were made only or primarily by Haeckel's clerical adversaries, or that they can only have been religiously motivated attacks on evolution. "As drawings of embryos, the well-known figures of Haeckel here reproduced are totally valueless," the *American Naturalist* commented.[48] Thus, Minot's stern condemnation of Haeckel was part of a larger controversy fomented by Haeckel's aggressive rhetorical style and techniques of illustration.

Another important American biologist, Alexander Agassiz, was just as vocal as Minot in his criticisms of Haeckel. He followed in his father Louis's footsteps and became a professor of zoology at Harvard, as well as a director of the museum his father had established.[49] According to his obituary for the National Academy of Sciences, "there is hardly a scientific society which does not carry his name on its rolls."[50] His comments on Haeckel were just as sharp as Minot's: "A man so skilled in coarse invective, who has risen to such a height of intolerance . . . is proof against anything so tame as fact or argument. . . . He deliberately falsifies facts . . . manufactures with names and figures an archetype which never existed. . . . All the bitterness of his bigotry and dogmatism are poured forth. . . . Haeckel's claim to be recognized as a true and devoted student of nature will be forgotten. In its place he will gain, what he seems to seek, the front rank amongst scientific demagogues."[51] Like Haeckel, the junior Agassiz was a specialist in marine invertebrate zoology, but he was as temperamentally averse to speculation as Haeckel was disposed to it. In the conclusion of his book *The Revision of the Echini* (1874), he wrote that Darwin's "disciples cannot ask us to take as proved beyond question all the vagaries regarding this and that ancestor of the great animal kingdom, about which they talk with such sublime confidence."[52] It is reasonable to interpret this as an oblique reference to Haeckel, who was foremost among Darwin's "disciples," who drew up the first detailed phylogenetic histories ("vagaries regarding this and that ancestor of the great animal kingdom") and who spoke on the subject with "sublime confidence." He repeated his father's criticism of Haeckel's view of starfish (above), concluding that "unsupported as it is by any data, and simply thrown out as a theoretical hint, it need not detain us any longer."[53] He kept Darwinism at arm's length until at least 1880, when he argued, in a presentation to the American Association for the Advancement of Science, that the fossil record was too fragmentary to allow for a complete reconstruction of the evolutionary history of even one species—clearly implying the impossibility of attempting a reconstruction for *all* species, as Haeckel had done in his *General Morphology* (1866) and again in *The Evolution of Man* (1874, trans. 1876). In almost his every comment on Haeckel, the junior Agassiz returned to the same theme: that Haeckel was a speculator, and speculation was not science.

The marine invertebrate zoologist William Keith Brooks was a student of Alexander Agassiz who went on to a career of professional distinction, chairing the biology department at Johns Hopkins University for fourteen years (1894–1908).[54] Like the younger Agassiz and Charles Sedgwick Minot, Brooks was hostile to Haeckel, dismissing monism in a review of *Monism as Connecting Religion and Science* (a precursor to Haeckel's wildly successful *The Riddle of the Universe*), titled "The Tyranny of the Monistic Creed." "Science is justified by works, not by faith," he wrote for *Science*, "and when Haeckel says 'Credo' and not 'Scio' we need not discuss the value of his belief."[55] In the same vein, he rebuked Haeckel with an approving quote from Bacon (second only to Newton in the pantheon of nineteenth-century empiricism): "Whoever, unable to doubt, and eager to affirm, shall establish principles proved, as he believes, . . . and according to the unmoved truth of these, shall reject or receive others, . . . he shall exchange things for words, reason for insanity, the world for a fable, and shall be unable to interpret."[56] Indeed, according to Brooks, "the dogmatism of the monistic creed" had "obstructed scientific progress."[57] As to Haeckel's insistence that the logically consistent Darwinian had to be a materialist and an atheist, "the only logical necessity is that when our knowledge ends we should confess ignorance."[58] Brooks made many of the same points in *Foundations of Zoölogy* (1899), a series of popular lectures: again the approving quotes from Bacon, the insistence that he would "refuse to give any weight to the 'faith' of pious evolutionists," as well as the scorn for Haeckel's evolutionary religion: "They who agree with Haeckel may worship stones, if they see fit, but they seem to me to fail as completely as any South Sea islander to understand the nature of scientific evidence."[59]

The American ichthyologist David Starr Jordan (1851–1931) gave Brooks's *Foundations of Zoölogy* a glowing review in *Science*.[60] In addition to writing numerous scientific publications, Jordan was president of Indiana University

(1885–1891), then of Stanford (1891–1913), and also of the American Association for the Advancement of Science (1909–1910).[61] Like Haeckel, he was a successful popular science writer, firmly persuaded of evolution and not very sympathetic to Christianity.[62] Yet he rejected monism as speculative and unscientific, for it required one to "conceive that all categories at bottom are one, matter and force, sense and spirit, object and subject, Nature and God. This fundamental conception of monism has never been made really intelligible because it can be stated in no terms of human experience. There is no way known to us by which we can expose it to scientific tests."[63] He especially took Haeckel, "the most conspicuous of the scientific apostles of monism," to task for insisting that abiogenesis *must* have occurred.[64] This was important to Haeckel for religious reasons, as we have seen. In his *General Morphology* (1866), Haeckel had argued that extremely simple proto-organisms, which he called *Monera*, must have formed the bridge between living and inert matter in the very earliest history of life on earth. It was in this context that Jordan rebuked Haeckel: "First among these postulates of monism, or 'articles of faith' in Haeckel's scheme, comes the 'essential unity of organic and inorganic nature, the former having been evolved from the latter only at a relatively recent period.' . . . We know nothing whatever of how life began. . . . That monism demands spontaneous generation in itself proves nothing. . . . Science stops where the facts stop, and speculation cannot safely proceed any farther."[65] Haeckel insisted that his evolutionary religion would "effect at last the truly beatific union of religion and science, so painfully longed after by so many to-day." For Jordan, there could be no such union, for "the warfare of science [i.e., with religion] is . . . a conflict of tendencies in the human mind."[66] Thus, "the same spirit that leads the orthodox Gladstone to reject natural selection because it 'relieves God of the labour of creation,' causes the heterodox Haeckel to condemn Weismann's theories of heredity, not because they are at variance with facts, but because such questions are settled once for all by the great philosophic *dictum* of monism."[67] It was less *what* Haeckel thought than *how* he thought that Jordan found objectionable.

The American entomologist Vernon Kellogg joined Jordan at Stanford in 1894 and remained a professor there until 1920. Like Haeckel, Kellogg was a Darwinian and an author of popular science books, but in *Darwinism To-Day* (1907) he had only a few critical words to spare for the world's leading popularizer of Darwinism. Indeed, he blamed Haeckel for the receding fortunes of Darwinism viz. Neo-Lamarckism at that time: "The modern reaction and revolt against Darwinism is chiefly due to the activity and attitude taken by certain of its over-ardent friends [such as] the evolution champion Haeckel. . . . As a matter of fact, however, biologists generally are agreed that Haeckel's daring speculations and reckless progress in advance of positions grounded on observed fact have been, in a way, always reprehensible and dangerous to the fair fame of biological science."[68] As to Haeckel's proudest accomplishment, the "biogenetic law," it was "mostly wrong; and what is right in it is mostly so covered up by the wrong part, that few biologists longer have any confidence in discovering the right."[69]

The "biogenetic law" was the idea that the embryological development of the individual organism recapitulates the evolutionary history of the species, in a progression from a single, simple cell through more complex forms of organization. Humans, for example, were supposed to pass through a fish stage early in their embryological development, because we have fish among our evolutionary ancestors. Haeckel had argued that this could be seen in the growth of structures called "gill-slits," which appear in the early development of the fetus but disappear as it matures, just as humans have lost their gills now that we have passed beyond the fish stage of our evolution.[70] In this way, a series of vague and temporary anatomical resemblances were supposed to establish facts of evolutionary history stretching back hundreds of millions of years. For Haeckel, this type of capacious reasoning was routine. For American biologists, it was an irritant. "Plain, unadorned statement of facts and description of things as they are can not be looked for in his [Haeckel's] books," Kellogg warned the readers of *Science*.[71]

Haeckel was a few months shy of eighty when the First World War broke out, but as an ardent German nationalist he was willing to do his part. The German army's treatment of Belgian civilians in the early days of the war provoked furious denunciations in the Allied countries. In reply, some of Germany's leading intellectuals issued an "Appeal to the Civilized World," protesting the justice of Germany's cause and conduct. Haeckel was one of the signatories, as was the German Darwinist August Weismann (1834–1914).[72] In 1916 Haeckel published his last popular science book, *Eternity: World-War Thoughts on Life and Death, Religion, and the Theory of Evolution*, which included fresh polemics against "Christian dollarism . . . which unfortunately has become the highest expression of 'Christian love' in Columbania [Haeckel's neologism for the United States]."[73]

This can hardly have endeared him to Kellogg, who visited Belgium on a humanitarian relief mission during the war and had thus been in a position to see firsthand the conduct Haeckel defended. In *Headquarters Nights* (1917), he attributed the war to the social Darwinism of Germany's officer corps. "The creed of the *Allmacht* of a natural selection based on violent and fatal competitive struggle is the gospel of the German intellectuals; all else is illusion and anathema."[74] By that time Haeckel had been the world's leading popularizer of Darwinism for about four decades, and he had always insisted that it provided the scientific basis for an urgently needed reform of government. Thus, although Kellogg did not single out Haeckel by name (he had gone out of his way not to name names, he explained)[75] and notwithstanding that Haeckel could not reasonably be accused of militarism (which he had denounced as senseless, dysgenic slaughter in *History of Creation* and *Riddle of the Universe*),[76] a denunciation of German social Darwinism while Haeckel was still living could hardly fail to include him. Anyone who wanted to blame the war on German Darwinism, hence on Haeckel, could do so on Kellogg's authority after 1917.

PERIPHERAL SUPPORTERS

American biologists had hard words for Haeckel and his evolutionary religion. Could he count even one friend among them? We have already encountered some evidence of sympathy: Minot's admission in 1877 that "praises [of Haeckel] which the writer personally has heard, were bestowed for the most part by young persons,"[77] and Alphaeus Packard's defense in a postscript to this same article. Kellogg commented in a 1910 review of Haeckel's *Evolution of Man* that "'Der Haeckelismus in der Zoologie' is a subject that will not down wherever biologists come together. And its discussion usually leads to a going apart"—again conceding that Haeckel had his supporters.[78] But they do not seem to have defended him in print with anything like the energy of his detractors. The German American biologists Ernst Mayr (1904–2005) and Richard Goldschmidt (1878–1958) have left evidence of Haeckel's formative influence, but Mayr would not emigrate to the United States until 1932, Goldschmidt until 1935.[79]

Haeckel also had admirers among avowed freethinkers—the German American editor Paul Carus (1852–1919), "the great agnostic" Robert Ingersoll (1833–1899), and the American sociologist Lester Frank Ward (1841–1913) all wrote approvingly about Haeckel and his ideas. But the American ichnologist Carroll Lane Fenton was the only American biologist I have discovered who zealously advocated for Haeckel in print.[80] Fenton was a graduate student at the University of Chicago when he wrote *Haeckel's Monistic Philosophy* (1924), a mass-market paperback and part of a series of "little blue books" intended to disseminate Western literary and scientific culture generally, and free thought and socialism in particular, to working-class audiences.[81] In it, Fenton wrote that he had "such sympathy for [Haeckel's] ideas that I could not write of them other than with favorable prejudice," but he, too, confessed that Haeckel often became "unduly dogmatic" and that "we have no right to be so sure of our stand as was Haeckel."[82] Nevertheless, Fenton was Haeckel's champion, sternly rebuking "our bands of protestant devil-hunters who maintain such cordial relations with legislatures and boards of regents."[83] The Scopes trial began in July of the following year.

TRANSNATIONAL CONTEXTS

Much of the space Haeckel's monism might have occupied, under different circumstances, had already been claimed by Herbert Spencer and his disciple, the American philosopher John Fiske (1842–1901). Fiske reinterpreted Spencer for an American audience, stressing those elements of Spencer's system that were amenable to Christianity, and even believed Spencer had strengthened religion by placing it on a firm, scientific basis.[84] This was diametrically opposed to Haeckel's monism, which stressed the incompatibility between science and Christianity. Spencer's evolutionism had another important advantage over Haeckel's in the United States: it presented evolution and Gilded Age laissez-faire economics as the biological and social aspects of an underlying, universal law of progress. American readers were thus gratified to find both their religious and political commitments reaffirmed in Spencer's evolutionism, where Haeckel conceded nothing to the former, cared little for the latter, and had no American disciple to compare with Fiske. His most popular book, *The Riddle of the Universe*, was translated by Joseph McCabe (1867–1955), an Englishman and a staunch rationalist with little interest in harmonizing schemes.

With Spencer's synthetic philosophy as the dominant form of evolutionism in the United States, and with Haeckel's reputation tarnished through a combination of aggressive polemics and uncertain conclusions, American biologists had little to gain by allying themselves with his monistic program. By the time Spencer's influence began to recede, German belligerence in international affairs leading up to the First World War, and then the war itself, made a public stand in favor of Haeckel even more difficult, and by the time this too had passed, Haeckel had long since faded from the public imagination.

Other national contexts contribute to our understanding of Haeckel's poor reception in the United States. In midcentury Europe, Haeckel was one of a chorus of outspoken advocates of atheistic materialism, including the renegade theologian David Friedrich Strauss (1808–1874), the physician Ludwig Buchner (1824–1899), the popular science writer Karl Vogt (1817–1895), the Dutch physiologist Jacob Moleschott (1822–1893), the Irish physicist John Tyndall (1820–1893), and, of course, the political philosopher Karl Marx (1818–1883).[85] Thus, both Haeckel and sympathetic European readers might well have felt that they were playing an active part in one of the great issues of their time—the progress of free thought and the pursuit of knowledge, which had to struggle valiantly against a repressive and backward orthodoxy. European conservatives—traumatized by the history of past revolutions (i.e., of 1798 and 1848) and fearing future ones (as would eventually occur, in 1871 and 1917), and typically regarding the Christian religion as an indispensable foundation of the social order—interpreted materialism as a threat and denounced it at every opportunity.[86] Atheism, materialism, science, and progress could thus appear as a package, often (though not always or necessarily) with ominous hints of revolutionary violence.

Haeckel, a salaried and really quite conservative university professor, had no interest in revolution, but authors cannot always control how others understand them.[87] Lenin, for example, championed Haeckel's materialism and atheism in one of his political pamphlets precisely because he understood these as part of a more general intellectual package, the whole aim of which was revolution.[88] For the same reason, tsarist authorities repeatedly censored Haeckel's books, and Russian biologists distanced themselves from him publicly even while allowing him to set the agenda for their research.[89] When the Bolsheviks came to power, they tried (unsuccessfully) to incorporate Haeckel's thought into official party doctrine.[90]

In Britain, Huxley and Darwin kept Haeckel at a polite distance for similar reasons—the charge that Darwinism was atheism and materialism was a formidable obstacle to the dissemination of their views, and Haeckel was shouting from the rooftops that this was exactly what it meant, apparently oblivious to the ominous revolutionary implications, or the way in which they powerfully (if not quite logically) undermined Darwinism. For Haeckel, to be a Darwinian and to not be an atheist and a materialist demonstrated either lack of intelligence or lack of courage. He must have been disappointed when Huxley began using the term "agnosticism" to distance Darwinism from atheism and materialism, arguing that science could not answer metaphysical questions, where Haeckel's entire point was that Darwin *had* answered them.[91] Nothing could better illustrate the distance between him and the British scientific establishment in the 1860s: Huxley and Darwin were controversial, and Haeckel was simply outrageous.[92]

In rapidly secularizing Sweden, Haeckel was well regarded by his scientific contemporaries.[93] In Germany, he was controversial. Indeed, it was a public spat with the Berlin physiologist (and fellow Darwinian) Emil Du Bois-Reymond that called forth his wildly successful *Riddle of the Universe.* Like Huxley, Du Bois-Reymond maintained that some questions were beyond the reach of science and hence not proper objects of knowledge. For Haeckel, this was anathema. Methodological dualism implied metaphysical dualism, and from there it was but a short step to theology, the supernatural, and, in short, orthodox religion. No retreat, tactical, diplomatic, or otherwise, from a strictly atheistic and materialistic monism could be contemplated—a bold stand in the land of Martin Luther and Immanuel Kant and where atheism and materialism had long been associated with French imperial domination.[94] As Richards has shown, Haeckel had many other formidable critics in Germany as well.[95] His ideas and the way they were evaluated in Europe during the nineteenth century cannot be understood apart from the more general discussion around atheism, materialism, and revolution.

Thus, when Haeckel's books arrived in the United States, filled with daring speculations and fiery anti-Christian rhetoric, but shorn of their original, European context, they were perceived not so much as threatening or exciting (as in Europe) but as strange and histrionic. Where was the repressive orthodoxy that Haeckel never tired of denouncing? Granted, there had been some clerical opposition to Darwin when he had first published, but it simply was not the case, during Haeckel's lifetime (1834–1919), that the American clergy was locked in a death-struggle against evolution or Darwinism.[96] Haeckel protested too much. And where Haeckel's evolutionary monism had natural (if unwanted) allies in Europe's energetic and well-organized revolutionary parties, there were no comparable movements in the United States. Instead, there were well-attended churches and Spencer's evolutionary philosophy—attractively reimagined and repackaged by John Fiske and disseminated by the tireless post-Protestant missionary of popular science, Edward L. Youmans.[97] Haeckel's monism never stood a chance with the American public, and in that case, why should American biologists associate themselves with such a controversial writer? The salient difference is that the legacy of the French Revolution polarized discussion in Europe but not in the United States. In Europe, Haeckel was liable to be misinterpreted as a radical. In the United States he was, as the *Atlantic* commented in a review of his *Wonders of Life* (1904), "totally at sea as to what may reasonably be said in a popular book."[98]

Without a doubt, Haeckel was read outside of Europe and the United States, including by intellectuals such as the Chinese soldier-scholar Yan Fu (1854–1921),[99] the Indian mathematician Ināyatullāh Mashriqī (1888–1963),[100] and the Indian theologian Swami Vivekananda (1863–1902).[101] However, the intellectual response outside the West cannot be adequately characterized on the basis of the scholarship currently available in the English language. It is a task for future scholarship.

CONCLUSION

American biologists frequently complained that Haeckel was too much a speculator and too little a careful researcher. But he was capable of the kind of limited, empirically precise research they valued.[102] The difference between them was not one of capacity but of emphasis. As Haeckel explained in his *History of Creation*, Darwinism was "only a small fragment of a far more comprehensive doctrine—a part of the universal Theory of Development, which embraces in its vast range the whole domain of human knowledge. . . . Its general importance cannot be overestimated."[103] The point of studying medusae and radiolarians was not merely to establish narrow facts of zoology but to provide the empirical basis for a second Reformation, this time completing Martin Luther's work in repudiating the whole of Christian dogma and superstition, rather than only, as Haeckel thought, a part of it.[104] Empirical research was the beginning of the journey, not the end. This larger vision of science's religious mandate, as well as Haeckel's intellectual roots in German romanticism, authorized ambitious and far-reaching claims.[105] The important thing, as he saw it, was not to establish every pettifogging fact beyond a reasonable doubt but to press on boldly in pursuit of truth and progress.

For American biologists, this was reckless. Scientists might offer an occasional opinion on non-scientific subjects, but their true calling was empirical research in pursuit of knowledge of nature, not evangelism on behalf of a new religion. One can detect a growing distance between biology and theology in the popular science writing of the time, from Louis Agassiz's romantic, nondenominational theism and Gray's Presbyterian orthodoxy in the 1860s and 1870s to Brooks's, Jordan's, and Kellogg's reticent skepticism near the turn of the century, but Haeckel's bold heresies still lay far outside the mainstream of the science-religion discussion in the United States.

In *The Tragic Sense of Life*, Robert Richards reminds us that "one must recognize that religious beliefs, harbored by Haeckel's scientific contemporaries, often crawled beneath objections having a different surface character."[106] We have encountered eight such critics among American biologists, and this does not adequately characterize any of them—except, perhaps, Louis Agassiz, though it seems unjust to so briefly dismiss the views of one of the most accomplished naturalists of his time. As for the other biologists, the most overtly religious (Gray and LeConte) supported evolution, while the ones who scolded Haeckel in the most unambiguous language (Minot, Alexander Agassiz, Kellogg) were evolutionists with no obvious religious axe to grind. They simply found Haeckel unattractive because he failed to calibrate the certainty of his convictions to the strength of the evidence on which they rested, and because he was pointlessly offensive in the promotion of his own religious opinions. Perhaps Richards is right to argue that the controversy surrounding evolution and religion explains much of the polarized commentary Haeckel provoked in nineteenth-century Europe, but it doesn't seem to explain it on the other side of the Atlantic. To cast Haeckel as the champion of *science* and his critics as champions of *religion* is to overlook both Haeckel's own overtly religious agenda and the substantial scientific credentials of his critics.

Haeckel saw himself as a champion of science and reason. In his view this of course meant a militant atheism, and he poured scorn on people who preferred a more moderate approach. "Unfortunately, consecutive thought is a rare phenomenon in nature," he wrote of these equivocators.[107] Nevertheless, he looked confidently forward to a time when, "in the twentieth century, the prevalent superstition gives place to rational knowledge and to a monistic conception of the unity of God and the world."[108] Science was an invincible sledgehammer that would "shatter the false structures of superstition."[109] American biologists saw something very different when they looked at Haeckel—a dogmatist, a speculator, and an investigator of questionable reliability. In their view, he failed to make the appropriate distinction between fact and hypothesis, and he was pointlessly dogmatic about the latter. As David Starr Jordan wrote, "every honest, intelligent, courageous scientific man, [Haeckel] tells us, so far as he is truthful, competent, and brave, shares the same belief [i.e., monism]. . . . Against such assumption science may protest."[110] American biologists did protest, as we have seen.

CODA

From Complexity to the Transnational

Bernard Lightman and Sarah Qidwai

This volume has aimed to move the study of the history of science and religion forward by suggesting that the next step in the development of the field should be shifting the emphasis from complexity to the transnational. We envision transnational histories as stepping stones toward the future construction of a global history of science and religion. The contributors have offered three ways to accomplish this goal. First, the chapters draw attention to the fact that intellectuals around the globe appropriated fairly diverse evolutionary theories. Some, such as the figures in Silva's chapter on Argentina, were reacting to the evolutionary possibilities to be found in French positivism; others, the East Asian intellectuals discussed in Fan and Lightman, for example, found British evolutionists, such as Chambers, Darwin, Spencer, and Huxley, to be more useful for their purposes; while still others, such as American biologists, as Halverson shows, Yalçınkaya's Muslim intellectuals in the Ottoman World, and Yoshida and Weldon's pro- and anti-religious Japanese figures, were responding to German evolutionists like Haeckel. There were several evolutionary streams circulating around the world. Evolutionary theory was diverse in a second sense as well—in terms of the scientific disciplines in which these theories could be found. Evolution was not limited to biology. Global intellectuals found it in astronomy, as Guimont and Schaffer show, and, as Gold points out, in geology, in archaeology, and in the study of human antiquity. Opening up our historical analyses to the diversity of evolutionary theory continues the trend in the history of science to move away from an overemphasis on Darwin and challenges our understanding of what counts as "science."

If the goal in understanding the diversity of evolutionary theory leads us to decentering Darwin, the aim in investigating the diversity of religious traditions has helped us in the ongoing efforts to decenter Christianity in the study of the history of science and religion. Whether it be Fernando and Qidwai's study of religion and evolution in South Asia or Wan's chapter on Kang Youwei's reformation of Confucianism, this volume has explored the multiple interactions between diverse religious traditions. The authors have considered Hinduism, Islam, Catholicism, Protestantism, Buddhism, Confucianism, and even forms of unbelief such as atheism, monism, and agnosticism. Moreover, following David Livingstone's lead, the chapters have emphasized the differences that existed within the same religious traditions. Yoshida and Weldon, for example, call attention to the many varieties of Buddhism and how that led to a divergence between believers on how to approach evolution, while Yalçınkaya explores the tensions that existed between materialistic-leaning Muslims and those who chose to revive esoteric traditions in Islamic thought.

The third way that this volume seeks to move the field forward has to do with more thoroughly integrating the scholarship on empire, colonialism, indigeneity, and race into our analysis of nineteenth-century evolution and religion. In certain contexts, the impact of evolutionary theory can be understood only as part of the racial conflicts between Indigenous peoples and settlers. As Stenhouse shows, in the case of New Zealand, Māori intellectuals and some Christian missionaries contested the use of evolution by settlers as a sanction for race war. Guimont discusses how evolution provided a scientific justification for imperialism as religious pluralism was secularized. In Gold's examination of the study of religion and human antiquity, the debates concerning the level of scientific knowledge possessed by the ancient races were of crucial relevance for both Europeanized and anticolonial reformers in Egypt. Similarly, as Schaffer asserts, the heated controversy over the validity of ancient astral knowledge within the Asiatic Society in Calcutta was closely connected to British views of Hinduism and India. The colonial dimensions of educational institutions in Australia are explored by Barnes.

In 2010, Ron Numbers challenged science and religion scholars with his piece titled "Simplifying Complexity: Patterns in the History of Science and Religion." Numbers observed that "outside a small circle of experts the turn to complexity has left most people yawning, or, worse yet, unconvinced." Instead of abandoning complexity and "retreating to uncomplicated master-narratives," he recommended that we "search for mid-scale patterns, whether epistemic or social, demographic or geographical, theological or scientific."[1] In the rest of the chapter Numbers offered five mid-scale generalizations, which he labeled naturalization, privatization, secularization, globalization, and radicalization. When it came to globalization, Numbers admitted that he and others had tried, but failed, to discover a pattern to the way that science and religion varied from place to place. "How much more can we say," Numbers asked, after acknowledging that the "various European experiences" of science and religion "have even less in common with the Hindus of India, the Muslims of Iran, and the atheists of China."[2] *Evolutionary Theories and Religious Traditions: National, Transnational, and Global Perspectives, 1800–1920* is an attempt to respond to Numbers's question "how much more can we say?" and to problematize his attempt to simplify complexity. Whatever their topic, the contributors approach the study of science and religion using a multiplicity of scales, from the local, to the national, to the transnational.

This volume is also, in some ways, a response to Peter Harrison's magisterial book, *The Territories of Science and Religion* (2015), in which he attempts to offer a coherent, large-scale generalization in the face of Numbers's suggestion that we limit ourselves to the mid-scale. Harrison's big picture is grounded in the principle that science and religion, which began as interior virtues, were increasingly exteriorized after the Protestant Reformation. Only when exteriorization reached a peak in the nineteenth century was it possible for two bodies of knowledge now labeled "science" and "religion" to be opposed.[3] Harrison's brilliant study of the changing relationship between science and religion from the ancient Greeks to the present pays little attention to Hinduism, Confucianism, Buddhism, Islam, and other non-Western religious traditions. Harrison self-consciously limits himself to the Western tradition.[4] It is too early to expect a big picture of the global history of science and religion to emerge at this point. In pushing for a more transnational perspective, we are encouraging scholars who work in the field to move in the direction of integrating other religious traditions into their accounts so that we might, in the future, begin to catch glimpses of a big picture like Harrison's that includes the history of sciences and religions across the globe. Only then will "Western science and religion" and "Eastern science and religion" come into clearer focus, since both developed, especially in the nineteenth century, in tandem, rather than in isolation. We must turn our attention to their subtle interaction if we wish to move forward.

However, while this volume pushes the field into new and exciting territories, we recognize there are still other issues to expand upon that we have not covered. Although we have strived for geographical breadth, we could not cover every region of the world. The Russian Orthodox Church and the appropriation of Darwinism in Russia, for example, is one area that we do not deal with that requires more research. While some papers did discuss a variety of responses in various religions, we need a more nuanced approach among religious denominations—for example, the differences in Sunni and Shi'a responses to evolution in British India or even more coverage of other religions such as Baha'i. Furthermore, we have to pay far more attention to the role of women in the historical relationship between science and religion around the world. While imperialism is a strong focus, there are also a number of areas of critical theory—critical race theory, feminist and queer theory, and postcolonial theory—where even more connections could be drawn.[5] There is a pressing need for more work on North American Indigenous perspectives relating to evolution and human origin. No doubt we have overlooked other facets of the topic. We have not tried to be comprehensive or exhaustive. Rather, we have attempted to build on the previous work by Numbers, Brooke, Sivasundaram, Fehige, Brown, and Harrison in the hopes of inspiring researchers to push transnational perspectives even further as we work toward a global history of science and religion.

NOTES

INTRODUCTION

The authors are grateful to John Stenhouse, Meira Gold, Fa-ti Fan, and Simon Schaffer for their helpful suggestions on how to strengthen this introduction significantly.

1. Barry K. Gills and William Thomson, eds., *Globalization and Global History* (London: Routledge, 2006); Bruce Mazlish, *The New Global History* (London: Routledge, 2006).

2. Sebastian Conrad, *What Is Global History?* (Princeton, NJ: Princeton University Press, 2016); Jürgen Osterhammel, *The Transformation of the World: A Global History of the Nineteenth Century* (Princeton, NJ: Princeton University Press, 2014).

3. Michael Bergunder, "'Religion' and 'Science' within a Global Religious History," *Aries* 16, no. 1 (2016): 86–141, 96.

4. Sujit Sivasundaram, ed., "Focus: Global Histories of Science," *Isis* 101, no. 1 (March 2010): 95–158.

5. The "Decolonizing Science in Asia" field report calls for "decolonization" as a historiographical and methodological stand. This is a developing area of research. See Prakash Kumar, Projit Mukharji, and Amit Prasad, "Decolonizing Science in Asia," *Verge: Studies in Global Asias* 4, no. 1 (2018): 24–43.

6. Hugh Richard Slotten, introduction to *Modern Science in National, Transnational, and Global Context*, ed. Hugh Richard Slotten, Ronald L. Numbers, and David N. Livingstone, vol. 8 of *The Cambridge History of Science*, ed. David C. Lindberg and Ronald L. Numbers (Cambridge: Cambridge University Press, 2020), 1–6.

7. Slotten, introduction, 1.

8. Slotten, introduction, 3.

9. Pratik Chakrabarti and Michael Worboys, "Science and Imperialism since 1870," in Slotten, Numbers, and Livingstone, *Modern Science in National, Transnational, and Global Context*, 11.

10. We are not suggesting the identification of the global with the imperial, and we have tried to avoid this problem by including studies of apparently non-colonial or informally colonized cases in such sites as Spain, Argentina, China, Japan, Türkiye, Egypt, and the United States.

11. Andrew Goss, ed., *The Routledge Handbook of Science and Empire* (London: Routledge, 2021).

12. Those scholars who adopt the complexity principle refuse to break down the relationship between science and religion into simple narratives of conflict or harmony. They follow the lead of John Brooke, who insisted that empirical analysis is required in each historical context. See John Hedley Brooke, *Science and Religion: Some Historical Perspectives* (Cambridge: Cambridge University Press, 1991). For an overview of the move from conflict to complexity see Bernard Lightman, introduction to *Rethinking History, Science, and Religion: An Exploration of Conflict and the Complexity Principle*, ed. Bernard Lightman (Pittsburgh: University of Pittsburgh Press, 2019), 3–7.

13. John Hedley Brooke and Ronald L. Numbers, eds., *Science and Religion Around the World* (New York: Oxford University Press, 2011).

14. Sujit Sivasundaram, "A Global History of Science and Religion," in *Science and Religion: New Historical Perspectives*, ed. Thomas Dixon, G.N. Cantor, and Stephen Pumfrey (Cambridge: Cambridge University Press, 2010), 177–97.

15. Sivasundaram, "Global History," 177.

16. Yiftach Fehige, introduction to *Science and Religion: East and West*, ed. Yiftach Fehige (London: Routledge, 2016), 20.

17. It is important to note that the modern Western category of religion was established in the eighteenth century. Like science, religion is not a single entity, and colonialism played a big role in its framing. See Peter Harrison, "'Science' and 'Religion': Constructing the Boundaries," *Journal of Religion* 86, no. 1 (January 2006): 81–106.

18. Peter Harrison, *The Territories of Science and Religion* (Chicago: University of Chicago Press, 2015), 147.

19. David N. Livingstone, *Dealing with Darwin: Place, Politics, and Rhetoric in Religious Engagements with Evolution* (Baltimore: Johns Hopkins University Press, 2014), 12, 25–26.

20. C. Mackenzie Brown, ed., *Asian Religious Responses to Darwinism: Evolutionary Theories in Middle Eastern, South Asian, and East Asian Cultural Contexts*, Sophia Studies in Cross-Cultural Philosophy of Traditions and Cultures, vol. 33 (Cham, Switzerland: Springer, 2020), 3.

21. See Myrna Perez Sheldon, Ahmed Ragab, and Terence Keel, eds., *Critical Approaches to Science and Religion* (New York: Columbia University Press, 2023).

1: BRITISH ORIENTALISM ON HISTORIES OF RELIGION AND ASTRAL SCIENCES IN NORTHERN INDIA

Thanks for generous help to Nader El-Bizri, Eric Gurevitch, Bernie Lightman, Minakshi Menon, Sarah Qidwai, Suman Seth, and Jonathan Smith.

Epigraphs: John Mitchell, *Essay on the Best Means of Civilizing the Subjects of the British Empire in India and of Diffusing the Light of the Christian Religion throughout the Eastern World* (Edinburgh: Blackwood, 1805), 3; Jörg Fisch, "A Pamphlet War on Christian Missions in India, 1807–1809," *Journal of Asian History* 19 (1985): 68.

[Alexander Hamilton], "Sanscrit and Greek—Sanscrit Poetry," *Edinburgh Review* 33 (1820): 442 (Hamilton's emphasis); Rosane Rocher, *Alexander Hamilton: A Chapter in the Early History of Sanskrit Philology* (New Haven, CT: American Oriental Society, 1968), 117.

1. Michael S. Dodson, *Orientalism, Empire and National Culture* (Basingstoke, UK: Palgrave Macmillan, 2007), 62–71; Dhruv Raina, "Disciplinary Boundaries and Civilisational Encounter," *Studies in History* 17 (2001): 224–25, 242.

2. Burke to Robertson, 1777, in *Correspondence of Edmund Burke*, ed. Thomas W. Copeland (Cambridge: Cambridge University Press, 1958–1978), 3:350–51.

3. Dipesh Chakrabarty, *Provincializing Europe: Postcolonial Thought and Historical Difference*, new ed. (Princeton, NJ: Princeton University Press, 2008), 8.

4. Dalhousie University Archives, MS-2-726, James Dinwiddie Fonds, Journal B69 (May 2, 1803).

5. Sisir Kumar Das, *Sahibs and Munshis: An Account of the College of Fort William* (New Delhi: Orion, 1960), 5–7; David Kopf, *British Orientalism and the Bengal Renaissance* (Los Angeles: University of California Press, 1969), 47; Kapil Raj, *Relocating Modern Science* (Delhi: Permanent Black, 2006), 146–53.

6. Rosane Rocher and Ludo Rocher, *The Making of Western Indology: Henry Thomas Colebrooke and the East India Company* (Abingdon, UK: Routledge, 2012), 60–65; Kopf, *British Orientalism*, 87–89.

7. Thomas Colebrooke, *The Life of H.T. Colebrooke* (London: Trübner, 1873), 22, 244–53; William J. Ashworth, "The Calculating Eye," *British Journal for the History of Science* 27 (1994): 419–20.

8. Dinwiddie Fonds, Journal B69 (May 7, 1803); John Bentley, *Universal Tables of Simple Interest, Batta and Exchange* (Calcutta: Stuart and Cooper, 1787); *Calcutta Gazette*, August 11, 1791, p. 2; *East India Kalendar for the Year 1791* (London: Debrett, 1791), 51.

9. "Bentley's Hindu Astronomy," *Asiatic Journal* 21 (1826): 207–8; John Bentley, "Remarks on the Principal Aeras and Dates of the Ancient Hindus," *Asiatick Researches* 5 (1798): 315; John Bentley, "On the Antiquity of the Suryá Siddhánta, and the Formation of the Astronomical Cycles therein Contained," *Asiatick Researches* 6 (1799): 576.

10. Colebrooke, *Life of Colebrooke*, 250–1; Henry Colebrooke, "Hindu Astronomy," *Asiatic Journal* 21 (1826): 366 (Colebrooke's emphasis).

11. Fisch, "Pamphlet War," 28–30; Robert Frykenberg, "New Light on the Vellore Mutiny," in *East India Company Studies*, ed. Kenneth Ballhatchet and John Harrison (London: SOAS, 1986), 209.

12. Thomas Motte, "Narrative of a Journey to the Diamond Mines at Sumbhulpoor," *Asiatic Miscellany* 2 (1786): 12; Henry Morris, *Life of Charles Grant* (London: John Murray, 1904), 105, 110; Geoffrey Oddie, *Imagined Hinduism: British Protestant Missionary Constructions of Hinduism 1793–1900* (New Delhi: Sage, 2006), 71.

13. Parliamentary Register 35 (1793): 583–84; Ainslee Embree, *Charles Grant and British Rule in India* (New York: Columbia University Press, 1962), 151–55.

14. C.H. Philips, *The East India Company 1784–1834* (Manchester: Manchester University Press, 1961), 125n3; Kopf, *British Orientalism*, 129.

15. Thomas Maurice, *History of Hindostan*, vol. 1 (London: Bulmer, 1795), xxii–iii; Nigel Leask, *British Romantic Writers and the East* (Cambridge: Cambridge University Press, 1992), 105–6.

16. Thomas Maurice, *Memoirs of the Author of Indian Antiquities, Part 1* (London: Rivington 1821), 102 (Maurice's emphasis); Thomas R. Trautmann, "Does India Have History? Does History Have India?," *Comparative Studies in Society and History* 54 (2012): 179–80.

17. Thomas Maurice, *Observations on the Remains of Ancient Egyptian Grandeur and Superstition* (London: John Murray, 1818), v; Jed Z. Buchwald and Diane Greco Josefowicz, *The Zodiac of Paris* (Princeton, NJ: Princeton University Press, 2010), 99–115.

18. David Pingree, *Jyotiḥśāstra: Astral and Mathematical Literature* (Wiesbaden, Germany: Otto Harrassowitz, 1981), 12–13; Clemency Montelle and Kim Plofker, *Sanskrit Astronomical Tables* (Cham, Switzerland: Springer, 2018), 13; Christopher Minkowski, "Competing Cosmologies in Early Modern Indian Astronomy," in *Ketuprakasa*, ed. Charles Burnett, Jan Hogendijk and Kim Plofker (Leiden: Brill, 2004), 349–85.

19. William Jones, "On the Antiquity of the Indian Zodiac," *Asiatick Researches* 2 (1790): 290; C.A. Bayly, *Empire and Information* (Cambridge: Cambridge University Press, 1996), 247–48; Sumathi Ramaswamy, *Terrestrial Lessons* (Chicago: Chicago University Press, 2017), 26–29.

20. Romila Thapar, "Imagined Religious Communities? Ancient History and the Modern Search for a Hindu Identity," *Modern Asian Studies* 23 (1989): 218; Raj, *Relocating Modern Science*, 126–28; Oddie, *Imagined Hinduism*, 99.

21. Rosane Rocher, "Weaving Knowledge: Sir William Jones and the Indian Pandits," in *Objects of Enquiry: Life, Contributions and Influences of Sir William Jones*, ed. Garland Cannon and Kevin Brine (New York: New York University Press, 1995), 60–61; Raj, *Relocating Modern Science*, 134–38.

22. John Shore, *Memoirs of the Life, Writings and Correspondence of William Jones*, 2nd ed. (London: Hatchard, 1806), 2:238.

23. Maconochie for Dundas, March 15, 1788, National Library of Scotland MS 546, fol. 16v.

24. Mīr Muḥammad Ḥusayn, al-Ijābah ʿalā taʿlīq al-Jawnbūrī, British Library MS Delhi Arabic 1928, fols. 130–41; *Second Report of the Calcutta School-Book Society's Proceedings* (Calcutta: School & Mission Presses, 1819), 4; "Letter from Three Hindoos Engaged in Translating Ferguson's Astronomy," *Asiatic Journal* 15 (January 1823): 63; S.M. Razaullah Ansari, "Modern Astronomy in Indo-Persian Sources," in *Transfer of Modern Science and Technology to the Muslim World*, ed. Ekmeleddin İhsanoğlu (Istanbul: IRCICA, 1992), 135–36.

25. David Brewster, ed., *Ferguson's Astronomy Explained upon Sir Isaac Newton's Principles* (Edinburgh: Ballantyne, 1811), 1:4–7.

26. Thomas Paine, *The Age of Reason*, 2nd ed. (London: Eaton, 1795), 45–46; Dinwiddie Fonds, Journal B33 (October 11, 1796).

27. Ashworth, "Calculating Eye," 426; Rocher and Rocher, *Western Indology*, 27–31; Bentley, *Universal Tables*.

28. Raj, *Relocating Modern Science*, 131–34; Javed Majeed, *Ungoverned Imaginings: James Mill's "The History of British India and Orientalism"* (Oxford: Clarendon Press, 1992), 16–17, 42–43.

29. Jones to Chapman, 1785, in Garland Cannon, ed., *Letters of William Jones* (Oxford: Clarendon Press, 1970), 667; Colebrooke, *Life of Colebrooke*, 26; George Colebrooke to Joseph Banks, May 19, 1787, National Library of Australia MS 9 Banks Papers, series 3, item 38, 269–71; Rocher and Rocher, *Western Indology*, 17.

30. William Jones, "Second Anniversary Discourse," *Asiatick Researches* 1 (1788), 412; and "On the Philosophy of the Asiaticks," *Asiatick Researches* 4 (1795): 176; Henry Colebrooke, "On the Vedas," *Asiatick Researches* 8 (1805): 461–63.

31. Henry Colebrooke, "Of the Religious Ceremonies of the Hindus," *Asiatic Researches* 7 (1801): 279; Colebrooke, "On the Vedas," 465–67, 474–75; Kopf, *British Orientalism*, 40–41; Rocher and Rocher, *Western Indology*, 79.

32. Guillaume Le Gentil, *Voyage dans les mers de l'Inde* (Bern, Switzerland: Libraires Associés, 1780), 1:315–32; Sylvia Murr, *L'Inde Philosophique entre Bossuet et Voltaire*, 2 vols. (Paris: Adrien Maisonneuve, 1987), 2:177; Dhruv Raina, "'Becoming All Things to All': French Jesuit Scientists and the Construction of the Antiquity of the Sciences of India," in *L'Inde des Lumières*, ed. Marie Fourcade and Ines Županov (Paris: EHESS, 2013), 347–49; Michel Danino, "Le Gentil à Pondichéry: De Vénus à l'Inde," *Synergies Inde* 7 (2016): 39–41.

33. Ayeen Akbery (Calcutta: n.p., 1783), trans. Francis Gladwin, 1:327–28; Bayly, *Empire and Information*, 23.

34. Anthony Grafton, "Dating History: The Renaissance and the Reformation of Chronology," *Daedalus* 132, no. 2 (Spring 2003): 80–81.

35. Quintin Craufurd, *Sketches Chiefly Relating to the History, Religion, Learning and Manners of the Hindoos* (London: Cadell, 1790), 237; John Playfair, "Remarks on the Astronomy of the Brahmins," *Transactions of the Royal Society of Edinburgh* 2 (1790): 189.

36. Jean-Sylvain Bailly, *Traité de l'Astronomie Indienne et Orientale* (Paris: Debure, 1787), xvii; Edwin Burrows Smith, "Jean-Sylvain Bailly: Astronomer, Mystic, Revolutionary," *Transactions of the American Philosophical Society* 44 (1954): 455–59; Dhruv Raina, "Betwixt Jesuit and Enlightenment Historiography: Jean-Sylvain Bailly's History of Indian Astronomy," *Revue d'Histoire des Mathématiques* 9 (2003): 261–67; Pingree, *Jyotiḥśāstra*, 21–23.

37. Bailly, *Traité de l'Astronomie Indienne*, lii–liii and 159; Jean-Sylvain Bailly, *Essai*

sur les Fables (Paris: Guillaume de Bure, 1797), 4; Smith, "Bailly," 480–81; Raina, "Betwixt Jesuit and Enlightenment Historiography," 264.

38. "Rapport des Commissaires Nommés par l'Académie des Sciences, 23 Décembre 1785," and Laplace to Bailly, February 22, 1787, *Journal des Sçavans* (June 1787): 327, 330; Smith, "Bailly," 497.

39. [Guillaume Dubois de Rochefort], "Traité de l'Astronomie Indienne et Orientale par M. Bailly," *Journal des Sçavans* (June 1787): 324.

40. Raj, *Relocating Modern Science*, 110–11; Jack Morrell, "Professors Robison and Playfair and the Theophobia Gallica: Natural Philosophy, Religion and Politics in Edinburgh, 1789–1815," *Notes and Records of the Royal Society of London* 26 (1971): 45–46.

41. Edmund Burke, "Speech on the Opening of Impeachment, 16 February 1788," in *Writings and Speeches of Edmund Burke*, ed. P.J. Marshall and William B. Todd (Oxford: Oxford University Press, 1991), 6:345; Partha Chatterjee, *The Black Hole of Empire* (Princeton, NJ: Princeton University Press, 2012), 59–61.

42. Jeffrey Smitten, *Life of William Robertson* (Edinburgh: Edinburgh University Press, 2017), 226–27; Maconochie to Dundas, February 1801, National Library of Scotland MS 1074, fol. 78.

43. Robertson to Smith, June 14, 1759, in E.C. Mossner and I.S. Ross, eds., *Correspondence of Adam Smith* (Oxford: Oxford University Press, 1987), 40; Adam Smith, *Essays on Philosophical Subjects*, ed. Joseph Black and James Hutton (London: Cadell and Davies, 1795), 25, 44.

44. Dennis R. Dean, "Hutton on Religion and Geology," *Annals of Science* 32 (1975): 190–92; Morrell, "Robison and Playfair," 52–53; [Thomas Smith], "Astronomy of the Hindus," *Calcutta Review* 1 (May 1844): 264.

45. Jane Rendall, "Scottish Orientalism: From Robertson to Mill," *Historical Journal* 25 (1982): 46–49; Stewart J. Brown, "William Robertson, Early Orientalism and the Historical Disquisition on India of 1791," *Scottish Historical Review* 88 (2009): 304–8.

46. [Henry Brougham], "Memoir of Allan Lord Meadowbank," *Law Review* 2 (1845): 73–76; Edinburgh University Library, Maconochie papers, MS Coll-1137, no. 14; *Transactions of the Royal Society of Edinburgh* 2 (1790): 25; Maconochie to Abercromby, March 15, 1788, National Library of Scotland MS 546, fol. 15.

47. Maconochie to Dundas, 1783, National Archives of Scotland, Melville Papers, GD 51-3-617/2, fols. 406–9; Dharampal, *The Beautiful Tree* (Goa: Other India Press, 2000), 14–15.

48. Maconochie for Dundas, March 15, 1788, National Library of Scotland MS 546, fols. 16–17, copied in British Library MS Add 29233 fols. 237–38, misattributed to Reuben Burrow in R.H. Phillimore, *Historical Records of the Survey of India*, vol. 1 (Dehradun: Geodetic Branch, Survey of India, 1945), 156.

49. "Memorial Respecting the Astronomy of India," January 30, 1790, National Library of Scotland, Melville Papers, MS Acc.12001/1.

50. Playfair to William Robertson the younger, May 9, 1774, National Library of Scotland MS Add 3942, fol. 150.

51. Playfair, "Remarks," 152, 160; Thomas R. Trautmann, *Aryans and British India* (Berkeley: University of California Press, 1997), 84–88.

52. Playfair, "Remarks," 138–39, 153, 168–69.

53. Dundas to Robertson, June 27, 1791, National Library of Scotland, MS 3944, fol. 65; Ross to Robertson, July 9, 1792, National Library of Scotland, MS 3944, fol. 106; Brown, "William Robertson," 301.

54. Brown, "William Robertson," 306–8; Geoffrey Carnall, "Robertson and Contemporary Images of India," in *William Robertson and the Expansion of Empire*, ed. Stewart Brown (Cambridge: Cambridge University Press, 1997), 214–15; Silvia Sebastiani, "William Robertson, entre l'Amérique et l'Inde," in Fourcade and Županov, *L'Inde des Lumières*, 116–19.

55. National Library of Scotland, MS 3967, fols. 31, 35; William Robertson, *Historical Disquisition Concerning the Knowledge which the Ancients Had of India* (London: Strahan and Cadell, 1791), 303, 309, 360.

56. National Library of Scotland, MS 3967, fol. 36; Robertson, *Historical Disquisition*, 310–11.

57. Samuel Davis, "On the Astronomical Computations of the Hindus," *Asiatick Researches* 2 (1790): 225–26; Yale University Beinecke Library, William Jones notebook, MS Osborn c400, 111; Jones to Davis, February 28, 1789, in Cannon, *Letters*, 830; Rocher, "Weaving Knowledge," 64.

58. Davis, "Astronomical Computations," 238, 249, 273, 287; Bentley, "Remarks," 329.

59. Playfair to Robertson, May 12, 1792, National Library of Scotland, MS 3944, fol.103; William Robertson, *Historical Disquisition Concerning the Knowledge which the Ancients Had of India*, 2nd ed. (London: Strahan and Cadell, 1794), 439–41.

60. John Playfair, "Observations on the Trigonometrical Tables of the Brahmins," *Transactions of the Royal Society of Edinburgh* 4 (1798): 103–6; Dhruv Raina, "Contextualizing Playfair and Colebrooke on Proof and Demonstration in the Indian Mathematical Tradition (1780–1820)," in *History of Mathematical Proof in Ancient Traditions*, ed. Karine Chemla (Cambridge: Cambridge University Press, 2012), 238–39.

61. Jones to Davis, November 8, 1789, and October 27, 1790, in Cannon, *Letters*, 852, 876.

62. William Jones, "Supplement to the Essay on Indian Chronology," *Asiatick Researches* 2 (1790): 393; Jones to Davis, March 21 and April 11, 1790, in Cannon, *Letters*, 858, 862.

63. Jones to Davis, February 21, 1790; March 21, 1790; September 16, 1791, in Cannon, *Letters*, 857, 859, 887–88.

64. Jones to Davis, January 6, 1791, and October 11, 1791, in Cannon, *Letters*, 879, 889.

65. Jones to Hastings, October 20, 1791, in Cannon, *Letters*, 900; William Jones,

"On Asiatick History, Civil and Natural," *Asiatick Researches* 4 (1795): 9–10; Trautmann, *Aryans and British India*, 41–46.

66. John Playfair, "Questions and Remarks on the Astronomy of the Hindus," *Asiatick Researches* 4 (1795): 161–63; Jones to Davis, December 27, 1791, and September 14, 1793, in Cannon, *Letters*, 905 and 917; Joydeep Sen, *Astronomy in India, 1784–1876* (London: Pickering and Chatto, 2014), 20–21.

67. William Jones, "Philosophy of the Asiaticks," 178–79.

68. Simon Schaffer, "The Asiatic Enlightenments of British Astronomy," in *The Brokered World: Go-Betweens and Global Intelligence, 1770–1820*, ed. Simon Schaffer, Lissa Roberts, Kapil Raj, and James Delbourgo (Sagamore Beach, MA: Science History Publications, 2009), 79–83, 87–88; Dinwiddie Fonds, Journal B32 (September 22, 1796); William Hunter, "Some Account of the Astronomical Labours of Jayasinha, Rajah of Ambere, or Jayanagar," *Asiatick Researches* 5 (1798): 210–11.

69. Sanjay Subrahmanyam, "Traces of the Ancient in India: Notes on Two Possible Narratives," in *World Antiquarianism: Comparative Perspectives*, ed. Alan Schnapp, L. von Falkenhausen, P.N. Miller, and T. Murray (Los Angeles: Getty Foundation, 2013), 376.

70. "Account of the Life and Character of Tofuzzel Hussein Khan," *Asiatic Annual Review for 1803*, "Characters": 4; Dinwiddie Fonds, Journal B32 (September 3, 1796); [James Prinsep], "Historical View of Hindu Astronomy," *Quarterly Oriental Magazine* 5 (June 1826): 233–34.

71. C.A. Bayly, "Orientalists, Informants and Critics in Benares, 1790–1860," in *Perspectives of Mutual Encounters in South Asian History 1760–1860*, ed. Jamal Malik (Leiden: Brill, 2000), 110.

72. Duncan to Cornwallis, January 1, 1792, in *Bengal Past and Present* 8 (1914): 131–33; Ramaswamy, *Terrestrial Lessons*, 125–29; Dodson, *Orientalism*, 155–56.

73. Rocher and Rocher, *Western Indology*, 47–48, 65, 70.

74. Colebrooke, "Hindu Astronomy," 360, 364; John Bentley, "On the Hindu Systems of Astronomy Connected with Ancient and Modern History," *Asiatick Researches* 8 (1805): 198, 200, 243.

75. Thomas Aufrecht, *Catalogue of Sanskrit Manuscripts in the Library of Trinity College Cambridge* (Cambridge: Deighton Bell, 1869), v–vi, 40–46, 74–75; Dinwiddie Fonds, Journal B52 (January 21, 1799; January 26, 1799; February 3, 1799).

76. John Bentley, *Historical View of the Hindu Astronomy* (London: Smith, Elder, 1825), iv–v, 186, 194–95; Colebrooke, "Hindu Astronomy," 364.

77. Georges Cuvier, *Recherches sur les Ossemens Fossiles de Quadrupèdes*, vol. 1 (Paris: Deterville, 1812), 1, 100–102, 108; Rachel Laudan, "Georges Cuvier on the Oriental Origins of Science," *Nineteenth-Century French Studies* 17 (1988–1989): 24–25.

78. Bentley, *Historical View*, 251; Buchwald and Josefowicz, *Zodiac of Paris*, 217–18.

79. Pierre Simon Laplace, *Exposition du Système du Monde*, vol. 2 (Paris: Cercle Social, 1796), 210; Pierre Simon Laplace, *Exposition du Système du Monde*, 2nd ed. (Paris: Crapelet, 1798), 294.

80. Delambre to Craufurd, July 21, 1814, in Quintin Craufurd, *Researches Concerning the Laws, Theology, Learning, Commerce, etc. of Ancient and Modern India*, 2 vols. (London: Cadell and Davies, 1817), 2:352–53; Raina, "Disciplinary Boundaries," 227.

81. Bentley, *Historical View*, xxiv–xxvii; Dinwiddie Fonds, Lecture Notes C-14.

82. Bentley, "Remarks on the Principal Aeras," 315–17.

83. Bentley, "On the Antiquity," 538, 541–45, 556.

84. Bentley, "On the Antiquity," 559, 572–75.

85. Bentley MSS, Add.Ms.a.617, Trinity College Cambridge, "Hindu Astronomy, Ancient and Modern, Historically Arranged," 169 (Bentley's emphasis); Bentley, "On the Hindu Systems," 203, 239.

86. Rocher, *Hamilton*, 11, 65–66; [Alexander Hamilton], "On the Antiquity of the Surya Siddhanta by Mr J. Bentley," *Asiatic Annual Register for 1803*, "Account of Books," 33; [Alexander Hamilton], "*Asiatic Researches* vol. VI," *Edinburgh Review* 1 (October 1802): 42–43.

87. [Alexander Hamilton], "*Asiatic Researches* vol. VIII," *Edinburgh Review* 12 (April 1808): 42; Bentley, "On the Hindu Systems," 197; Sen, *Astronomy in India*, 30–31.

88. Bentley, "On the Hindu Systems," 240.

89. Bentley, "On the Hindu Systems," 203, 244.

90. Bentley, *Historical View*, 158, 169, 182, 208.

91. Bentley, *Historical View*, 76, 79, 97; [Prinsep], "Historical View," 262–63.

92. Joshua Marshman, *Elements of Chinese Grammar* (Serampore, India: Mission Press, 1814), 154–55; Thomas Stamford Raffles, *History of Java*, vol. 1 (London: Black, Parbury, and Allen, 1817), 418; Bentley, "Hindu Astronomy," 167.

93. Bentley, *Historical View*, 109–11; Bentley, "Hindu Astronomy," 157.

94. Macintosh to Butler, January 16, 1809, in *Memoirs of the Life of Sir James Macintosh*, ed. Robert Macintosh, vol. 1, 2nd ed. (London: Moxon, 1836), 447; [Hamilton], "*Asiatic Researches* vol. VIII," 49; Bentley, *Historical View*, xxiii; Rocher, *Alexander Hamilton*, 90–91.

95. [John Playfair], "Laplace's System of the World," *Edinburgh Review* 15 (January 1810): 416; [John Playfair], "Cuvier on the Theory of the Earth," *Edinburgh Review* 22 (January 1814): 472–75.

96. [John Playfair], "Algebra and Arithmetic of the Hindus," *Edinburgh Review* 29 (November 1817): 161–63; Raina, "Contextualizing Playfair and Colebrooke," 238–46; Rocher and Rocher, *Western Indology*, 133–35; Bentley MSS, Add.Ms.a.617, Trinity College Cambridge, "Translation of Part of the Beej Ganita or Algebra of the Hindus."

97. [John Playfair], "Bentley on the Indian Astronomy," *Edinburgh Review* 10 (July 1807): 461, 450.

98. [Playfair], "Bentley on the Indian Astronomy," 460.

99. [Playfair], "Bentley on the Indian Astronomy," 465–66.

100. [Hamilton], "Sanscrit Poetry," 442; [Prinsep], "Historical View," 259–60; Colebrooke, *Life of Colebrooke*, 245–46.

101. James Mill, *History of British India*, vol. 1 (London: Baldwin, Cradock, and Joy, 1817), 396–400; Trautmann, *Aryans and British India*, 117–24; Majeed, *Ungoverned Imaginings*, 128; Dodson, *Orientalism*, 69; [James Mill], "Hindu Astronomy," *Westminster Review* 2 (July 1824): 279.

102. Raj, *Relocating Modern Science*, 138.

2: DEBATING EVOLUTION AND RELIGION IN NINETEENTH-CENTURY SOUTH ASIA

1. Lynn A. Phelps and Edwin Cohen, "The Wilberforce-Huxley Debate," *Western Journal of Communication* 37, no. 1 (1973): 56–64, https://doi.org/10.1080/10570317309373768.

2. See for example John Hedley Brooke, *Science and Religion: Some Historical Perspectives* (Cambridge: Cambridge University Press, 1991), 45; Richard England, "Censoring Huxley and Wilberforce: A New Source for the Meeting That the *Athenaeum* 'Wisely Softened Down,'" *Notes and Records* 71, no. 4 (December 20, 2017): 371–84, 372; Nanna Katrine Lüders Kaalund, "Oxford Serialized: Revisiting the Huxley–Wilberforce Debate through the Periodical Press," *History of Science* 52, no. 4 (December 2014): 429–53, 429; Ian Hesketh, *Of Apes and Ancestors: Evolution, Christianity, and the Oxford Debate* (Toronto: University of Toronto Press, 2009), 6.

3. Janet Browne, *Darwin's Origin of Species: A Biography* (London: Atlantic Books, 2006), 1.

4. J. Vernon Jensen, "Return to the Wilberforce–Huxley Debate," *British Journal for the History of Science* 21, no. 2 (June 1988): 161–79, https://doi.org/10.1017/S0007087400024742.

5. For example, Richard England examines how an account of the debate was omitted from the July 1860 edition of the *Athenaeum* but printed in the *Oxford Chronicle*. See England, "Censoring Huxley and Wilberforce."

6. On the relationship of these figures to the history of evolutionary thought, see the following articles: Duncan M. Porter, "On the Road to the *Origin* with Darwin, Hooker, and Gray," *Journal of the History of Biology* 26, no. 1 (1993): 1–38; Joel S. Schwartz, "Charles Darwin's Debt to Malthus and Edward Blyth," *Journal of the History of Biology* 7, no. 2 (1974): 301–18; Barbara G. Beddall, "'Notes for Mr. Darwin': Letters to Charles Darwin from Edward Blyth at Calcutta: A Study in the Process of Discovery," *Journal of the History of Biology* 6 (1973): 69–95, https://doi.org/10.1007/bf00137299; Christine Brandon-Jones, "Edward Blyth, Charles Darwin, and the Animal Trade in Nineteenth-Century India and Britain," *Journal of the History of Biology* 30, no. 2 (1997): 145–78; Ernst Haeckel, *A Visit to Ceylon*, trans. Clara Bell (Boston: Bradlee Whidden, 1889); Robert C. Stauffer, "Haeckel, Darwin, and Ecology," *Quarterly Review of Biology* 32, no. 2 (1957): 138–44; Robert J. Richards, "The Linguistic Creation of Man: Charles Darwin, August Schleicher, Ernst Haeckel, and

the Missing Link in Nineteenth-Century Evolutionary Theory," in *Experimenting in Tongues: Studies in Science and Language*, ed. Matthias Dörries (Stanford, CA: Stanford University Press, 2002), 21–48; Georgy S. Levit and Uwe Hossfeld, "Ernst Haeckel in the History of Biology," *Current Biology* 29, no. 24 (2019): 1276–84.

7. See Donald S. Lopez Jr.'s careful treatment of the event in *Buddhism and Science: A Guide for the Perplexed* (Chicago: University of Chicago Press, 2008), 39–73.

8. Kitsiri Malalgoda, *Buddhism in Sinhalese Society, 1750–1900: A Study of Religious Revival and Change* (Berkeley: University of California Press, 1976), 229.

9. Malalgoda, *Buddhism*.

10. Dhruv Raina and S. Irfan Habib, "The Moral Legitimation of Modern Science: Bhadralok Reflections on Theories of Evolution," *Social Studies of Science* 26, no. 1 (1996): 9–42, 10.

11. Patrick Peebles, *The History of Sri Lanka* (Westport, CT: Greenwood, 2006), 53.

12. See Anne Blackburn's account of the central place of "interreligious debate culture" including not only new texts from Europe but also older authoritative Pali texts. Anne Blackburn, *Locations of Buddhism: Colonialism and Modernity in Sri Lanka* (Chicago: University of Chicago Press, 2010), 6. On debate in Buddhism in this period see Blackburn, *Locations of Buddhism*, 6; Richard Gombrich, *Theravada Buddhism: A Social History from Ancient Benares to Modern Colombo*, 2nd ed. (London: Routledge, 2006); Elizabeth J. Harris, *Theravada Buddhism and the British Encounter: Religious, Missionary and Colonial Experience in Nineteenth-Century Sri Lanka* (London: Routledge, 2006); Ananda Tissa Kumara and Nandana Weerarathna, eds., *Kristiyani Prajnaptiyata Piliturak saha Kristiyani Prajnapitiya* by Bentota Atthadassa (Colombo: S. Godage and Brothers, 2005); R.F. Young and G.P.V. Somaratne, *Vain Debates: The Buddhist-Christian Controversies of Nineteenth-Century Ceylon* (Vienna: Sammlung De Nobili, 1996).

13. On *ijtihad* see Wael B. Hallaq's classic essay, "Was the Gate of Ijtihad Closed?," *International Journal of Middle East Studies* 16, no. 1 (1984): 3–41. Alternatively, we might draw on intellectual histories around Qur'anic exegesis or *tafsir*. See Andreas Gorke and Johanna Pink, *Tafsir and Islamic Intellectual History: Exploring the Boundaries of a Genre* (Oxford: Oxford University Press, 2015).

14. For example, see Marwa Elshakry, *Reading Darwin in Arabic* (Chicago: University of Chicago Press, 2013).

15. See Sarah A. Qidwai, "Darwinism and the History of Evolutionary Biology: What Is Excluded?," in *Beyond the Darwinian Revolution: Historicizing Evolution from the Past to the Present*, ed. Ian Hesketh (Pittsburgh: University of Pittsburgh Press, 2022), 137–50.

16. Mackenzie Brown, *Hindu Perspectives on Evolution: Darwin, Dharma, and Design* (London: Routledge, 2012), 5.

17. Mackenzie Brown, ed., *Asian Religious Responses to Darwinism: Evolutionary*

Theories in Middle Eastern, South Asian, and East Asian Cultural Contexts (Cham, Switzerland: Springer, 2020).

18. Luzia Savary, *Evolution, Race and Public Spheres in India* (New York: Routledge, 2019), 24.

19. On the continued use of "South Asia" see, for example, Pratik Chakrabarti, "Introduction: States of Healing," *South Asian History and Culture* 4, no. 1 (January 1, 2013): 1–8, https://doi.org/10.1080/19472498.2012.750451; Asif A. Siddiqi, "Technology in the South Asian Imaginary," *History and Technology* 31, no. 4 (October 2, 2015): 341–49, https://doi.org/10.1080/07341512.2016.1142632; Assa Doron and Alex Broom, "Health and Healing in South Asia," *South Asian History and Culture* 1, no. 2 (March 25, 2010): 193–98, https://doi.org/10.1080/19472491003590577; Lisa Alette Brooks, Anthony Cerulli, and Victoria Sheldon, eds., "Special Issue: Medicines and Memories in South Asia," *Asian Medicine* 15, no. 1 (November 2020).

20. For critiques of area studies framings, see Martin W. Lewis and Karen Wigen, *The Myth of Continents: A Critique of Metageography* (Berkeley: University of California Press, 1997); Willem van Schendel, "Geographies of Knowing, Geographies of Ignorance: Jumping Scale in Southeast Asia," *Environment and Planning D: Society and Space* 20, no. 6 (December 2002): 647–68, https://doi.org/10.1068/d16s; Siddiqi, "Technology in the South Asian Imaginary," 341; On alternative framing as "Southern Asia" see Anne M. Blackburn and R. Michael Feener, "Sufis and *Saṅgha* in Motion: Toward a Comparative Study of Religious Orders and Networks in Southern Asia," in *Buddhist and Islamic Orders in Southern Asia*, ed. R. Michael Feener and Anne M. Blackburn (Honolulu: University of Hawai'i Press, 2018), 8.

21. Here a kind of doublespeak operates: in the history of science literature, geographic regions of study—"China," "India," "South Asia," and "Turkey"—are naturalized as static and unchanging, inscribed in stone, rather than culturally construed and in flux. By contrast, when describing scientific practitioners or theories, historians of science are entirely alert to translation, flux, movement, and flows.

22. Steven Kemper, *Rescued from the Nation: Anagarika Dharmapala and the Buddhist World* (Chicago: University of Chicago Press, 2015), 18.

23. Kemper, *Rescued from the Nation*, 8.

24. For example, Anushka Kahandagama, "Buddhist Revivalist Movement under British Rule in the Nineteenth Century Sri Lanka: 'Sciences' of Enlightenment and Buddhism" (MPhil diss., South Asian University, Delhi, 2019).

25. See Su Lin Lewis, 'Between Orientalism and Nationalism: The Learned Society and the Making of 'Southeast Asia,'" *Modern Intellectual History* 10, no. 2 (2013): 353–74; Shwe Zan Aung, "Buddhism and Science," *Journal of the Burma Research Society* 8, no. 2 (1918): 99–106; Shwe Zan Aung, "The World Is Round," *Journal of the Burma Research Society* 7, no. 2 (1917): 184–86. Lewis's analysis reads philosopher Shwe Zan Aung in relation to English-education literate societies in colonized cities like Rangoon. The trends fit well with similar themes in Ceylon.

26. See Hallaq, "Was the Gate of Ijtihad Closed?"; and Gorke and Pink, *Tafsir and Islamic Intellectual History.*

27. Charu Singh, "Science in the Vernacular? Translation, Terminology, and Lexicography in the *Hindi Scientific Glossary* (1906)," in "Indigenous Knowledges and Colonial Sciences in South Asia," special issue, *South Asian History and Culture* 13, no. 1 (2022): 63–86, https://doi.org/10.1080/19472498.2021.2001200.

28. See Avril Ann Powell, *Muslims and Missionaries in Pre-Mutiny India* (New York: Routledge, 1995), 235–71.

29. Shoaib Ahmed Malik, *Islam and Evolution: Al-Ghazālī and the Modern Evolutionary Paradigm*, Routledge Science and Religion Series (London: Routledge, 2021).

30. Barbara Daly Metcalf, *Islam in South Asia in Practice* (Princeton, NJ: Princeton University Press, 2009), 1.

31. Elshakry, *Reading Darwin*, 120.

32. Margrit Pernau, "Fluid Temporalities: Saiyid Ahmad Khan and the Concept of Modernity," *History and Theory* 58, no. 4 (December 2019): 107–31; Savary, *Evolution, Race and Public Spheres*; Sarah Qidwai, "Darwin or Design: Sir Sayyid Ahmad Khan's Views on Human Evolution," in *The Cambridge Companion to Sayyid Ahmad Khan*, ed. Yasmin Saikia and Raisur Rahim (Cambridge: Cambridge University Press, 2019), 214–32.

33. Sayyid Ahmad Khan, *The Mahomedan Commentary on the Holy Bible Genesis I–XI Part II* (Allygurh, 1865), 66.

34. Sayyid Ahmad Khan, "Adna Halat se Aala Halat par Insaan ki Taraqqi" ("The Stages of Human Development from an Inferior to Superior State"), reprinted in *Maqalat-e-Sir Sayyid*, ed. Muhammad Ismail Panipati (Lahore: Majlis-e Taraqqi-e Adab, 1962).

35. In an earlier publication Qidwai has sketched out how this article does not fit into the historiographical landscape of the history of evolutionary biology. There are other responses as well to Sayyid Ahmad's words, and we have to unpack them further. See Qidwai, "Darwin or Design."

36. See Muhammad Al-Ghazali, *Socio-Political Thought of Shah Wali Allah* (Delhi: Adam Publisher, 2004).

37. Al-Ghazali, *Socio-Political Thought of Shah Wali Allah*, 51–76.

38. "Proceedings," in Shan Muhammad, *The Aligarh Movement: Basic Documents, 1864–1898* (Meerut, India: Meenakshi Prakashan, 1978), 21–23.

39. Stephan Popp, "Muhammad Iqbal—Reconstructing Islam along Occidental Lines of Thought," *Interdisciplinary Journal for Religion and Transformation in Contemporary Society* 5, no. 1 (2019): 201–29.

40. Muhammad Iqbal and Sheikh M. Saeed, *The Reconstruction of Religious Thought in Islam* (Stanford, CA: Stanford University Press, 2013).

41. Iqbal and Saeed, *Reconstruction of Religious Thought*, 23–49.

42. Iqbal and Saeed, *Reconstruction of Religious Thought*, 23–49.

43. Souleymane Bachir Diagne , *Postcolonial Bergson*, trans. Lindsay Turner (New York: Fordham University Press, 2019).

44. Anushka Kahandagama makes this connection in her MPhil dissertation, "Buddhist Revivalist Movement," 8.

45. Ananda Guruge, *Return to Righteousness* (Colombo: Sri Lanka Department of Culture, 1965); David Karunaratne, අනගාරික ධර්මපාලතුමා [*Anagarika Dharmapala*] (Kandy, Sri Lanka: Buddhist Publication Society, 1964); T.S. Dharmabandhu, සිංහල වීරයෝ [*Sinhala heroes*] (Colombo: M.D. Gunasena, 1973); Kemper, *Rescued from the Nation*.

46. See Harshana Rambukwella's treatment of Sinhala nationalistic scholarship on Dharmapāla in *The Politics and Poetics of Authenticity: A Cultural Genealogy of Sinhala Nationalism* (London: UCL Press, 2018). See also Sarath Amunugama, *The Lion's Roar: Anagarika Dharmapala and the Making of Modern Buddhism* (Oxford: Oxford University Press, 2019); Sandagomi Coperahewa ed., *Anagarika Dharmapala and India–Sri Lanka Relations* (Colombo, Sri Lanka: Centre for Contemporary Indian Studies, 2015); Bhadrajee Hewage, *A Name for Every Chapter: Anagarika Dharmapala and Ceylonese Buddhist Revivalism* (London: Sulochana Publishing, 2020).

47. Historians of Buddhism, however, have been very interested in Dharmapāla's views on science. These debates have not made their way into the global history of science, and perhaps this says something about disciplinary bounding between the study of religions and history of science. See Lopez, *Buddhism and Science*; and David L. McMahan, *The Making of Buddhist Modernism* (Oxford: Oxford University Press, 2009).

48. See, for instance, his modes of denigrating Islamic and Christian approaches to evolution in Karunaratne, අනගාරික ධර්මපාලතුමා. This was, after all, the period that saw world religions emerging as equivalent "faiths."

49. The Parliament ran September 11–27, 1893; for more detail on the event see Eric Jozef Ziolkowski, *Museum of Faiths: Histories and Legacies of the 1983 World Parliament of Religions* (Charlottesville: University of Virginia Press, 1993), 25, 44.

50. C.A. Bayly, *The Birth of the Modern World, 1780–1914* (Malden, MA: Blackwell, 2004), 20; qtd. in Kemper, *Rescued from the Nation*, 424.

51. Marwa Elshakry, "Global Darwin: Eastern Enchantment," *Nature* 461 (October 29, 2009): 1200–1201.

52. Elshakry, "Global Darwin," 1201.

53. Dharmapāla's ideas echoed those of Sōen Shaku, a Japanese Zen priest who gave a speech at the same Parliament in 1893 titled "The Law of Cause and Effect, as Taught by the Buddha."

54. McMahan, *Buddhist Modernism*, 64.

55. See Brown's treatment of Hindus attempting to grapple with phrenology in *Hindu Perspectives on Evolution*, 64.

56. Kemper, *Rescued from the Nation*, 13.

57. Brown, *Asian Religious Responses to Darwinism*, 9.

58. Kemper, *Rescued from the Nation*, 425; for a more nuanced perspective, Anne Blackburn describes Dharmapāla's changing notions of Buddhism as "locative pluralism," shifting based on who his audience was. Blackburn, *Locations of Buddhism*, 209–10.

59. Stephen Prothero, "Henry Steel Olcott and 'Protestant Buddhism,'" *Journal of the American Academy of Religion* 63, no. 2 (Summer 1995): 289, 297.

60. McMahan, *Buddhist Modernism*, 70.

61. McMahan, *Buddhist Modernism*, 122.

62. Victor Sōgen Hori, "D.T. Suzuki and the Invention of Tradition," *Eastern Buddhist* 47, no. 2 (2016): 41–81.

63. McMahan, *Buddhist Modernism*, 6.

64. For more see Sujit Sivasundaram, *Islanded: Britain, Sri Lanka and the Bounds of an Indian Ocean Colony* (New Delhi: Oxford University Press, 2014; first published by the University of Chicago Press, 2013), 334.

65. Michael Roberts, "For Humanity. For the Sinhalese. Dharmapala as Crusading Bosat," *Journal of Asian Studies* 56 (1997): 1006–32.

66. McMahan, *Buddhist Modernism*, 65.

67. Savary, *Evolution, Race and Public Spheres*.

68. Anne Blackburn speculates that these anonymous contributors might have been Buddhist monks. *Locations of Buddhism*, 84n30.

69. Savary, *Evolution, Race and Public Spheres*, 3.

70. In response, *Morning Star*, the bilingual evangelical Jaffna newspaper, believed "there is not a single geologist, or biologist, or botanist anywhere who is not known out of his own district, who does not believe in the evolution of species." The paper cited from the publications of American naturalist James Dwight Dana to muse on the nature of scientific proof: "What is *proved?* That for which sufficient evidence has been presented to require a reasonable man to believe it." Anon., "Evolution," *Morning Star* 57, no. 11 (May 27, 1897).

71. "Both Darwin and Huxley may be cited as having expressed themselves strongly that there was no contradiction between science and religion," the paper explained to its readers. See also Brown's comments on the reception of Spencer in Japan and China in *Asian Religious Responses to Darwin*, 4.

72. Jaffna College was started in 1823 at the Batticaloa seminary. The aim of his paper in 1912 was to "trace the ascent of man's body from the lower animal conditions, and the evolution of his mind, and lastly . . . the forces that were at work in the evolution of the human race to higher conditions." See J.V. Chelliah, *A Century of English Education: The Story of Batticotta Seminary and Jaffna College*, rev. ed. (Vaddukoddai: Jaffna College, 1984).

73. Drummond's *Lowell Lectures on the Ascent of Man* were published in 1894 and sold 10,000 copies within a year. James R. Moore, "Evangelicals and Evolution: Henry Drummond, Herbert Spencer, and the Naturalisation of the Spiritual

World," *Scottish Journal of Theology* 38, no. 3 (August 1985): 383–418, https://doi.org/10.1017/S0036930600041028.

3: EVOLUTION IN COLONIAL AUSTRALIA

1. Ann Mozley, "Evolution and the Climate of Opinion in Australia, 1840–76," *Victorian Studies* 10 (1967): 411–30; Barry W. Butcher, "Darwin's Australian Correspondents: Deference and Collaboration in Colonial Science," in *Nature in Its Greatest Extent: Western Science in the Pacific*, ed. Roy MacLeod and Philip F. Rehbock (Honolulu: University of Hawai'i Press, 1988), 139–57; Colin Finney, *Paradise Revealed: Natural History in Nineteenth-Century Australia* (Melbourne: Museum of Victoria, 1993), 97–113, 131–44; Tom Frame, *Evolution in the Antipodes: Charles Darwin and Australia* (Sydney: University of New South Wales Press, 2009), 91–124; Ann Moyal, with Robert E. Marks, "The Scientists and Darwin's *The Origin of Species* in Nineteenth Century Australia: A Re-evaluation," *Journal and Proceedings of the Royal Society of New South Wales* 152 (2019): 5–26.

2. David Wade Chambers, "Does Distance Tyrannize Science?," in *International Science and National Scientific Identity: Australia between Britain and America*, ed. R.W. Home and S.G. Kohlstedt (Dordrecht: Kluwer Academic Publishers, 1991), 19–38; Barry W. Butcher, "Darwin Down Under: Science, Religion, and Evolution in Australia," in *Disseminating Darwinism: The Role of Place, Race, Religion, and Gender*, ed. Ronald L. Numbers and John Stenhouse (Cambridge: Cambridge University Press, 1999), 40–41.

3. Ronald L. Numbers and John Stenhouse, "Introduction," in Numbers and Stenhouse, *Disseminating Darwinism*, 1–2.

4. Finney, *Paradise Revealed*.

5. Amy Way, "Natural Selection and the 'Antiquity of Man': Intellectual Impacts in the Australian Colonies," *Studies in History and Philosophy of Science* 90 (2021): 310–11; Moyal, with Marks, "Scientists," 5–6.

6. R.J.W. Selleck, *The Shop: The University of Melbourne, 1850–1939* (Carlton: Melbourne University Press, 2003), 31.

7. Clifford Turney, Ursula Bygott, and Peter Chippendale, *Australia's First: A History of the University of Sydney*, vol. 1 (Sydney: University of Sydney in association with Hale and Iremonger, 1991), 124–25; Selleck, *Shop*, 55.

8. Stephen Joseph Pascoe, "The Social and Spatial Construction of Student Housing: The University of Melbourne in an Age of Expansion" (MA diss., University of Melbourne, 2011).

9. Selleck, *Shop*, 31.

10. Mueller's anti-Darwinism was, however, less public than has often been realized: A.M. Lucas, "Ferdinand von Mueller's Interactions with Charles Darwin and His Response to Darwinism," *Archives of Natural History* 37, no. 1 (2010): 102–30.

11. Butcher, "Darwin's Australian Correspondents." Butcher later repudiated this analysis, writing in 1999 that it had come to him to seem "a naive attempt at model

building," though it is difficult from Butcher's wording to pinpoint exactly what was objectionable about his earlier treatment; possibly the issue was a suggestion that pro-evolutionary "heterodoxies" had "laid the groundwork to some extent for the eventual triumph of Darwinism" from the 1880s; Butcher, "Darwin Down Under," 40. A version of the offending analysis also appeared as a chapter of Barry Butcher, "Darwinism and Australia, 1836–1914" (PhD diss., University of Melbourne, 1992), 44–73. Other scholars have found Butcher's chapter subtle and useful: Chambers, "Does Distance Tyrannize Science?," 34, 37; Moyal, with Marks, "Scientists," 17. See also Finney, *Paradise Revealed*, 108–13.

12. Richard Davies Hanson, *Law in Nature &c, &c, &c.: Four Papers Read before the Adelaide Philosophical Society* (Adelaide: Adelaide Philosophical Society, 1864); Finney, *Paradise Revealed*, 98, 104–5; Moyal, with Marks, "Scientists," 17–18; Walter W. Phillips, "Religious Response to Darwin in Australia in the Nineteenth Century," *Journal of Australian Studies* 14, no. 26 (1990): 37–51.

13. Butcher, "Darwin's Australian Correspondents," 140–41.

14. Ann Mozley Moyal, "Sir Richard Owen and His Influence on Australian Zoological and Palaeontological Science," *Records of the Australian Academy of Science* 3, no. 2 (1975): 41–56.

15. David N. Livingstone, *Dealing with Darwin: Place, Politics, and Rhetoric in Religious Engagements with Evolution* (Baltimore: Johns Hopkins University Press, 2014), 2.

16. For an overview, see Alan Lester, "Imperial Circuits and Networks: Geographies of the British Empire," *History Compass* 4, no. 1 (2006): 124–41.

17. Tony Ballantyne, "Rereading the Archive and Opening Up the Nation-State: Colonial Knowledge in South Asia (and Beyond)," in *After the Imperial Turn: Thinking With and Through the Nation*, ed. Antoinette Burton (New York: Duke University Press, 2003), 113.

18. Tamson Pietsch, *Empire of Scholars: Universities, Networks and the British Academic World* (Manchester: Manchester University Press, 2013), 3.

19. Lorraine Daston, "The Moral Economy of Science," *Osiris* 10 (1995): 2–24; Rob Boddice, *The Science of Sympathy* (Urbana: University of Illinois Press, 2016), 13–16; Rob Boddice, *The History of Emotions* (Manchester: Manchester University Press, 2018), 194–201. These models differ from classic accounts of "moral economy" as describing the organizing principles of non-market economies in subsistence societies. See E.P. Thompson, "The Moral Economy of the English Crowd in the Eighteenth Century," *Past and Present* 50 (1971): 76–136; James C. Scott, *The Moral Economy of the Peasant: Rebellion and Subsistence in Southeast Asia* (New Haven, Yale University Press, 1976).

20. Daston, "Moral Economy," 4.

21. Cf. Johanna Siméant, "Three Bodies of Moral Economy: The Diffusion of a Concept," *Journal of Global Ethics* 11, no. 2 (2015): 169–70; Norbert Götz, "'Moral Economy': Its Conceptual History and Analytical Prospects," *Journal of Global Ethics* 11, no. 2 (2015): 156–58.

22. Daston, "Moral Economy," 6.

23. Daston, "Moral Economy," 23–24; Boddice, *Science of Sympathy*, 16; Boddice, *History of Emotions*, 201. On Daston's subsequent development of the concept, since reframed as "epistemic virtues," see Lorraine Daston and Peter Galison, *Objectivity* (New York: Zone Books, 2007), 39–42; Paul White, "Introduction [to 'Focus: The Emotional Economy of Science']," *Isis* 100, no. 4 (2009): 796.

24. Warwick Anderson, *The Cultivation of Whiteness: Science, Health and Racial Destiny in Australia* (Carlton South: Melbourne University Press, 2002), 2, 11–70.

25. Turney, Bygott, and Chippendale, *Australia's First*, 15–20.

26. Catholic Education Commission NSW and Macquarie University, *REPstats*, 2018, accessed May 21, 2021, http://www.repstats.info/.

27. Turney, Bygott, and Chippendale, *Australia's First*, 25.

28. Turney, Bygott, and Chippendale, *Australia's First*, 4–15, 53–55.

29. Rowan Williams, "Secularism, Faith and Freedom," in *The New Visibility of Religion: Studies in Religion and Cultural Hermeneutics*, ed. Graham Ward and Michael Hoelzl (London: Continuum, 2008), 45–56. Williams's use of the concepts of positive and negative liberty draws on Isaiah Berlin's classic distinction. My thanks to John Stenhouse for suggesting the pertinence of Williams's categories to the present argument.

30. Turney, Bygott, and Chippendale, *Australia's First*, 4–15, 27–28, 38–56, 71–92.

31. G.F.C., "Andrew's in Retrospect," *Magazine of St Andrew's College* 1, no. 36 (1939): 20–22; M.E. Eyers, "Sidere collegium idem mutato," *Johnsman* (1967): 5–6.

32. Turney, Bygott, and Chippendale, *Australia's First*, 60, 64–71; Darwin to Charles Lyell, [December 10, 1859], Darwin Correspondence Project, "Letter no. 2575," accessed August 30, 2021, https://www.darwinproject.ac.uk/letter/DCP-LETT-2575.xml.

33. Edward Hamilton, Charles Nicholson, and E. Deas Thomson to George Biddell Airy, Sir John Herschel, Henry Malden, and Henry Denison, May 7, 1851, and *Papers relating to the University of Sydney, and to the University College, Sydney, New South Wales* (London: Richard Taylor, 1851), both in the Papers of George Biddell Airy (hereafter Airy Papers), Cambridge University Library, Royal Greenwich Observatory Archives, RGO 6/337, 7–12, 209–26. I have used the microfilm duplicate of these papers held by the National Library of Australia, Canberra.

34. *Papers relating to the University of Sydney*, 7.

35. These passages were not reproduced in *Papers relating to the University of Sydney* but are quoted in Turney, Bygott, and Chippendale, *Australia's First*, 61.

36. *Papers relating to the University of Sydney*, 33.

37. Turney, Bygott, and Chippendale, *Australia's First*, 65–70.

38. Paul Cooper, "The Philanthropic and Religious Views of Professor John Smith," *Church Heritage* 20, no. 4 (2018): 230–63.

39. "Testimonials in Favour of John Smith," testimonial from George Dickie, November 22, 1851, Airy Papers, RGO 6/337, 274.

40. James McCosh and George Dickie, *Typical Forms and Special Ends in Creation* (Edinburgh: Thomas Constable, 1856).

41. "Testimonials in Favour of John Smith," testimonials from William MacGillivray, November 26, 1851, and John Croumbie Brown, November 26, 1851, both in Airy Papers, RGO 6/337, 274, 276.

42. Geoffrey Serle, *The Golden Age: A History of the Colony of Victoria 1851–1861* (Melbourne: Melbourne University Press, 1963), 382.

43. Serle, *Golden Age*, 336.

44. Serle, *Golden Age*, 336, 343.

45. Selleck, *Shop*, 32.

46. Philip Raymont, "An Australian Hybrid: Australia's Universities and Their Colleges," *History of Education Review* 30, no. 2 (2001): 73–85.

47. Selleck, *Shop*, 17, 22–24.

48. Selleck, *Shop*, 22.

49. Raymont, "Australian Hybrid," 77.

50. *An Act to Incorporate and Endow the University of Melbourne, 1853* (Victoria), ss. VIII, XII. On the working out of these provisions once the colleges were established, see Selleck, *Shop*, 131–34.

51. Selleck, *Shop*, 29.

52. *Papers relating to the University of Melbourne, Victoria, (Late Port Phillip,) Australia* (London: Bradbury and Evans, 1854), copy in Airy Papers, RGO 6/335, 21–32. The *Papers* are summarized in Selleck, *Shop*, 30–32.

53. *Papers relating to the University of Melbourne*, 16–19.

54. Herschel to Airy, July 12, 1854; Airy to Herschel, July 15, 1854; J.A.C. Oudemans to Airy, June 20, 1854; Airy to Oudemans, July 4, 1854; selection committee notes on the application of Stevenson Macadam for the natural science professorship, University of Melbourne, n.d. [May–June 1854]; all in Airy Papers, RGO 6/335, 59–62, 118–19, 200.

55. Malden to Airy, May 5, 1854, and Airy to Malden, May 6, 1854, both in Airy Papers, RGO 6/335, 80–82.

56. G.C. Fendley, "McCoy, Sir Frederick (1817–1899)," *Australian Dictionary of Biography* (hereafter *ADB*), accessed May 19, 2021, https://adb.anu.edu.au/biography/mccoy-sir-frederick-4069/text6491; Selleck, *Shop*, 34–35; Herschel to Barry, August 18, 1854, University of Melbourne Council Papers, University of Melbourne Archives (hereafter UMA) 1962.0019, Unit 2, 76–77.

57. *Papers relating to the University of Melbourne*, 20–21.

58. "Opening of the University of Melbourne," *The Argus* (Melbourne), April 14, 1855, 4–5.

59. For overviews, see Selleck, *Shop*, 32–40, 50.

60. Herschel to Barry, August 18, 1854.

61. Selleck, *Shop*, 34, 131–32; J.A. La Nauze, "Hearn, William Edward (1826–1888)," *ADB*, accessed May 21, 2021, https://adb.anu.edu.au/biography/hearn-william-edward

-3743/text5893; "Wilson, William Parkinson (1826–1874)," *ADB*, accessed May 21, 2021, https://adb.anu.edu.au/biography/wilson-william-parkinson-4870/text8143.

62. Selleck, *Shop*, 36–37; "City and Suburbs," *The Advocate* (Melbourne), July 15, 1899, 18; "Prominent Topics," *Advocate*, September 2, 1899, 12; "Church and Organ," *Melbourne Punch*, September 7, 1899, 4.

63. G.C. Fendley, "Irving, Martin Howy (1831–1912)," *ADB*, accessed May 21, 2021, https://adb.anu.edu.au/biography/irving-martin-howy-3840/text6099.

64. Selleck, *Shop*, 71–75.

65. A.L. Kenny, "Halford the Man," *Medical Journal of Australia*, February 1, 1941, 130.

66. Daston, "Moral Economy," 4.

67. Barry W. Butcher, "Gorilla Warfare in Melbourne: Halford, Huxley and 'Man's Place in Nature,'" in *Australian Science in the Making*, ed. R.W. Home (Cambridge: Cambridge University Press, 1988), 153–69; Selleck, *Shop*, 104–9; Finney, *Paradise Revealed*, 99–103, 106–8.

68. T.H. Huxley, "On the Physical Basis of Life," *Fortnightly Review* 5, no. 26, n.s. (1869): 129–45, reprinted as *Protoplasm: The Physical Basis of Life* (Melbourne: Peter C. Alcock, 1869); Frederick McCoy, *The Order and Plan of Creation: The Substance of Two Lectures Delivered in Connection with the Early Closing Association* (Melbourne: Samuel Mullen, 1870).

69. Turney, Bygott, and Chippendale, *Australia's First*, 61–63, 77–78, 257–71.

70. "Testimonials in Favour of John Smith, M.D., Fordyce Lecturer on Agriculture, and Assistant in the Chemical Laboratory of Marischal College, Aberdeen," testimonials from the Senatus Academicus of Marischal College and University, November 22, 1851, Thomas Clark, November 21, 1851, MacGillivray, November 26, 1851, and Peter Redfern, November 25, 1851, all in Airy Papers, RGO 6/337, 271–72, 274–75.

71. "Philosophical Society," *Sydney Morning Herald*, November 12, 1863, 8.

72. On the "Essay" and *Plutology*, see Butcher, "Darwinism and Australia," 124–42, 154–55; J.A. La Nauze, "Hearn on Natural Religion: An Unpublished Manuscript," *Historical Studies: Australia and New Zealand* 12, no. 45 (1965): 119–22; J.A. La Nauze, *Political Economy in Australia: Historical Studies* (Carlton: Melbourne University Press, 1949), 45–97. The "Essay," which remains unpublished, is in University of Melbourne Library Special Collections, 210 HEAR.

73. Robert M. Young, *Darwin's Metaphor: Nature's Place in Victorian Culture* (Cambridge: Cambridge University Press, 1985), 126–63.

74. Butcher, "Darwinism and Australia," 130, 141.

75. Gregory C.G. Moore, "Selling Plutology: Correspondence Relating to the Failure of Australia's First Economics Text," *History of Economics Review* 35, no. 1 (2002): 64–77.

76. Butcher, "Gorilla Warfare."

77. Leonard Huxley, *Life and Letters of Thomas Henry Huxley*, 3 vols. (London: Macmillan, 1900), 1:76–81, 93.

78. T.G. Evans to Edward F. A'Beckett, draft telegram, n.d. [ca. 1872–1874], Registrar's Department Correspondence, University of Adelaide Archives, Series 169, 20/37.

79. University of Adelaide, *Annual Report, 1874–5* (January 31, 1876).

80. Phillips, "Religious Response."

81. *The Adelaide University: Addresses Delivered at the Inauguration* (Adelaide: W.C. Cox, 1876), 9–34, 22n.

82. Butcher, "Darwin's Australian Correspondents," 156; D.J. Mulvaney and J.H. Calaby, *"So Much That Is New": Baldwin Spencer, 1860–1929, a Biography* (Carlton: Melbourne University Press, 1985), 68–70, 78–82; University of Melbourne Council Papers, UMA 1962.0019, Unit 1, volume on appointments of Professors Laurie, Masson, Spencer, and Lyle, 225, 247–49, 259–62; Huxley, *Life and Letters*, 2:148–49.

83. Finney, *Paradise Revealed*, 113, 136–44.

84. "Science," *Melbourne University Review* 3, no. 1 (1887): 28–32.

85. Turney, Bygott, and Chippendale, *Australia's First*, 256–71.

86. Edward Meyrick, "Charles Darwin," *Sydney University Review* 3 (July 1882): 244–53, 245.

87. For example, Frederic Wood Jones, *Habit and Heritage* (London: Kegan Paul, Trench, Trubner and Co., 1943).

88. Mark Francis, "Social Darwinism and the Construction of Institutionalised Racism in Australia," *Journal of Australian Studies* 20, nos. 50–51 (1996): 90–105. Similar points are made by Anderson, *Cultivation of Whiteness*, 3.

89. Barry W. Butcher, "Darwinism, Social Darwinism, and the Australian Aborigines: A Reevaluation," in *Darwin's Laboratory: Evolutionary Theory and Natural History in the Pacific*, ed. Roy MacLeod and Philip F. Rehbock (Honolulu: University of Hawai'i Press, 1994), 371–94.

90. Joel Barnes, "Australian Universities and Atlantic Slavery," *History Australia* 20, no. 1 (2023): 49–50.

91. Francis, "Social Darwinism," 94–97.

92. Melbourne Centre for the Study of Higher Education, "Indigenous History of the University of Melbourne Project," accessed October 12, 2022, https://melbourne-cshe.unimelb.edu.au/research/research-projects/policy-and-management-in-higher-education/indigenous-history-project.

4: AS ABOVE, SO BELOW

With thanks to Meira Gold, David Halperin, Daniel Halverson, Bernard V. Lightman, Justin Mullis, Jaume Navarro, Roberto Sirvent, Sarah A. Qidwai, Frederick V. Simmons, and Horace A. Smith for discussions that contributed toward this work.

1. Qur'an 1:2, 16:8, and 41:9; see Alan Jones, ed., *The Qur'an* (Edinburgh: Edinburgh University Press, 2007), 23, 250, 437–38; Muhammed Jamaluddin El-Fandy,

On Cosmic Verses in the Quran, trans. Ismail Cashmiry, rev. Shawki Sukkary (Cairo: Supreme Council for Islamic Affairs, 1967), 33–38; Michael J. Crowe, *The Extraterrestrial Life Debate 1750–1900: The Idea of a Plurality of Worlds from Kant to Lowell* (Cambridge: Cambridge University Press, 1986), 202–15.

2. Walter Stephens, "Strategies of Interspecies Communication, 1100–2000," in *Conversations with Angels: Essays Towards a History of Spiritual Communication, 1100–1700*, ed. Joad Raymond (London: Palgrave Macmillan, 2011), 25–48.

3. Simon Schaffer, "The Nebular Hypothesis and the Science of Progress," in *History, Humanity, and Evolution: Essays for John C. Greene*, ed. James R. Moore (Cambridge: Cambridge University Press, 1989), 142–53.

4. Jules Verne, *From the Earth to the Moon* (New York: Hurst and Co., 1909 [1865]), 15.

5. Verne, *From the Earth*, 12.

6. Verne, *From the Earth*, 56–57.

7. Jean Chesneaux, "Jules Verne's Image of the United States," trans. Frances Chew, *Yale French Studies* 43 (1969): 111–27.

8. Verne, *From the Earth*, 57.

9. H.G. Wells, *The First Men in the Moon* (New York: Airmont, 1965 [1901]), 65–66.

10. Wells, *First Men*, 148–58.

11. Matthew Dimmock, *Mythologies of the Prophet Muhammad in Early Modern English Culture* (Cambridge: Cambridge University Press, 2013), 118.

12. Wells, *First Men*, 135.

13. Wells, *First Men*, 156–60.

14. Wells, *First Men*, 129–31.

15. H.G. Wells, *The War of the Worlds* (New York: Modern Library, 2002 [1898]), 40.

16. Wells, *War*, 5, 169.

17. Wells, *War*, 3–6; William Makepeace Thackeray, *The Works of William Makepeace Thackeray*, vol. 13, *Ballads and Miscellanies* (New York: Harper and Brothers, 1899), 531–33; Richard Fulton, "The Sudan Sensation of 1898," *Victorian Periodicals Review* 42, no. 1 (Spring 2009): 37–63.

18. Paul A. Cantor and Peter Hufnagel, "The Empire of the Future: Imperialism and Modernism in H.G. Wells," *Studies in the Novel* 38, no. 1 (Spring 2006): 36–56.

19. H.G. Wells, "In the Avu Observatory," in *The Stolen Bacillus and Other Incidents* (London: Methuen, 1895), 36–52.

20. Jack Williamson, "The Evolution of the Martians," in *The War of the Worlds: Fresh Perspectives on the H.G. Wells Classic*, ed. Glenn Yeffeth (Dallas: BenBella Books, 2005), 189–98; Michael Livingston, "'The Tripods of Vulcan and Mars: Homer, Darwin, and the Fighting Machines of H.G. Wells's *The War of the Worlds*," *The Wellsian* 32 (2009): 58–60; Elana Gomel, "Shapes of the Past and the Future: Darwin and the Narratology of Time Travel," *Narrative* 17, no. 3 (Oct. 2009): 334–52; H.G. Wells

and Lisa M. Lane, "The Theory of Evolution: A Recovered Article," *The Wellsian* 42 (2019): 69–73.

21. Catherine L. Newell, *Destined for the Stars: Faith, the Future, and America's Final Frontier* (Pittsburgh: University of Pittsburgh Press, 2019). This is of course not to say that the European empires did not also lay rhetorical frameworks for space exploration; see David J. Sivier, "Voyaging to the Heart of Darkness: The Slave Trade and the 1841 Niger Expedition as Precursors to Modern Interplanetary Colonization," *Journal of the British Interplanetary Society* 56, no. 1/2 (Jan. 2003): 56–65.

22. Edward Everett Hale, "The Brick Moon, Part I," *Atlantic Monthly* 24, no. 144 (Oct. 1869): 451–61; Stephen Baxter, "From the Brick Moon to Island One: Space Habitats in Science Fiction Before O'Neill's *High Frontier*," *Journal of the British Interplanetary Society* 72, no. 9/10 (Sep./Oct. 2019): 297.

23. Edward Everett Hale, "The Brick Moon, Part III," *Atlantic Monthly* 24, no. 146 (Dec. 1869): 681.

24. Hale, "Part III," 685.

25. Hale, "Part III," 687.

26. Konstantin Tsiolkovsky, "The Aims of Astronautics," trans. X. Danko, in *The Call of the Cosmos*, ed. V. Dutt (Moscow: Foreign Languages Publishing House, 1960), 370–72.

27. Asif A. Siddiqi, *The Red Rockets' Glare: Spaceflight and the Russian Imagination, 1857–1957* (Cambridge: Cambridge University Press, 2010), 43–73.

28. Albert Einstein, "Why Socialism?," *Monthly Review* 1, no. 1 (May 1949): 9.

29. E.P. Thompson, "Time, Work-Discipline, and Industrial Capitalism," *Past & Present* 38 (Dec. 1967): 56–97.

30. Peter Galison, *Einstein's Clock's, Poincaré's Maps: Empires of Time* (New York: W.W. Norton, 2003), 303–22.

31. Bernard Lightman, *Victorian Popularizers of Science: Designing Nature for New Audiences* (Chicago: University of Chicago Press, 2007), 400, 403.

32. Schaffer, "Nebular Hypothesis," 132.

33. Schaffer, "Nebular Hypothesis."

34. Schaffer, "Nebular Hypothesis," 145; Verne, *From the Earth*, 14–15.

35. Joseph T. Ross, "Hegel, Analogy, and Extraterrestrial Life," in *Astrobiology, History, and Society: Life Beyond Earth and the Impact of Discovery*, ed. Douglas A. Vakoch (New York: Springer, 2013), 86–101.

36. Louis-Auguste Blanqui, *Eternity by the Stars: An Astronomical Hypothesis*, trans. and ed. Frank Chouraqui (New York: Contra Mundum Press, 2013 [1872]), 78–112.

37. Catherine Gallagher, *Telling It Like It Wasn't: The Counterfactual Imagination in History and Fiction* (Chicago: University of Chicago Press, 2018), 66–73.

38. John Hedley Brooke, "Natural Theology and the Plurality of Worlds: Observations on the Brewster-Whewell Debate," *Annals of Science* 34, no. 3 (1977): 221–86; Lightman, *Victorian Popularizers*, 307–17.

39. Karl Marx and Friedrich Engels, *The German Ideology*, ed. C.J. Arthur (New York: International Publishers, 1970), 77, 108–9.

40. Friedrich Engels, *Anti-Dühring: Landmarks of Scientific Socialism* (Bolton, ON: Leftist Public Domain Project, 2020 [1878]), 65–74; Friedrich Engels, *Dialectics of Nature*, ed. J.B.S. Haldane (New York: International Publishers, 1940 [1883]), 271–78.

41. Keith A. Spencer, "Fear of a Capitalist Planet," *Jacobin*, April 15, 2015, https://www.jacobinmag.com/2015/04/aliens-extraterrestrials-active-seti/; McKenzie Wark, *Molecular Red: Theory for the Anthropocene* (New York: Verso Books, 2015), 11–12, 53–55; Rupert Wildt, "Note on the Surface Temperature of Venus," *Astrophysical Journal* 91 (Mar. 1940): 266–68.

42. Spencer, "Fear."

43. Stephen Baxter, "H.G. Wells' Enduring Mythos of Mars," in Yeffeth, *War of the Worlds*, 181–88.

44. Percival Lowell, *The Soul of the Far East* (Boston: Houghton, Mifflin, 1888), 164.

45. Lowell, *Soul of the Far East*, 162–93.

46. Percival Lowell, *Occult Japan, or The Way of the Gods* (Cambridge, MA: Riverside Press, 1894), 323.

47. Percival Lowell, *Mars and Its Canals* (New York: Macmillan, 1906), 6–7.

48. K. Maria D. Lane, *Geographies of Mars: Seeing and Knowing the Red Planet* (Chicago: University of Chicago Press, 2011), 141–216; Crowe, *Extraterrestrial*, 480–546; Jenn Boneza, "Scientists Escorted Off of Mauna Kea," KHON2 News, July 16, 2019, https://www.khon2.com/local-news-2/no-arrests-have-been-made-in-the-controversial-tmt-project-with-day-two-of-construction-to-begin/. For a general overview of early science fiction's links to European and American imperialism, see John Rieder, *Colonialism and the Emergence of Science Fiction* (Middletown, CT: Wesleyan University Press, 2008), 34–60; Patricia Kerslake, *Science Fiction and Empire* (Liverpool: Liverpool University Press, 2011), 63–83.

49. Edgar Rice Burroughs, *A Princess of Mars* (New York: Grosset and Dunlap, 1917), 1–34.

50. Edgar Rice Burroughs, *The Gods of Mars* (Chicago: A.C. McClurg and Co., 1918), 133–46.

51. Lowell, *Mars and Its Canals*, 17.

52. Martin Brückner, "Popular Map Genres in American Literature," in *Literature and Cartography: Theories, Histories, Genres*, ed. Anders Engberg-Pedersen (Cambridge, MA: MIT Press, 2017), 346–49; Bernard Loing, "Space and Time in Wells and Jules Verne," *The Wellsian* 27 (2004): 3–13.

53. Linda Dryden, "H.G. Wells and Joseph Conrad: A Literary Friendship," *The Wellsian* 28 (2005): 2–13.

54. Joseph Conrad, *Heart of Darkness* (London: Norton, 1988 [1899]), 29.

55. C.S. Lewis, *Out of the Silent Planet* (Quebec City: Samizdat, 2015 [1938]), 83–88; C.S. Lewis, *Perelandra* (Quebec City: Samizdat, 2015 [1943]), 174–75.

56. Griffith Taylor, "The Evolution and Distribution of Race, Culture, and Language," *Geographical Review* 11, no. 1 (Jan. 1921): 54–119; Michael F. Robinson, *The Lost White Tribe: Explorers, Scientists, and the Theory that Changed a Continent* (Oxford: Oxford University Press, 2016), 147–62; Emily Kern, "Out of Asia: A Global History of the Scientific Search for the Origins of Humankind, 1800–1965" (PhD thesis, Princeton University, 2018), 44–57.

57. Taylor, "Evolution," 82–83.

58. Svante Arrhenius, trans. H. Borns, *Worlds in the Making: The Evolution of the Universe* (New York: Harper and Bros., 1908), 212–30; Stéphane Tirard, "The Relationship Between the Origins of Life on Earth and the Possibility of Life on Other Planets: A Nineteenth-Century Perspective," in *Astrobiology, History, and Society: Life Beyond Earth and the Impact of Discovery*, ed. Douglas A. Vakoch (New York: Springer, 2013), 103–13; Crosbie Smith, *The Science of Energy: A Cultural History of Energy Physics in Victorian Britain* (Chicago: University of Chicago Press, 1998), 289, 303.

59. Anthony F. Aveni, "The Star of Bethlehem," *Archaeology* 51, no. 6 (Nov./Dec. 1998): 36.

60. Newell, *Destined*, 182–204.

61. C.S. Lewis, "Will We Lose God in Outer Space?," *Christian Herald* 81 (Apr. 1958): 19, 74–76.

62. Jason Colavito, "George Knapp: Christian Fundamentalists in the Pentagon Shut Down Government Paranormal and UFO Probes Due to Demon Fears," *Jason Colavito Blog*, October 17, 2018, http://www.jasoncolavito.com/blog/george-knapp-christian-fundamentalists-in-the-pentagon-shut-down-government-paranormal-and-ufo-probes-due-to-demon-fears.

63. Junno Arocho Esteves, "Seen and Unseen: The Role of Faith, Reason in the Search for Alien Life," *Catholic Review*, June 21, 2021, https://catholicreview.org/seen-and-unseen-the-role-of-faith-reason-in-the-search-for-alien-life/.

64. Galya Diment, "Introduction: The Wells Effect," in *H.G. Wells and All Things Russian*, ed. Galya Diment (New York: Anthem Press, 2019), 1–10; Patrick Parrinder, "On Lenin's Supposed Response to *The Time Machine*," *Science Fiction Studies* 45, no. 2 (Jul. 2018): 407–8.

65. Ekaterina Yudina, "Dibs on the Red Planet: The Bolsheviks and Mars in the Russian Literature of the Early Twentieth Century," in *Visions of Mars: Essays on the Red Planet in Fiction and Science*, ed. Howard V. Hendrix, George Slusser, and Eric S. Rabkin (Jefferson, NC: McFarland, 2011), 51–55; George Slusser, "The Martians Among Us: Wells and the Strugatskys," in Hendrix, Slusser, and Rabkin, *Visions of Mars*, 56–72; Robert Crossley, *Imagining Mars: A Literary History* (Middletown, CT: Wesleyan University Press, 2011), 174–76; Richard Stites, "Fantasy and Revolution: Alexander Bogdanov and the Origins of Bolshevik Science Fiction," in *Red Star: The First Bolshevik Utopia*, ed. Loren R. Graham and Richard Stites (Bloomington: Indiana University Press, 1984), 1–16; Siddiqi, *Red Rockets' Glare*, 19–21, 99–103.

66. Vladimir Lenin, "Letter 168: To His Mother," Marxists.org, 2011, accessed March 22, 2017, https://www.marxists.org/archive/lenin/works/1908/jul/00mau.htm.

67. Spencer, "Fear."

5: EVOLUTION IN TIMES OF POLITICAL TRANSFORMATIONS

1. Jaume Navarro, "Draper in Spain: The Conflicting Circulation of the Conflict Thesis," *Zygon: Journal of Religion and Science* 54 (2019): 1107–24; and Diego Núñez, introduction to *Historia de los Conflictos entre la Religión y la Ciencia*, by John W. Draper (Barcelona: Alta Fulla/Mundo Científico, 1987), 7–39.

2. Vicente Cacho Viu, *La Institución Libre de Enseñanza* (Madrid: Sociedad Estatal de Conmemoraciones Culturales, 2010).

3. Nicolás Salmerón, prologue to *Historia de los Conflictos entre la Religión y la Ciencia*, by John W. Draper (Barcelona: Alta Fulla/Mundo Científico, 1987), lxxii.

4. Jaume Navarro, "Building a Nation: Spanish Engineers in the Science-and-Religion Narratives," in Jaume Navarro and Kostas Tampakis, eds., *Science, Religion and Nationalism* (London: Routledge, 2024).

5. See James A. Secord, *Victorian Sensation: The Extraordinary Publication, Reception and Secret Authorship of Vestiges of the Natural History of Creation* (Chicago: University of Chicago Press, 2000); Bernard Lightman, *Victorian Popularizers of Science: Designing Nature for New Audiences* (Chicago: University of Chicago Press, 2007).

6. Nicolás Masson de Morvilliers, "Espagne," in *Encyclopédie Méthodique ou par ordre des Matières: Géographie moderne*, vol. 1. (París: Pandoucke, 1782), 554–68.

7. José Luis Peset, "La Polémica de la Ciencia Española" (public conference at the Fundación Juan March, April 29, 2010), accessed October 19, 2021, https://www.march.es/conferencias/anteriores/voz.aspx?p1=22669.

8. José Luis Mora García, "A [casi] Siglo y Medio de la Polémica de la Ciencia Española: Su Noble e Involuntario 'Instigador': Gumersindo de Azcárate," in *La Ciencia Española*, ed. Victor Navarro Brotons and Gerardo Bolado (Madrid: Fundación Ignacio Larramendi, 2019), lvii–lxxvi. See also Gonzalo Capellán, *Gumersindo de Azcárate: Biografía intelectual* (Valladolid: Junta de Castilla y León, 2005).

9. Gumersindo de Azcárate, *El Self-government y la Monarquía Doctrinaria* (Madrid: Librerías de A. de San Martín, 1877).

10. The Constitution of 1869 had, for the first time, deprived the Catholic Church of its exclusive role in the country. The 1876 Constitution placed the church back in its traditional role and Spain became again a confessional state.

11. Marcelino Menéndez Pelayo, "Mr Masson Redivivo," *Revista Europea*, July 30, 1876.

12. Panentheism, a term coined by Krause, tries to be a middle way between traditional theism and pantheism. In panentheism, God and the world are deeply interrelated but preserving their separate essences. See John Culp, "Panentheism," *Stanford Encyclopedia of Philosophy* (Fall 2020 ed.), ed. Edward N. Zalta, https://plato.stanford.edu/archives/fall2020/entries/panentheism/.

13. On Krausism, see Diego Nuñez, *La Metalidad Positiva en España* (Madrid: Ediciones de la UAM, 1975).

14. On Perojo, see María Dolores Díaz Regadera et al., eds., *Artículos Filosóficos y Políticos de José del Perojo (1875–1908)* (Madrid: Ediciones de la Universidad Autónoma, 2003).

15. José del Perojo, *Ensayos sobre el Movimiento Intelectual en Alemania: Primera Serie* (Madrid: Imprenta de Medina y Navarro, 1875).

16. The last instance of a reissuing and study was the 2019 volume by Navarro Brotons et al., *La Ciencia Española.*

17. An exception to the norm, but who did not take part in the 1870s debates, was the mathematician José Echegaray.

18. Carlos Darwin, *El Origen del Hombre* (Barcelona: Editorial Renaixença, 1876).

19. Thomas Glick, *Darwin en España* (Barcelona: Ediciones Península, 1982); Francisco Pelayo, *Ciencia y Creencia en España durante el Siglo XIX* (Madrid: Ediciones del CSIC, 1999); Francisco Pelayo, "El impacto del darwinismo en la sociedad española del siglo XIX," *Hispania Nova* 14 (2015): 310–29; José Sala Català, *Ideología y Ciencia Biológica en España entre 1860 y 1881: La Difusión de un Paradigma* (Madrid: CSIC, 1987).

20. Juan Vilanova, *Origen, Naturaleza y Antigüedad del Hombre* (Madrid: Imp. de la Cía. de impresores y Libreros del Reino, 1872); Juan Vilanova, *Compendio de Geología* (Madrid: Imprenta de Alejandro Gómez Fuentenebro, 1872); Juan Vilanova, *La Creación: Historia Natural* (Barcelona: Montaner y Simón, 1872–1876); Juan Vilanova, *Atlas Geográfico Universal* (Madrid: Astort Hermanos, 1877).

21. This was the nickname the writer Benito Perez Galdós gave him.

22. Alberto Gomis Blanco and Jaume Josa Llorca, *Bibliografía Crítica Ilustrada de las Obras de Darwin en España (1857–2008)* (Madrid: CSIC, 2009).

23. Francisco Sunyer Capdevila, *Dios* (Barcelona: López Editor, 1969), 67.

24. The article was anonymous, but according to Pelayo, it is easily attributable to Tubino.

25. Francisco María Tubino, "La Ciencia del Hombre según las más Recientes e Importantes Publicaciones," *Revista Contemporánea* 3 (1877): 414.

26. Pelayo, *Ciencia y Creencia*, n240.

27. Antonio Machado, "Darwinismo," *Revista Mensual de Filosofía, Literatura y Ciencias* 4 (1872): 524–25.

28. Serrano Fatigati, *La Evolución en la Naturaleza* (Madrid: I. de Aribau, 1874), 35.

29. Juan Vilanova, "Importancia de la Paleontología," *Revista Europea* 3 (1875): 485.

30. Miguel Mir, *Harmonía entre la Ciencia y la Fe* (Madrid: Impr. Manuel Tello, 1881), 283.

31. Pelayo, *Ciencia y Creencia*, 279.

32. "El Darwinismo ante la Crítica," *El Consultor de los Párrocos* 2 (1873): 19.

33. Nicolás Antonio, "La Controversia Católica," *La Controversia* 5 (1891): 675–76.

34. José Salamero Martínez, "Nuestra Regla de Doctrina y de Conducta," *La Controversia* 1 (1887): 5.

35. Salamero Martínez, "Nuestra Regla," 6.

36. "Movimiento Científico," *La Controversia* 1 (1887): 137.

37. "Movimiento Científico," 522.

38. "Movimiento Científico," *La Controversia* 2 (1888): 569.

39. José María Menéndez de la Pola, "Un Defensor de Darwin," *La Controversia* 3 (1889): 817–20.

40. Pilar Díaz Sánchez, "Los Ecos del Darwinismo en España a través de la Literatura: Escritores y Escritoras," *Investigaciones Feministas* 0 (2009): 183–203.

41. Emilia Pardo Bazán, "Reflexiones Científicas contra el Darwinismo I," *La Ciencia Cristiana* 4 (1877): 289.

42. Pardo Bazán, "Reflexiones I," 291.

43. Pardo Bazán, "Reflexiones I," 291.

44. Emilia Pardo Bazán, "Reflexiones Científicas contra el Darwinismo VIII," *La Ciencia Cristiana* 5 (1877): 487.

45. Emilia Pardo Bazán, "Reflexiones Científicas contra el Darwinismo V," *La Ciencia Cristiana* 5 (1877): 405.

46. Pardo Bazán, "Reflexiones V," 407.

47. Juan Manuel Orti y Lara, "Los Puntos Negros de la Ciencia Moderna I," *La Ciencia Cristiana* 1 (1877): 20.

48. Orti y Lara, "Los Puntos Negros," 113.

49. Alberto Vázquez Bragado, "Ciencia Cristiana frente a Ciencia Moderna en la España de la segunda mitad del siglo XIX: Un estudio de la Revista La Ciencia Cristiana," *Llull* 33 (2010): 123–48.

50. Félix Sardà y Salvany, *El Liberalismo es Pecado* (Barcelona: L y T. Católica, 1884).

51. Francesc Xavier Vall i Solaz, "Representacions Visuals Catalanes del Dawinisme durant el segle XIX," *Actes d'Història de la Ciència i de la Tècnica* 8 (2015): 85–136.

52. *El Museo Universal*, May 17 and 31, and June 14 and 28, 1863. Reproductions from Charles Henry Bennett in *Illustrated Times*, between May 2 and October 10, 1863.

53. "Teoría Filosófica," *La Hormiga de Oro*, November 22, 1884, 748–49.

54. "Darwinismo," *La Hormiga de Oro*, June 15, 1892, 263.

55. *La Chispa: Semanario Católico Casi Humorístico Ilustrado* 11, July 10, 1890, 16.

56. *La Chispa: Semanario Católico Casi Humorístico Ilustrado* 38, January 15, 1891, 7.

57. *La Chispa: Semanario Católico Casi Humorístico Ilustrado* 24, October 9, 1890, 5; and 38, January 15, 1891, 5.

58. *La Chispa: Semanario Católico Casi Humorístico Ilustrado* 1, April 24, 1890, 2.

59. *Lo Burinot* 37, November 10, 1880, 4.

60. *Lo Burinot* 51, December 17, 1880, 4.

61. Solange Hibbs, *Iglesia, Prensa y Sociedad en España (1868–1904)* (Alicante: Instituto de Cultura Científica, 1995).

62. *Almanach de La Campana de Gracia* (Barcelona: Lopez Editor, 1878), 4.

63. *El Motín*, April 10, 1881.

64. *El Motín*, October 9, 1887.

65. Agustí Nieto-Galan, "A Republican Natural History in Spain around 1900: Odón de Buen (1863–1945) and His Audiences," *Historical Studies in the Natural Sciences* 42 (2012): 159–89.

66. Jaume Navarro, "Science contra Science: The Battle for Legitimate Knowledge in the Spanish Catholic Journals in the Early Twentieth Century," in *So What's New About Scholasticism? How Neo-Thomism Helped Shape the Twentieth Century*, ed. Rajesh Heynickx and Stéphane Symons (Berlin: De Gruyer, 2018), 205–26.

6: ARGENTINE POSITIVISM ON EVOLUTION AND RELIGION IN THE LATE NINETEENTH CENTURY

1. Miguel de Asúa, "Jesuit Science in the Missions of Paraguay and Río de la Plata," in *Latin American Perspectives on Science and Religion*, ed. Ignacio Silva (London: Pickering and Chatto, 2014), 71–84; de Asúa, "Darwin among the Pagans: Secularisation and the Reception of the Theory of Evolution in Buenos Aires," *Science and Christian Belief* 31 (2019): 4–25; de Asúa, "Three Centuries of Scientific Culture and Catholicism in Argentina," in *Rethinking History, Science, and Religion*, ed. Bernard Lightman (Pittsburgh: University of Pittsburgh Press, 2019), 37–49. The volume I edited, *Latin American Perspectives on Science and Religion*, includes three chapters devoted to historical explorations of the relations between science and religion, including Héctor Velázquez Fernández's "Reception, Debate and Rejection of Darwinism in the Second Half of the Nineteenth Century in Latin America," which is also an example of the first strategy I mention here.

2. Thomas Glick, Miguel Ángel Puig-Samper, and Rosaura Ruiz Gutiérrez, eds., *The Reception of Darwinism in the Iberian World: Spain, Spanish America and Brazil* (Boston: Springer, 2001).

3. Bernard Lightman, ed., *Global Spencerism* (Leiden: Brill, 2016).

4. Héctor Aldo Palma, *Las Huellas de Darwin en la Argentina* (Buenos Aires: Teseo-Press, 2016), 156. I have translated this and all subsequent quotations originally in Spanish or Portuguese in this chapter.

5. See Adriana Novoa, "The Rise and Fall of Spencer's Evolutionary Ideas in Argentina, 1870–1910," in Lightman, *Global Spencerism*, 173–91.

6. Marcelo Montserrat, "The Evolutionist Mentality in Argentina: An Ideology of Progress," in Glick, Puig-Samper, and Ruiz Gutiérrez, *Reception of Darwinism in the Iberian World*, 11.

7. José Manuel Estrada, *El Génesis de nuestra raza* (Buenos Aires: Imprenta de La Bolsa, 1862), 46–47.

8. Estrada, *El Génesis*, 49.

9. Both quotations are taken from Palma, *Las Huellas de Darwin*, 155.

10. See de Asúa, "Darwin among the Pagans," 7.

11. See his 1880 *La Antigüedad del Hombre en el Plata* (La Plata: Taller de Impresiones Oficiales, 1915).

12. Florentino Ameghino, *Los Cuatro Infinitos* (Buenos Aires: Ediciones Selectas América, 1919), 293.

13. Ameghino, *Los Cuatro Infinitos*, 296–97.

14. de Asúa, "Darwin among the Pagans," 13.

15. Reproduced in Florentino Ameghino, *Obras completas y correspondencia científica*, 24 vols. (La Plata: Taller de Impresiones Oficiales, 1913), 23:505–9.

16. As quoted in Pablo Perazzi, "Muerte y Transfiguración del Sabio: La batalla por el Legado de Florentino Ameghino," in *Historias de Arqueología Sudamericana*, ed. J. Nastri and L. Menezes Ferreira (Buenos Aires: Fundación de Historia Natural Félix de Azara, 2010), 123–36, 130.

17. Florentino Ameghino, *Les questions concernant Torigine des Hominiens d'aprés les derniers travaux et les derniéres découvertes (textos francés y castellano)*, in *Obras completas*, 19:933.

18. Ameghino, *La Antigüedad*, 150.

19. Ameghino, *Filogenia*, in *Obras completas*, 4:386.

20. Ameghino, *Filogenia*, 263.

21. Ameghino, *Filogenia*, 264.

22. Ameghino, *Visión y Realidad*, in *Obras completas*, 10:26.

23. Ameghino, *Diario de un Naturalista*, in *Obras completas*, 2:48.

24. For more details on the links made between the nebular hypothesis and evolutionary theory, see Ronald L. Numbers, *Creation by Natural Law: Laplace's Nebular Hypothesis in American Thought* (Seattle: University of Washington Press, 1977); Simon Schaffer, "The Nebular Hypothesis and the Science of Progress," in *History, Humanity, and Evolution: Essays for John C. Greene*, ed. James R. Moore (Cambridge: Cambridge University Press, 1989), 131–64.

25. Ameghino, *Filogenia*, 389. See Gray's parallel text on these ideas: "The nebular hypothesis—a natural consequence of the theory of gravitation and of the subsequent progress of physical and astronomical discovery—has been denounced as atheistical even down to our own day. But it is now largely adopted by the most theistical natural philosophers as a tenable and perhaps sufficient hypothesis, and where not accepted is no longer objected to, so far as we know, on philosophical or religious grounds." Asa Gray, *Darwiniana*, ed. Anderson Hunter Dupree (Cambridge, MA: Belknap Press, 1963), 113.

26. *Funeral cívico de homenaje a la memoria del sabio naturalista Dr. D. Florentino Ameghino en La Plata* (La Plata: Taller de Impresiones Oficiales, 1911), 17.

27. José Ingenieros, "La Santidad Moral," in *Funeral cívico de homenaje*, 66.

28. Alberto Caturelli, *Historia de la Filosofía en la Argentina, 1600–2000* (Buenos Aires: Ciudad Argentina, 2001), 428.

29. "D. Pedro Scalabrini," *El Monitor de la Educación Común* 521 (1916): 210.

30. Julio R. Contreras et al., *Pedro Scalabrini (1848–1916): Educador y Naturalista* (Buenos Aires: Fundación de Historia Natural Félix de Azara, 2019), 57.

31. Pedro Scalabrini, "Filosofía Didáctica," *El Monitor de la Educación Común* 509 (1915): 293–98, 296. It is rather remarkable that while most of these names are *creators* of different philosophies (rational, moral, methodical, mathematical, zoological, palaeontological, etc.), no one less than Newton and Darwin are called *discoverers* "of universal attraction" and "of natural selection in the struggle for existence," respectively.

32. Pedro Scalabrini, *Materialismo, Darwinismo, Positivismo: Diferencias y Semejanzas* (Paraná, Brazil, 1888), 216.

33. See Pedro Scalabrini, "Calendario positivista: Algunas observaciones (Continuación)," *La Escuela Positiva* 2, no. 23 (1896): 1152–55. For more details on these thoughts, see Malvina T. Antonietta de Gabardini, *Revista La Escuela Positiva: Corrientes 1895–1899; Introducción e índices* (Resistencia: Gómez Lestani, 1995), 36.

34. Scalabrini, *Materialismo*, 231.

35. Scalabrini, *Materialismo*, 185.

36. Scalabrini, *Materialismo*, 228.

37. Scalabrini, *Materialismo*, 228.

38. Scalabrini, *Materialismo*, 186.

39. Scalabrini, *Materialismo*, 230.

40. Scalabrini, *Materialismo*, 186.

41. Pedro Scalabrini, "Museo Escolar Argentino," *El Monitor de la Educación Común* 243 (1894): 635–39, 639.

42. Pedro Scalabrini, "Conciliación Relativa," *La Escuela Positiva* 2, no. 17 (1896): 848–52. See Gabardini, *Revista La Escuela Positiva*, 86. Due to the COVID-19 pandemic, I was not able to consult this journal in full, not being able to study the details of this conciliation, mentioned in Gabardini's work.

43. Alfredo Ferreira, "Catolicismo y Positivismo," *La Escuela Positiva* 1, no. 3 (1895): 111–14, 112.

44. Ferreira, "Catolicismo y Positivismo," 113.

45. José Manuel Estrada, "Sustraendos," in *Obras Completas de Juan Manuel Estrada*, vol. 11 (Buenos Aires: Compañía Sud Americana de Billetes de Banco, 1904), 298.

46. Alfredo Ferreira, "Una Ética Química," *El Monitor de la Educación Común* 24 (1916): 280–84, 281. Ferreira mitigates his claims by the end of his paper, stating that sociology will become the science that will explain and dictate morality in a near future, but only because recurring to chemistry would be too burdensome, sourcing the very first causes of morality rather than proximate causes.

47. Estrada, "Sustraendos," 298.

48. Estrada, "Sustraendos," 297.

49. José Manuel Estrada, *La Política Liberal Bajo la Tiranía de Rosas* (Buenos Aires: Librería del Colegio, De Pedro Igón y Cía, 1897), 90 (emphasis added).

50. Estrada, "Sustraendos," 297.

51. Estrada, "Estudios Clásicos," in *Obras Completas*, 11:378.

52. Estrada, "Filosofía," in *Obras Completas*, 11:444.

53. Estrada, "Problemas Argentinos," in *Obras Completas*, 11:52.

54. Estrada, "Estudios Clásicos," 379.

55. Estrada, "Estudios Clásicos," 378.

56. Estrada, "Segunda Enseñanza," in *Obras Completas*, 11:410.

57. Estrada, "Gobierno Escolar," in *Obras Completas*, 11:159.

58. Estrada, "Por qué pedimos libertad," in *Obras Completas*, 11:439–40.

59. Estrada, "Patronato," in *Obras Completas*, 11:136.

60. Estrada, "Por qué pedimos libertad," 441.

61. Estrada, *La Política Liberal*, 234.

62. Estrada, "La Familia y la Escuela," in *Obras Completas*, 11:229–30.

63. Manuel D. Pizarro, *Miscelánea II* (Córdoba: La Minerva, 1899), 195–96.

64. Pizarro, *Miscelánea II*, 267.

65. Manuel D. Pizarro, *Miscelánea IV* (Córdoba: La Minerva, 1902), 96.

66. Manuel D. Pizarro, *Miscelánea I* (Córdoba: La Minerva, 1898), 239.

67. Pizarro, *Miscelánea I*, 238.

68. Pizarro, *Miscelánea II*, 269.

69. Pizarro, *Miscelánea I*, 255 (emphasis added).

70. See José M. Blanco S.J., *La Evolución Antropológica y Ameghino* (Buenos Aires: Alfa y Omega, 1916). For context and a detailed discussion of Blanco's papers on this matter, see Miguel de Asúa, "Los artículos del P. José María Blanco S.I. en la revista Estudios sobre la evolución y las teorías antropológicas de Ameghino," *Stromata* 65, nos. 3–4 (2009): 313–35.

71. Gabino Barreda, *Oración Cívica* (Mexico: Cuadernos de Cultura Latinoamericana, UNAM, 1979), 10.

72. Barreda, *Oración*, 10.

73. Barreda, *Oración*, 10.

74. Barreda, *Oración*, 1.

75. All quotations in this paragraph are taken from Laura Ibarra García, "El Positivismo de Gabino Barreda. Un Estudio desde la Teoría Histórico-Genética," *Acta Sociológica* 60 (2013): 11–38, 25.

76. Barreda, as quoted in the editorial "La Síntesis de Augusto Comte," *Revista Positiva* 47 (1904): 542.

77. Ladislau Netto, *Observaciones sobre la Teoría de la Evolución leídas en la Sociedad Científica Argentina* (Buenos Aires: Imprenta de La Nación, 1882).

78. Netto, *Observaciones*.

79. Regina Cândida Ellero Gualtieri, "O Evolucionismo na Produção Científica do Museu Nacional do Rio de Janeiro (1876–1915)," in *A recepção do Darwinismo no Brasil*, ed. Heloisa Maria Bertol Domingues, Magali Romer Sá, and Thomas Glick (Rio de Janeiro: Editora Fiocruz, 2003), 45–96, 62.

80. Juan Enrique Lagarrigue, *Carta al Señor Don J. Alfredo Ferreira* (Santiago de Chile: Imprenta y Librería Ercilla, 1900), 6.

7: PREHISTORIC AND PRIMEVAL PASTS

1. Peter Bowler, *The Invention of Progress: The Victorians and the Past* (Oxford: Blackwell, 1989); A. Bowdoin Van Riper, *Men among the Mammoths: Victorian Science and the Discovery of Human Prehistory* (Chicago: University of Chicago Press, 1993); Martin J.S. Rudwick, *Bursting the Limits of Time: The Reconstruction of Geohistory in the Age of Revolution* (Chicago: University of Chicago Press, 2005); Matthew R. Goodrum, "The Idea of Human Prehistory: The Natural Sciences, the Human Sciences, and the Problem of Human Origins in Victorian Britain," *History and Philosophy of the Life Sciences* 34, no. 1/2 (2012): 117–45; Clive Gamble, *Making Deep Time: Zeal, Perseverance, and the Time Revolution of 1859* (Oxford: Oxford University Press, 2021).

2. Efram Sera-Shriar, "Introduction: From the Beginning," in *Historicizing Humans: Deep Time, Evolution, and Race in Nineteenth Century British Sciences*, ed. Efram Sera-Shriar (Pittsburgh: University of Pittsburgh Press, 2018), 1–13.

3. Marwa Elshakry, "Histories of Egyptology in Egypt: Some Thoughts," in *Histories of Egyptology: Interdisciplinary Measures*, ed. William Carruthers (New York: Routledge, 2015), 185.

4. Mogens Trolle Larsen, "Orientalism and Near Eastern Archaeology," in *Domination and Resistance*, ed. Daniel Miller, Michael Rowlands, and Chris Tilley (London: Routledge, 1995), 228–38; Allison Mickel, "Orientalism and Origins: The Search for Firsts in the 'Cradle of Civilisation,'" in *Interrogating Human Origins: Decolonisation and the Deep Human Past* (London: Routledge, 2019), 139–56; Pratik Chakrabarti, *Inscriptions of Nature: Geology and the Naturalization of Antiquity* (Baltimore: Johns Hopkins University Press, 2020), esp. 55–83.

5. David Gange, *Dialogues with the Dead: Egyptology in British Culture and Religion, 1822–1922* (Oxford: Oxford University Press, 2013), 24–31, 96–120.

6. Elliott Colla, *Conflicted Antiquities: Egyptology, Egyptomania, Egyptian Modernity* (Durham, NC: Duke University Press, 2007), 103.

7. Marwa Elshakry, *Reading Darwin in Arabic, 1860–1950* (Chicago: University of Chicago Press, 2013), 35, 86–91.

8. Adam Mestyan, *Primordial History, Print Capitalism, and Egyptology in Nineteenth-Century Cairo* (Cairo: Institut Français D'Archéologie Orientale, 2021), 36–39.

9. Translated in Mestyan, *Primordial History*, 79.

10. Colla, *Conflicted Antiquities*, 121–55.

11. On Barak, *On Time: Technology and Temporality in Modern Egypt* (Berkeley: University of California Press, 2013); Jennifer L. Derr, *The Lived Nile: Environment, Disease, and Material Colonial Economy in Egypt* (Stanford, CA: Stanford University Press, 2019), 30–31.

12. Daniel A. Stolz, *The Lighthouse and the Observatory: Islam, Science, and Empire in*

Late Ottoman Egypt, Science in History (Cambridge: University of Cambridge Press, 2018), 43; Derr, *Lived Nile*, 30, 99.

13. Stolz, *Lighthouse and the Observatory*, 43, 88.

14. Donald M. Reid, *Whose Pharaohs? Archaeology, Museums, and Egyptian National Identity from Napoleon to World War I* (Berkeley: University of California Press, 2002), 61; F. Robert Hunter, *Egypt under the Khedives, 1805–1879: From Household Government to Modern Bureaucracy* (Cairo: American University in Cairo Press, 1999), 92–93.

15. Meira Gold, "Ancient Egypt and the Geological Antiquity of Man, 1847–1863," *History of Science* 57, no. 2 (2019): 194–230.

16. Joseph Hekekyan to Leonard Horner, December 3, 1857, 37459.648, Hekekyan MSS. British Library, London.

17. Kevin van Bladel, *The Arabic Hermes: From Pagan Sage to Prophet of Science* (Oxford: Oxford University Press, 2009); Christian H. Bull, *The Tradition of Hermes Trismegistus: The Egyptian Priestly Figure as a Teacher of Hellenized Wisdom* (Leiden: Brill, 2018).

18. Joseph Hekekyan Bey, *A Treatise on the Chronology of Siriadic Monuments* (London: Taylor and Francis, 1863), vi.

19. Quoted in Darrell Dykstra, "Joseph Hekekyan and the Egyptian School in Paris," *Armenian Review* 35, no. 2 (1982): 165.

20. Pratik Chakrabarti, "Situating the Empire in History of Science," in *The Routledge Handbook of Science and Empire*, ed. Andrew Goss (London: Routledge, 2021), 13–14.

21. Chakrabarti, *Inscriptions of Nature*, 2.

22. Daniel Stolz, "Another Empire: Science in the Ottoman Lands," in Goss, *Routledge Handbook of Science and Empire* (London: Routledge, 2021), 178.

23. Ahmed Abdel-Rahim Mustafa, "The Hekekyan Papers," in *Political and Social Change in Modern Egypt: Historical Studies from the Ottoman Conquest to the United Arab Republic*, ed. Peter Malcolm Holt (London: Oxford University Press, 1968), 69; David G. Jeffreys, "Joseph Hekekyan at Heliopolis," in *Studies on Ancient Egypt in Honour of H.S. Smith*, ed. Anthony Leahy and William John Tait (London: Egypt Exploration Society, 1999), 158; Gold, "Ancient Egypt," 204–7.

24. Reid, *Whose Pharaohs?*, 22, 59–63; Colla, *Conflicted Antiquities*.

25. Elshakry, *Reading Darwin*, 59–62.

26. Mirjam Brusius, "Towards a History of Preservation Practices: Archaeology, Heritage, and the History of Science," *International Journal of Middle East Studies* 47, no. 3 (2015): 574–79; Mirjam Brusius, "Hitting Two Birds with One Stone: An Afterword on Archaeology and the History of Science," *History of Science* 55, no. 3 (2017): 383–91; William Carruthers and Stéphane Van Damme, "Disassembling Archaeology, Reassembling the Modern World," *History of Science* 55, no. 3 (2017): 255–72.

27. For "semicolonial" Egypt, see Timothy Mitchell, *Colonising Egypt* (Berkeley: University of California Press, 1991); Juan Cole, *Colonialism and Revolution in the Middle East: Social and Cultural Origins of Egypt's 'Urabi Movement* (Princeton, NJ:

Princeton University Press, 1993); Hunter, *Egypt under the Khedives*; Khaled Fahmy, *All the Pasha's Men: Mehmed Ali, His Army and the Making of Modern Egypt* (Cairo: American University in Cairo Press, 2002); Barak, *On Time*; On Barak, *Powering Empire: How Coal Made the Middle East and Sparked Global Carbonization* (Oakland: University of California Press, 2020).

28. Eve M. Troutt Powell, *A Different Shade of Colonialism: Egypt, Great Britain, and the Mastery of the Sudan* (Berkeley: University of California Press, 2003), 7; Omnia S. El Shakry, *The Great Social Laboratory: Subjects of Knowledge in Colonial and Postcolonial Egypt* (Stanford, CA: Stanford University Press, 2007), 3; Stolz, *Lighthouse and the Observatory*, 10.

29. Hunter, *Egypt under the Khedives*, 92–93, 111–17; F. Robert Hunter, "Egypt's High Officials in Transition from a Turkish to a Modern Administrative Elite, 1849–1879," *Middle Eastern Studies* 19, no. 3 (1983): 277–300.

30. Reid, *Whose Pharaohs?*, 273–74.

31. Journal Entry, 37448.179–184, Hekekyan MSS; Paul Sedra, *From Mission to Modernity: Evangelicals, Reformers and Education in Nineteenth-Century Egypt* (New York: Palgrave Macmillan, 2011), 63–66.

32. Mustafa, "Hekekyan Papers"; Hunter, "Egypt's High Officials," 285–86; Dykstra, "Egyptian School"; Reid, *Whose Pharaohs?*, 59–63; Jeffreys, "Joseph Hekekyan at Heliopolis"; Sedra, *From Mission to Modernity*, 63–83; Timothy Mitchell, *Rule of Experts: Egypt, Techno-Politics, Modernity* (Berkeley: University of California Press, 2002), 68–69.

33. Paul Crozet, "The Teaching of Descriptive Geometry in Egypt," in *Descriptive Geometry: The Spread of a Polytechnic Art*, ed. Évelyne Barbin, Marta Menghini, and Klaus Volkert (Cham, Switzerland: Springer, 2019), 365–67.

34. Dykstra, "Egyptian School," 174–75.

35. Sedra, *From Mission to Modernity*, 67, 80–82.

36. Paul Crozet, *Les Sciences Modernes en Égypte: Transfert et Appropriation (1805–1902)* (Paris: Geuthner, 2008); El Shakry, *Great Social Laboratory*; Elshakry, *Reading Darwin*; Stolz, *Lighthouse and the Observatory*.

37. Hunter, *Egypt under the Khedives*, 112–15; Mustafa, "Hekekyan Papers," 70.

38. Leonard Horner, "An Account of Some Recent Researches Near Cairo . . . ," *Philosophical Transactions of the Royal Society of London* 148 (1858): 53–92.

39. For details of this long-distance collaboration, see Gold, "Ancient Egypt."

40. Hekekyan to Horner, November 16, 1857, 37459.590–97, Hekekyan MSS.

41. "Memoir on the Roda Nilometer," 37459.598, Hekekyan MSS.

42. "Memoir on the Roda Nilometer," 37459.643, Hekekyan MSS.

43. "Memoir on the Roda Nilometer," 37459.613–14, Hekekyan MSS.

44. "Memoir on the Roda Nilometer," 37459.611–12, Hekekyan MSS.

45. Horner to Hekekyan, December 17, 1857, 37459.655, Hekekyan MSS.

46. Horner to Hekekyan, August 2, 1858, 37459.663, Hekekyan MSS.

47. Horner to Hekekyan, August 2, 1858, 37459.663, Hekekyan MSS.

48. Horner to Hekekyan, November 21, 1862, 37459.756–57, Hekekyan MSS.

49. Horner to Hekekyan, June 16, 1863, 37459.760, Hekekyan MSS.

50. Horner to Hekekyan, November 21, 1863, 37459.765–66, Hekekyan MSS.

51. Hekekyan to Horner, November 16, 1857, 37459.590, Hekekyan MSS.

52. Hekekyan to Horner, July 22, 1862, 37459.743–44, Hekekyan MSS.

53. Hekekyan Bey, *Siriadic Monuments*, xi–xxxvii.

54. Hekekyan Bey, *Siriadic Monuments*, xviv.

55. Hekekyan Bey, *Siriadic Monuments*, 13–14.

56. Simon Schaffer, "Oriental Metrology and the Politics of Antiquity in Nineteenth-Century Survey Sciences," *Science in Context* 30, no. 2 (2017): 177.

57. Zaraza Friedman, "Nilometer," in *Encyclopaedia of the History of Science, Technology, and Medicine in Non-Western Cultures*, ed. Helaine Selin (Berlin: Springer, 2008), 1752–53.

58. Schaffer, "Oriental Metrology," 184–85.

59. Schaffer, "Oriental Metrology," 182; Jed Z. Buchwald and Diane Greco Josefowicz, *The Zodiac of Paris: How an Improbable Controversy Over an Ancient Egyptian Artifact Provoked a Modern Debate between Religion and Science* (Princeton, NJ: Princeton University Press, 2010).

60. Pamela Pilbeam, *Saint-Simonians in Nineteenth-Century France: From Free Love to Algeria* (London: Palgrave Macmillan, 2013), 125–29.

61. Hekekyan Bey, *Siriadic Monuments*, 100.

62. Jehane Ragai and Gregg DeYoung, "Calendars in Egypt," in Selin, *Encyclopaedia of the History of Science, Technology, and Medicine*, 445–46.

63. Hekekyan Bey, *Siriadic Monuments*, 9.

64. Hekekyan Bey, *Siriadic Monuments*, 11.

65. Reginald Stuart Poole, *Horae Aegyptiacae* (London: John Murray, 1851).

66. Hekekyan Bey, *Siriadic Monuments*, 149.

67. Hekekyan Bey, *Siriadic Monuments*, 149.

68. François de Blois, "The 'Sabians' (*Sabi'un*) in Pre-Islamic Arabic," *Acta Orientalis* 56 (1995): 39–61; François de Blois, "Ṣābi'," in *Encylopedia of Islam*, 2nd ed. (Leiden: Koninklijke Brill NV, 2012), http://dx.doi.org/10.1163/1573-3912_islam_COM_0952; Jonathan Elukin, "Maimonides and the Rise and Fall of the Sabians: Explaining Mosaic Laws and the Limits of Scholarship," *Journal of the History of Ideas* 63, no. 4 (2002): 619–37.

69. Hekekyan Bey, *Siriadic Monuments*, xxxiii, 96.

70. Okasha El Daly, *Egyptology: The Missing Millennium; Ancient Egypt in Medieval Arabic Writings* (London: UCL Press, 2005), 31–37, 50–54, 109–14.

71. Gold, "Ancient Egypt," 216–18.

72. Hekekyan Bey, *Siriadic Monuments*, 8–9.

73. Bull, *Hermes Trismegistus*, 382.

74. Hekekyan Bey, *Siriadic Monuments*, xi–xxxvii.

75. Hekekyan Bey, *Siriadic Monuments*, xiii–xiv, xxx–xxxi, xxxiii.

76. Hekekyan Bey, *Siriadic Monuments*, xvi–xvii.

77. Louis Gardet, "'Ilm al-Kalām," in *Encyclopedia of Islam*, http://dx.doi.org/10.1163/1573-3912_islam_COM_0366; Elshakry, *Reading Darwin*, 154–56, 181–82.

78. Reid, *Whose Pharaohs?*, 29; Colla, *Conflicted Antiquities*, 74.

79. Colla, *Conflicted Antiquities*, 76–91.

80. Reid, *Whose Pharaohs?*, 30–31; van Bladel, *Arabic Hermes*, 128; El Daly, *Missing Millennium*, 113–15.

81. Reid, *Whose Pharaohs?*, 106–9; Colla, *Conflicted Antiquities*, 127–29; Mestyan, *Primordial History*, 40.

82. John W. Livingstone, "Western Science and Educational Reform in the Thought of Shaykh Rifaa Al-Tahtawi," *International Journal of Middle East Studies* 28, no. 4 (1996): 543–64.

83. Reid, *Whose Pharaohs?*, 54–63; Colla, *Conflicted Antiquities*, 116–20; Gold, "Ancient Egypt," 205.

84. Elshakry, "Histories of Egyptology," 191–93.

85. Mestyan, *Primordial History*, 40.

86. Reid, *Whose Pharaohs?*, 109–11.

87. Reid, *Whose Pharaohs?*, 111; Elshakry, "Histories of Egyptology," 192.

88. Reid, *Whose Pharaohs?*, 109–10.

89. Mestyan, *Primordial History*, 45–46.

90. Mestyan, *Primordial History*, 36–39.

91. Mestyan, *Primordial History*, 39, 43–44.

92. Gold, "Ancient Egypt," 205.

93. Reid, *Whose Pharaohs?*, 120–23.

94. Reid, *Whose Pharaohs?*, 96.

95. *Athenaeum*, November 14, 1863, p. 644.

96. "History and Biography," *Westminster Review* 25, no. 1 (Jan. 1864): 285–86.

97. Thierry Zarcone, "Freemasonry and Islam," in *Handbook of Freemasonry*, ed. Henrik Bogdan and Jan A.M. Snoek (Leiden: Brill, 2014), 150–52, 240–41.

98. Karim Wissa, "Freemasonry in Egypt 1798–1921: A Study in Cultural and Political Encounters," *British Journal of Middle Eastern Studies* 16, no. 2 (1989): 154–56.

99. A. Albert Kudsi-Zadeh, "Afghāni and Freemasonry in Egypt," *Journal of the American Oriental Society* 92, no. 1 (1972): 27; Jacob M. Landau, "Prolegomena to a Study of Secret Societies in Modern Egypt," *Middle Eastern Studies* 1, no. 2 (1965): 135–86.

100. Elshakry, *Reading Darwin*, 61–63, 73–75, 81–86, 339.

101. Sedra, *From Mission to Modernity*, 68.

102. Elshakry, *Reading Darwin*, 119–24.

103. Elshakry, *Reading Darwin*, 161–218, 175.

104. Cited in Colla, *Conflicted Antiquities*, 145.

105. Elshakry, *Reading Darwin*, 61–63, 73–75, 81–86, 339.

106. Stolz, *Lighthouse and the Observatory*, 77.

107. Mahmud Bey Falaki, *L'Âge et le but des pyramides lus dans sirius* (Alexandria: Imprimerie Française Mourès, 1865), 6.

108. Stolz, *Lighthouse and the Observatory*, 95–107; Schaffer, "Oriental Metrology," 199–205; Mestyan, *Primordial History*, 41.

109. Falaki, *L'Âge et le but des pyramides*, 13–19.

110. Stolz, *Lighthouse and the Observatory*, 96–107, esp. 103.

111. Simon Schaffer, "Metrology, Metrication, and Victorian Values," in *Victorian Science in Context*, ed. Bernard Lightman (Chicago: University of Chicago Press, 1997), 449–59.

112. Charles Piazzi Smyth, *Life and Work at the Great Pyramids* (Edinburgh: Edmonston and Douglas, 1867), 250.

113. Smyth, *Life and Work*, 252–55; Schaffer, "Oriental Metrology," 202.

114. Hunter, *Egypt under the Khedives*, 88.

115. Darrell Dykstra, "Pyramids, Prophets, and Progress: Ancient Egypt in the Writings of 'Alī Mubārak," *Journal of the American Oriental Society* 114, no. 1 (1994): 54–65; Reid, *Whose Pharaohs?*, 179–81; Colla, *Conflicted Antiquities*, 136–42.

116. Stolz, *Lighthouse and the Observatory*, 106.

117. A. Dufeu, *Decouverte de l'age et de la véritable destination des quatre pyramides de gizeh* (Paris: Ve A. Morel et Cie, 1873), preface.

118. Charles Piazzi Smyth, "The Pyramids of Egypt," *Quarterly Journal of Science* 10 (Oct. 1873): 514.

119. Erik Hornung, *The Secret Lore of Egypt: Its Impact on the West* (Ithaca, NY: Cornell University Press, 2001); Caroline Tully, "Egyptosophy in the British Museum: Florence Farr, the Egyptian Adept and the *Ka*," in *The Occult Imagination in Britain, 1875–1947*, ed. Christine Ferguson and Andrew Radford (London: Routledge, 2017), 131–45; Eleanor Dobson, *Victorian Alchemy: Science, Magic and Ancient Egypt* (London: UCL Press, 2022); Simon Magus, *Rider Haggard and the Imperial Occult: Hermetic Discourse and the Romantic Contiguity* (Leiden: Brill, 2022).

120. George Oliver, *The Antiquities of Freemasonry* (London: Richard Spencer, 1843); Albert G. Mackey, *The Symbolism of Freemasonry* (New York: Clark and Maynard, 1869); Joscelyn Godwin, *The Theosophical Enlightenment* (New York: State University of New York Press, 1994), 277–81; Hornung, *Secret Lore of Egypt*, 141–54.

121. James Bronwick, *Pyramid Facts and Fancies* (London: C. Kegan Paul, 1877), 219–20.

122. Smyth, *Life and Work*, 368.

123. W.M. Flinders Petrie, *The Pyramids and Temples of Gizeh* (London: Field and Tuer, 1883).

124. Meira Gold, "Victorian Egyptology and the Making of a Colonial Field Science, 1850–1906" (PhD diss., University of Cambridge, 2020).

125. Donald M. Reid, "Indigenous Egyptology: The Decolonization of a Profession?," *Journal of the American Oriental Society* 105, no. 2 (1985): 239; Colla, *Conflicted Antiquities*, 178–80, 211–16; Donald M. Reid, *Contesting Antiquity in Egypt:*

Archaeologies, Museums, and the Struggle for Identities from World War I to Nasser (Cairo: American University in Cairo Press, 2015).

8: EVOLUTION, RELIGION, AND RACIAL POLITICS IN NEW ZEALAND, 1814–1920

1. David N. Livingstone, *Dealing with Darwin: Place, Politics and Rhetoric in Religious Engagements with Evolution* (Baltimore: Johns Hopkins University Press, 2014), 1–2.

2. John Hedley Brooke, *Science and Religion: Some Historical Perspectives* (Cambridge: Cambridge University Press, 1991).

3. Ronald L. Numbers, "Review of *Science and Religion*, by John Hedley Brooke," *Metascience*, n.s., 1 (1992): 35–39; Ronald L. Numbers, "Simplifying Complexity: Patterns in the History of Science and Religion," in *Science and Religion: New Historical Perspectives*, ed. Thomas Dixon, Geoffrey Cantor, and Stephen Pumphrey (Cambridge: Cambridge University Press, 2011), 247–62.

4. Peter Harrison, "Conflict, Complexity, and Secularization in the History of Science and Religion," in *Rethinking History, Science and Religion: An Exploration of Conflict and the Complexity Principle*, ed. Bernard Lightman (Pittsburgh: University of Pittsburgh Press, 2019), 221–34, quotation on 224.

5. Vincent O'Malley, *The New Zealand Wars/Ngā Pakanga o Aotearoa* (Wellington: Bridget Williams Books, 2019).

6. Tony Ballantyne, *Entanglements of Empire: Missionaries, Māori and the Question of the Body* (Auckland: Auckland University Press, 2015); Hirini Kaa, *Te Hāhi Mihinare: The Māori Anglican Church* (Wellington: Bridget Williams Books, 2020), 18–50.

7. Malcolm Falloon, "The Māori Conversion and Four Early Converts" (PhD thesis, University of Otago, 2020), 53–80.

8. J.C. Prichard, *Researches into the Physical History of Mankind*, 3rd ed. (London: Sherwood, Gilbert and Piper, Paternoster Row, and J. and A. Arch, Cornhill, 1836); H. Augstein, *James Cowles Prichard's Anthropology: Remaking the Science of Man in Early Nineteenth Century Britain* (Leiden: Brill, 2016).

9. Aborigines Protection Society [APS], *First Annual Report* (London: William Ball, 1838); Zoe Laidlaw, "Aunt Anna's Report: The Buxton Women and the Aborigines Select Committee, 1835–37," *Journal of Imperial and Commonwealth History* 32, no. 2 (2006): 1–28.

10. Charles Swaisland, "The Aborigines Protection Society, 1837–1909," *Slavery and Abolition* 21, no. 2 (2000): 265–80.

11. APS, *First Annual Report*, 10.

12. Adrian Desmond and James R. Moore, *Darwin's Sacred Cause: Race, Slavery and the Quest for Human Origins* (New York: Houghton Mifflin Harcourt, 2009), 145.

13. Patricia Burns, *Fatal Success: A History of the New Zealand Company* (Auckland: Heinemann Reid, 1989), 96.

14. Claudia Orange, *The Treaty of Waitangi* (Wellington: Bridget Williams Books, 1987), 32–59.

15. Diane B. Paul, John Stenhouse, and Hamish G. Spencer, "The Two Faces of Robert FitzRoy, Captain of HMS Beagle and Governor of New Zealand," *Quarterly Review of Biology* 88, no. 3 (2013): 219–25.

16. Allan K. Davidson, ed., *A Controversial Churchman: Essays on George Selwyn, Bishop of New Zealand and Lichfield, and Sarah Selwyn* (Wellington: Bridget Williams Books, 2011).

17. See further, John Stenhouse, "Church and State in New Zealand, 1835–1870: Religion, Politics and Race," in *Church and State in Old and New Worlds*, ed. Hilary M. Carey and John Gascoigne (Leiden: Brill, 2011), 233–60.

18. R. Maunsell to Governor Grey, October 18, 1847, quoted in Grant Phillipson, *The Northern South Island*, Rangahaua Whanui series (Wellington: Waitangi Tribunal, 1995), 131.

19. Bernard Semmel, "The Philosophic Radicals and Colonialism," *Journal of Economic History* 21, no. 4 (1961): 513–25.

20. Hansard, Parliamentary Debates, 3rd series, vol. 82, July 1845, cols. 843–51 (J.A. Roebuck); S.A. Beaver, "John Arthur Roebuck (1802–1879)," in *Oxford Dictionary of National Biography* (Oxford: Oxford University Press, 2019), https://doi.org/10.1093/ref:odnb/23945.

21. William Fox, *The Six Colonies of New Zealand* (London: John Parker, 1851), 54, 64; Keith Sinclair and Raewyn Dalziel, "Fox, William," in *Dictionary of New Zealand Biography* (first published 1990), accessible at *Te Ara—the Encyclopedia of New Zealand*, accessed July 26, 2021, https://teara.govt.nz/en/biographies/1f15/fox-william.

22. Patrick Brantlinger, *Dark Vanishings: Discourse on the Extinction of Primitive Races, 1800–1930* (Ithaca, NY: Cornell University Press, 2003); Russell McGregor, *Imagined Destinies: Aboriginal Australians and the Doomed Race Theory* (Melbourne: Melbourne University Press, 1997).

23. Wiremu Tamihana Te Waharoa, *New Zealand Spectator* 5 (January 11, 1851), reprinted in *Colonial Intelligencer, or Aborigines' Friend* 3, no. 40 (1851): 243–51.

24. G.A. Selwyn to W. Gladstone, September 15, 1846, microfilm MS 426, Hocken Library, Dunedin.

25. Atholl Anderson, "Shortland, Edward," *Dictionary of New Zealand Biography* (first published 1990, updated December 2013), accessible at *Te Ara—the Encyclopedia of New Zealand*, accessed May 31, 2021, https://teara.govt.nz/en/biographies/1s11/shortland-edward.

26. Edward Shortland, *The Southern Districts of New Zealand* (London: Longman, Brown, Green, and Longmans, 1851), 40–41, 77.

27. G.A. Selwyn to H.J. Tancred, Auckland, April 28, 1860, *Appendices to the Journals of the House of Representatives* (*AJHR*), 1860, 1 E1 a.

28. *Great Britain Parliamentary Papers* (*GBPP*), (Shannon: Irish University Press, 1968–), [2798] vol. 41: 60–61.

29. *Hawkes Bay Herald*, November 10, 1860, p. 4; O. Hadfield, *The Second Year of One of England's Little Wars* (London: Williams and Norgate, 1861), 75–90.

30. *The Oracle of Reason* 1 (1841): 5; see further J. Stenhouse, "Imperialism, Atheism and Race: Charles Southwell, Old Corruption and the Maori," *Journal of British Studies* 44, no. 4 (2005): 754–74.

31. *Auckland Examiner*, December 11, 1856, and October 29, 1857.

32. *Auckland Examiner*, January 1, 1857.

33. Charles Southwell, "New Zealand, Its Missionaries and Its People," *Reasoner: Journal of Free Thought and Positive Philosophy*, March 1, 1857: 34–35, and March 8, 1857: 38.

34. Stenhouse, "Imperialism," 770–73.

35. See further Kenton Storey, "The Fourth Estate at War: New Zealand's Press during the Taranaki War, 1860–61," in *Tutu te puehu: New Perspectives on the New Zealand Wars*, ed. John Crawford and I.C. McGibbons (Wellington: Steele Roberts, 2018), 145–67.

36. Joseph Giles, "Waitara and the Native Question," *Southern Monthly Magazine* 1 (1863): 209–16, quotation on 215; Giles, "Our Colonization and Its Ethics," *Southern Monthly Magazine* 1 (1863): 547–52, quotation on 552.

37. A.S. Atkinson, Journal, September 10, 1862, and June 5, 1863, Alexander Turnbull Library, Wellington.

38. *Otago Witness*, March 27, 1869, p. 7; for similar views, see *Otago Daily Times*, April 5, 1862, p. 6.

39. *The Press*, November 29, 1862, p. 1.

40. *Lyttelton Times*, July 8, 1867, p. 3.

41. Darwin Correspondence Project, "Letter no. 2503," accessed February 9, 2021, https://www.darwinproject.ac.uk/letter/DCP-LETT-2503.xml.

42. Darwin Correspondence Project, "Letter no. 295," accessed February 4, 2021, https://www.darwinproject.ac.uk/letter/DCP-LETT-295.xml.

43. Charles Darwin, *The Descent of Man, and Selection in Relation to Sex*, vol. 1 (London: John Murray, 1871), 168–69.

44. Darwin Correspondence Project, "Letter no. 11745," accessed February 14, 2021, https://www.darwinproject.ac.uk/letter/DCP-LETT-11745.xml.

45. Darwin Correspondence Project, "Letter no. 13230," accessed February 14, 2021, https://www.darwinproject.ac.uk/letter/DCP-LETT-13230.xml.

46. J. Browne, *Charles Darwin: Voyaging* (London: Jonathon Cape, 1995), 390.

47. L. Bastings, "A History of the New Zealand Society, 1851–68: A Wellington Scientific Centenary," *Transactions and Proceedings of the Royal Society of New Zealand* 80 (1952): 359–65, quotation on 360.

48. George F. Bowen, "Inaugural Address of Governor Sir George Ferguson Bowen, G.C.M.G., to the New Zealand Institute, As its first President, August 4 1868," *Transactions and Proceedings of the New Zealand Institute* (*TPNZI*) 1 (1868): 3–9, quotation on 4.

49. Michael Hoare, *Beyond the "Filial Piety": Science History in New Zealand—A Critical Review of the Art*, Second Cook Lecture 1976 (Melbourne: Hawthorn Press, 1977), 10; Charles Fleming, *Science, Settlers and Scholars: The Centennial History of the Royal Society of New Zealand* (Wellington: Royal Society of New Zealand, 1987), 95; Ross Galbreath, *Scholars and Gentlemen Both: G.M. and Allan Thomson in New Zealand Science and Education* (Wellington: Royal Society of New Zealand, 2002), 21.

50. Francis Lucian Reid, "'The Democratic Politician Does Not Trouble Himself with Science': Class and Professionalisation in the New Zealand Institute, 1867–1903," *Tuhinga* 16 (2005): 21–31, quotation on 21; see further F. Reid, "Province of Science: James Hector and the New Zealand Institute, 1867–1903" (PhD thesis, University of Cambridge, 2007).

51. J.E. Martin, "Science and Species of Politician in the Nineteenth Century—Monkeys or Bullfrogs?," *Journal of the Royal Society of New Zealand* 47, no. 1 (2017): 27–32.

52. R.W. Home, "Australia, New Zealand and Oceania," in *The Cambridge History of Science*, vol. 8, *Modern Science in National, Transnational and Global Context*, ed. Hugh R. Slotten, Ronald L. Numbers, and David N. Livingstone (Cambridge: Cambridge University Press, 2020), 712–35, quotation on 718.

53. R.K. Dell, "Hector, James," *Dictionary of New Zealand Biography* (first published in 1990), accessible at *Te Ara—the Encyclopedia of New Zealand*, accessed March 17, 2021, https://teara.govt.nz/en/biographies/1h15/hector-james.

54. James Hector to J.D. Hooker, August 13, 1865, in Rowan Burns and Simon Nathan, *My Dear Hooker: Transcriptions of Letters from James Hector to Joseph Dalton Hooker between 1860 and 1898* (Geoscience Society of New Zealand miscellaneous publication 133B, 2012), 70–71.

55. See further John Stenhouse, "Creationists and Evolutionists in New Zealand, 1800–2010: Science, Religion, Politics and Race," *Almagest: International Journal for the History of Scientific Ideas* 12 (2021): 94–123, 100–103.

56. W.T.L. Travers, "On the Changes Effected in the Natural Features of a New Country by the Introduction of Civilized Races," *TPNZI* 2 (1869): 299–313, quotations on 308, 312, 313.

57. Livingstone, *Dealing with Darwin*, 12, 2.

58. Travers, "On the Changes," 308, 310, 312–13.

59. John Stenhouse, "'A Disappearing Race Before We Came Here': Doctor Alfred Kingcome Newman, the Dying Maori, and Victorian Scientific Racism," *New Zealand Journal of History* 30, no. 2 (1996): 124–40.

60. See further Stenhouse, "'Disappearing Race,'" 128–32.

61. Alfred Kingcome Newman, "A Study of the Causes Leading to the Extinction of the Maori," *TPNZI* 14 (1881): 459–77, quotations on 465–70.

62. Newman, "Study of the Causes," 474–77.

63. James Belich, *The New Zealand Wars and the Victorian Interpretation of Racial Conflict* (Auckland: Auckland University Press, 1986), 299.

64. Malcolm Nicolson, "Medicine and Racial Politics: Changing Images of the New Zealand Maori in the Nineteenth Century," in *Imperial Medicine and Indigenous Societies*, ed. David Arnold (Manchester: Manchester University Press, 1988), 66–104, quotation on 88.

65. *TPNZI* 14 (1881): 539–40; *New Zealand Times*, January 29, 1882, p. 3.

66. A.K. Newman, "A New Zealand Association of Science," *New Zealand Journal of Science* 1 (1882): 145–50.

67. Stenhouse, "'Disappearing Race,'" 136–37.

68. *Maori Messenger: Te Karere Maori*, November 30, 1860, p. 1.

69. Ross Galbreath, *Walter Buller: The Reluctant Conservationist* (Wellington: GP Books, 1989); *New Zealand Parliamentary Debates* (*NZPD*) 59 (1887): 122.

70. Walter Buller, "Address by the President," *TPNZI* 17 (1884): 443–46 (emphasis in original).

71. *TPNZI* 17 (1884): 445–46.

72. *Appendices to the Journals of the House of Representatives* (*AJHR*), 1891, G-1, xxix–xxx.

73. *NZPD* 78 (1892): 299.

74. Apirana Ngata, "The Past and Future of the Maori," *The Press*, December 3, 1894, p. 4; and December 19, 1894, p. 2.

75. M.P.K. Sorrenson, "Ngata, Apirana Turupa," in *Dictionary of New Zealand Biography* (first published 1996), accessible at *Te Ara—the Encyclopedia of New Zealand*, accessed July 20, 2021, https://teara.govt.nz/en/biographies/3n5/ngata-apirana-turupa.

76. *NZPD* 240 (1934): 1181–82.

77. See further Barbara Brookes, "'Aristocrats of Knowledge': Māori Anthropologists and the Survival of the 'Race,'" in *Eugenics at the Edges of Empire: New Zealand, Australia, Canada and South Africa*, ed. D.B. Paul, John Stenhouse, and Hamish G. Spencer (Cham, Switzerland: Palgrave Macmillan, 2018), 267–88.

78. M.P.K. Sorrenson, "Buck, Peter Henry," in *Dictionary of New Zealand Biography* (first published 1996, updated May 2002), accessible at *Te Ara—the Encyclopedia of New Zealand*, accessed July 14, 2021, https://teara.govt.nz/en/biographies/3b54/buck-peter-henry.

79. Te Rangi Hiroa, "The Passing of the Maori," *TPNZI* 55 (1924): 362–75 passim.

80. Hiroa, "Passing," 366–75; see further Helen M. Leach, "Sir Peter Buck and the Samoans," *New Zealand Journal of Archaeology* 15 (1993): 57–66.

9: EVOLUTION AND RELIGION IN TRANSNATIONAL CONTEXTS

We thank the following scholars for their input and suggestions in the writing of this chapter: Clinton Godart, Erik Hammerstrom, Ian Hesketh, Joachim Kurtz, Sean Lei, Kuan-yen Liu, Christine Luk, and Sarah Qidwai. We are especially grateful to Tomoko Kitagawa, who played the leading role in creating the section on Japan.

1. Peter Harrison, *The Territories of Science and Religion* (Chicago: University of Chicago Press, 2015), 99.

2. Harrison, *Territories*, 101.

3. Harrison, *Territories*, 147.

4. James Secord, "Global Darwin," in *Darwin*, ed. William Brown and Andrew C. Fabian (Cambridge: Cambridge University Press, 2010), 32.

5. The names of Japanese and Chinese historical figures are given in the original order, with surname first.

6. See, e.g., Fa-ti Fan, "Modernity, Region, and Technoscience: One Small Cheer for Asia as Method," *Cultural Sociology* (2016): 352–68; Fa-ti Fan, "Science in Cultural Borderlands: Methodological Reflections on the Study of Science, European Imperialism, and Cultural Encounter," *East Asian Science, Technology, and Society: An International Journal* 1, no. 2 (2007): 213–31.

7. Secord, "Global Darwin," 35.

8. Bernard Lightman, *Victorian Popularizers of Science: Designing Nature for New Audiences* (Chicago: University of Chicago Press, 2007, 378–88); Bernard Lightman, "The International Scientific Series and the Communication of Darwinism," *Journal of Cambridge Studies* 5 (2010): 27–38.

9. Examples of Huxley's books translated into different languages: *Evidence as to Man's Place in Nature* (French, German, Chinese, Japanese), *Lessons in Elementary Physiology* (Russian, French, German), *Manual of the Anatomy of Vertebrated Animals* (German, French), *Physiography* (French), *Hume* (French), *The Crayfish* (German), *Introductory Science Primer* (Spanish), and *Collected Essays* (volumes translated into French, Swedish, and Chinese).

10. Bernard Lightman, ed., *Global Spencerism: The Communication and Appropriation of a British Evolutionist* (Leiden: Brill, 2016).

11. Scholars have noticed Huxley's debt to German romanticism through Thomas Carlyle. See, for example, Frank Turner, "Victorian Scientific Naturalism and Thomas Carlyle," *Victorian Studies* 18 (1975): 325–43. But Huxley's conception of Eastern religions may have been mediated through his encounters with German romanticism. The connections between German romanticism and Eastern religions have been explored. See James Hodkinson and Jeffrey Morrison, eds., *Encounters with Islam in German Literature and Culture* (Rochester, NY: Camden House, 2009); Debra N. Prager, *Orienting the Self: The German Literary Encounter with the Eastern Other* (Rochester, NY: Camden House, 2014).

12. Henry R. Teddler, ed., *Descriptive Sociology: Or, Groups of Sociological Facts, Classified and Arranged by Herbert Spencer: Chinese*, comp. E.T.C. Werner (London: Williams and Norgate, 1910), iii.

13. Jonathan H. Turner, "Herbert Spencer's Sociological Legacy," in *Herbert Spencer: Legacies*, ed. Mark Francis and Michael W. Taylor (London: Routledge, 2015), 63.

14. Teddler, *Descriptive Sociology*, table 9.

15. Marjorie Wheeler-Barclay, *The Science of Religion in Britain, 1860–1915* (Charlottesville: University of Virginia Press, 2010).

16. For a particularly penetrating analysis of Tylor's historicizing of religion and his relationship to scientific naturalism, see Efram Sera-Shriar, "Historicizing Belief: E.B. Tylor, *Primitive Culture*, and the Evolution of Religion," in *Historicizing Humans: Deep Time, Evolution, and Race in Nineteenth-Century British Sciences*, ed. Efram Sera-Shriar (Pittsburgh: University of Pittsburgh Press, 2018), 68–112. Also relevant is the following comparison of Tylor's and Darwin's developmental anthropology in the early 1870s by the same author: Efram Sera-Shriar, *The Making of British Anthropology 1813–1871* (London: Pickering and Chatto, 2013), 147–76.

17. Herbert Spencer, "Last Words About Agnosticism and the Religion of Humanity," *Nineteenth Century* 16 (Nov. 1884): 827–30, 838.

18. Herbert Spencer, *The Principles of Sociology*, 3 vols. (New York: D. Appleton, 1897), 3:7, 12, 21.

19. Spencer, *Principles*, 3:49–51, 60, 150–51, 157, 173, 255–57.

20. T.H. Huxley, "On the Ethnology and Archaeology of India," in T.H. Huxley, *The Scientific Memoirs of Thomas Henry Huxley*, ed. Professor Michael Foster and Professor E. Ray Lankester, 4 vols. (London: Macmillan, 1901), 3:427–28.

21. Thomas H. Huxley, *Science and Hebrew Tradition* (London: Macmillan, 1893), 287–88, 308, 319–20, 347, 367, 371–72.

22. Huxley, *Science and Hebrew Tradition*, 181, 183, 185.

23. Thomas H. Huxley, *Science and Christian Tradition* (London: Macmillan, 1894), 234, 240. For an insightful discussion of how Muslim fatalism was condemned in debates on the relationship between science and religion, see M. Alper Yalçinkaya, "Kindred Fatalisms: Debating Science, Islam, and Free Will in the Darwinian Era," *Annals of Science* 79, no. 3 (2022): 364–85.

24. Thomas H. Huxley, *Evolution and Ethics and Other Essays* (London: Macmillan, 1894), 60–62.

25. Huxley, *Evolution and Ethics*, 91, 96.

26. Donald S. Lopez Jr., *Buddhism and Science: A Guide for the Perplexed* (Chicago: University of Chicago Press, 2008), 4.

27. Philip C. Almond, *The British Discovery of Buddhism* (Cambridge: Cambridge University Press, 1988), 34.

28. Huxley, *Evolution and Ethics*, 96.

29. Almond, *British Discovery of Buddhism*, 98–99.

30. Donald S. Lopez Jr., "Buddhism," in *Science and Religion Around the World*, ed. John Hedley Brooke and Ronald L. Numbers (Oxford: Oxford University Press, 2011), 216.

31. Lopez Jr., *Buddhism and Science*, 7.

32. Leonard Huxley, *Life and Letters of Thomas Henry Huxley*, 2 vols. (New York: D. Appleton, 1901), 2:375–76.

33. G. Clinton Godart, *Darwin, Dharma, and the Divine: Evolutionary Theory and Religion on Modern Japan* (Honolulu: University of Hawai'i Press, 2017), 17.

34. G. Clinton Godart, "'Philosophy' or 'Religion'? The Confrontation with Foreign Categories in Late Nineteenth Century Japan," *Journal of the History of Ideas* 69, no. 1 (2008): 71.

35. Isomae Jun'ichi, *Kindai nihon no shūkyō gensetsu to sono keifu: Shūkyō, kokka, Shinto* (Tokyo: Iwanami, 2003).

36. James R. Batholomew, *The Formation of Science in Japan* (New Haven, CT: Yale University Press, 1989); Kenkichiro Koizumi, "The Emergence of Japan's First Physicists, 1868–1900," *Historical Studies in the Physical Sciences* 6 (1975): iv–108; Takaaki Inuzuka, *Alexander Williamson: A Victorian Chemist in the Making of Modern Japan* (London: UCL Press, 2021).

37. The next to appear was Huxley's *On the Origin of Species: Or, The Causes of the Phenomena of Organic Nature* (1862) in 1879. The full-length translation of Darwin's *Origin* came out much later in 1896.

38. G. Clinton Godart, "Spencerism in Japan: Boom and Bust of a Theory," in Lightman, *Global Spencerism*, 59.

39. Traditionally, it was believed that Morse was the first person to lecture on evolutionary theory, but Ron Amundson claims there were two other lecturers at Tokyo medical school (*Tokyo igakkō*) and Dōshisha in Kyoto.

40. For a summary of the introduction of evolutionary theory, see Yulia Burenina, "Japanese Responses to Evolutionary Theory, with Particular Focus on Nichiren Buddhist," in *Asian Religious Responses to Darwinism*, ed. C. Mackenzie Brown (London: Springer, 2020), 340–43. The list of translations related to evolutionary theory is available in the journal *Historical English Studies in Japan* (*Eigakushi kenkyū*). Godart also provides the list of the translations of Spencer's work in his article in *Global Spencerism*, "Spencerism in Japan."

41. Godart, "Spencerism in Japan," 61.

42. Godart, "Spencerism in Japan," 62.

43. Douglas R. Howland, *Translating the West: Languages and Political Reason in Nineteenth-Century Japan* (Honolulu: University of Hawai'i Press, 2002), 173; Douglas Howland, "Herbert Spencer and Political Theory in Early Meiji Japan," in *Comparative Studies in Society and History* 42, no. 1 (2000): 67–68.

44. Setoguchi Akihisa, "Nihon ni okeru shinkaron no dōnyū," *Seibutsu kagaku* 56 (2004): 23–29.

45. Murakami Yōichirō, "Meiji shoki no seibutsu shinkaron jyuyō o megutte," *Hikakushisō kenkyū* 7 (1980): 2–3; Koyama Chūmaru, "Uchimura Kanzō ni okeru kagaku to shūkyō," *Hikakushisō kenkyū* 7 (1980): 7; Ōuchi Saburō, "Nihon kirisutokyō ni okeru shinkaron no mondai: Uchimura Kanzō o chūshin to shite," *Nihon bunka kenkyūjo hōkoku* 14 (1978): 7.

46. Godart, *Darwin, Dharma, and the Divine*, 13–14.

47. Yamashita Shigekazu, "Meiji jidai ni okeru Spencer no juyou," *Kokugakuin hōgaku* 12, no. 1 (1974): 85–88.

48. Saitō Shōji, "Meijiki ni okeru Spencer bunken o tazunete," *Shakai gaku ronsō* 24 (1962): 23–31.

49. Minakata's note in Spencer's *Three Essays*. Matsui Ryūgo, "Minakata Kumagusu zōshochū no Herbert Spencer chosaku ni mirareru kakikomi," *Kumagusu kenkyū* 9 (2015): 231, 244.

50. Minakata worked at the British Library from 1895 to 1898. Befitting the theme of international encounters of this chapter, he actually met and befriended Sun Yat-sen, the Chinese revolutionary, who was exiled in London during this period.

51. He wrote on various topics and published fifty-one articles in *Nature* throughout his life.

52. Herbert Spencer, *Principles of Sociology* (New York, 1896); quoted in Ryūgo Matsui, "Minzoku idō ni yoru takai kannen no hassei ni kansuru giron," *Gendai-Shiso* 47, no. 2 (2014): 250–59.

53. Herbert Spencer, *The Study of Sociology* (New York: D. Appleton, 1894).

54. Minakata's note in Spencer's *Study of Sociology*. Matsui, "Minakata Kumagusu zōshochū," 230.

55. The date of this note is estimated as before 1900. Minakata was still in England. Matsui Ryūgo, "Minakata Kumagusu zōshochū no Herbert Spencer chosaku ni mirareru kakikomi," in *Minakata Kumagusu: Fukugan no gakumon kōsō* (Tokyo: Keio University Press, 2016), 239.

56. Harada Ken'ichi, *Minakata Kumagusu: Shinkaron, seiji, sei* (Tokyo: Heibonsha, 2003), 35; Peter Szalay, "Minakata Kumagusu no ingaron to Herbert Spencer: Doki Hōryū ōfuku shokan ni mirareru giron o toushite," *Kumagusu kenkyū* 10 (2016): 189.

57. Szalay, "Minakata Kumagusu," 194.

58. Herbert Spencer, *First Principles* (Frankfurt: Outlook, 2020), 81.

59. Szalay, "Minakata Kumagusu," 197, 205.

60. Minakata Kumagusu, *Doki Hōryū ōfuku shokan* (Tokyo: Yasaka shobō, 1990), 314.

61. Minakata, *Doki Hōryū ōfuku shokan*, 235–38.

62. He taught there for about eight years starting in 1879. Godart, *Darwin, Dharma, and the Divine*, 33.

63. Yamashita Shigekazu, "Fenollosa and Spencer: An Examination of "Seitai Kaishinron (1880)," *Eigakushi kenkyū* 24 (1992): 1–13.

64. Yamashita, "Fenollosa and Spencer," 7.

65. Kiyozawa Manshi, *Kiyozawa Manshi zenshū*, ed. *Kengyō Nishimura and Haya Akegarasu*, vol. 1 (Kyoto: Hōzōkan, 1953), 204.

66. Bernat Marti-Oroval, "Kiyozawa Manshi no 'shūkyōshin' to jūkyū seiki Europe no shūkyō kigen tankyū: Herbert Spencer oyobi Max Müller tono kankei o chūshin ni," *Bukkyōgaku* 59 (2018): 46–47.

67. Marti-Oroval, "Kiyozawa Manshi no 'shūkyōshin' to jūkyū seiki," 48–49.

68. Yulia Burenina, "An Understanding of Evolutionary Theory in Modern Nichiren Buddhism: Focusing on the Thoughts of Honda Nissho and Tanaka Chigaku," *Matani ronshū* 7 (2013): 17–18.

69. Kishio Satomi, *Japanese Civilization, Its Significance and Realization: Nichirenism and the Japanese National Principles* (London: K. Paul, Trench, Trubner, 1923).

70. E.g., Jing Zhu, Mingjun Zhang, and Michael Weisberg, "Why Does the Chinese Public Accept Evolution?," *Studies in the History and Philosophy of Science*, part A, vol. 81 (June 2020): 116–24.

71. Benjamin Schwartz, *In Search of Wealth and Power: Yan Fu and the West* (Cambridge, MA: Harvard University Press, 1964); James Pusey, *China and Charles Darwin* (Cambridge, MA: Harvard University Asian Center, 1983); Yang Haiyan, "Knowledge Across Borders: The Early Communication of Evolution in China," in *The Circulation of Knowledge Between Britain, India and China*, ed. Bernard Lightman, Gordon McOuat, and Larry Stewart (Leiden: Brill, 2013), 179–208; Xiaoxin Jin, "The Evolution of Evolutionism in China, 1870–1930," *Isis* 111 (2020): 46–66; Xiaoxin Jin, "Translation and Transmutation: The Origin of Species in China," *British Journal for the History of Science* (2019): 117–41; Ke Zunke and Li Bin, "Spencer and Science Education in China," in Lightman, *Global Spencerism*, 78–102.

72. I borrow the term from Nelson Goodman, *Ways of Worldmaking* (Indianapolis: Hackett, 1978).

73. Hu Weiqing, "Jindai lai Hua chuanjiashi yu jinhualun," *Shijie zongjiao yanjiu* 3 (2001): 63–73; Benjamin Elman, *On Their Own Terms: Science in China, 1550–1900* (Cambridge, MA: Harvard University Press, 2005), chaps. 8 and 9.

74. Godart, *Darwin, Dharma, and the Divine*, chap. 1.

75. For biographical information of Yan and much more, see Max Ko-wu Huang, *Yan Fu and the Origins of Chinese Liberalism* (Hong Kong: Chinese University of Hong Kong, 2008).

76. Cao Juren, *Zhongguo xueshu sixiangshi suibi* (Beijing: Sanlian shuju, 2003).

77. Wang Tiangen, *Tianyanlun chuanbo yu Qingmo minchu de shehui dongyuan* (Heifei: Hefei Gongye Daxue chubanshe, 2006); Qian Li, "The Idea of Progress in Modern China: The Case of Yan Fu," in *Chinese Visions of Progress*, ed. Thomas Fröhlich and Axel Schneider (Leiden: Brill, 2020), 103–31; Hu Shi, *Sishi zishu* (Shanghai: Yadong tushuguan, 1933), 49–50.

78. Wang, *Tianyanlun chuanbo*, 122.

79. Sean Hsiang-lin Lei, "The Dawn of Science as Cultural Authority in China: *Tianyanlun* (On Heavenly Evolution) in the Post-1895 Debate over the Engagement with Western Civilization," *East Asian Science, Technology and Society* 16, no. 3 (2022): 408–32; Huang Kewu, "Hewei tianyan? Yan Fu tianyan zhi xue de neihan yu yiyi," *Zhongyang yanjiuyuan jindaishi yanjiu jikan* 85 (2014): 129–87; Shen Guowei, *Yan Fu yu kexue* (Nanjing: Fenghuang chubanshe, 2017); Fa-ti Fan, "Nature and Nation in Chinese Political Thought: The National Essence Circle in Early Twentieth-Century

China," in *The Moral Authority of Nature*, ed. Lorraine Daston and Fernando Vidal (Chicago: University of Chicago Press, 2004), 409–37.

80. Huxley, *Evolution and Ethics*.

81. Kuan-yan Liu, "Yan Fu's Daoist Reinterpretation of Evolutionism," in Brown, *Asian Religious Responses*, 287–318; Kuan-yan Liu, "Yan Fu's Xunzian-Confucian Translation of Thomas Huxley's Evolution and Ethics," in Brown, *Asian Religious Responses*, 257–86.

82. Shen Guowei, *Yan Fu yu Kexue*; Huang, "Hewei tianyan?"

83. Zheng Kuangmin, *Liang Qichao qimeng sixiang de dongxue beijing* (Shanghai: Shanghai shudian, 2009); Fa-ti Fan, "Redrawing the Map: Science in Twentieth-Century China," *Isis* 98 (2007): 524–38.

84. Kung-Chuan Hsiao, *A Modern China and a New World—K'ang Yu-wei, Reformer and Utopian, 1858–1927* (Seattle: University of Washington Press, 1975); Hao Chang, *Chinese Intellectuals in Crisis: The Search for Order and Meaning, 1890–1911* (Berkeley: University of California Press, 1987).

85. Zheng Kuangmin, *Liaing Qichao qimeng sixiang de dongxue beijing*, chap. 2.

86. Liang Qichao, "Tianyanxue shizu Daerwen zhi xueshuo ji qi zhuanlue," in *Yinbinshi quanji*, vol. 5, *Yinbinshi wenji zhi shisan* (Shanghai: Zhonghua shuju, 1936), 12–18. Originally published in *Xinmin congbao* (March 1902).

87. All quotes in this paragraph are from Liang Qichao, "Tianyanxue shizu Daerwen zhi xueshuo ji qi zhuanlue."

88. Ma Junwu, "Xinpai shenwuxue (ji tianyanxue) jia xiaoshi," in *Ma Junwu ji*, ed. Mo Shixiang (Wuhan: Huazhong Shifan Daxue chubanshe, 1991), 5–12, originally published in *Xinmin congbao* (May 1902). This piece marked the beginning of Ma's long-term translation project of Darwin's *Origin*, which he eventually finished in 1920.

89. E.g., Yang, "Knowledge Across Borders."

90. Tze-Ki Hon, "Zhang Zhidong's Proposal for Reform: A New Reading of the *Qianxue pian*," in *Rethinking the 1898 Reform Period: Political and Cultural Change in Late Qing China*, ed. Rebecca Karl and Peter Zarrow (Cambridge, MA: Harvard East Asian Center, 2002), 77–98; Vincent Goossaert, "1898: The Beginning of the End for Chinese Religion?," *Journal of Asian Studies* 65 (2006): 307–36.

91. Huang Kewu, "Minguo chunian Kongjiao de zhenlun," *Taiwan Shida lishi xuebao* 12 (1984): 197–223.

92. Erik Hammerstrom, *The Science of Chinese Buddhism: Early Twentieth-Century Engagement* (New York: Columbia University Press, 2015).

93. Tan Sitong, *Renxue*. The text is available on the Chinese Text Project: https://ctext.org/wiki.pl?if=en&chapter=402475 (no publication details are given).

94. Li-chuan Tai, "Cong Xujiahui Bowuyuan dao Zhendan Bowuyuan—Faguo Yesuhui shi zai jindai Zhongguo de ziranshi yanjiu huodong," *Zhongyang Yanjiuyuan Lishiyuyuan suo jikan* 82 (2013): 329–85.

95. Barry Brundell, "Catholic Church Politics and Evolution Theory, 1894–1902," *British Journal of the History of Science* 34 (2001): 81–95.

96. Joachim Kurtz, "Messenger of the Sacred Heart: Li Wenyu and the Jesuit Periodical Press in Late Qing Shanghai," in *From Woodblocks and the Internet*, ed. Cynthia Brokaw and Christopher Reed (Leiden: Brill, 2010), 79–109; Wang, *Tianyanlun chuanbo*, 214–57.

97. Quotes are from Li Wenyu, *Xu Liku* (Shanghai: Tusanwan Yinshuguan, 1915), 14, 15, 21.

98. Liang Qichao, "Jinhualun gemnin zhe Jide zhi xueshuo," in *Yinbingshi quanji*, vol. 5, *Yinbinshi wenji zhi shier* (Shanghai: Zhonghua shuju, 1936), 78–86.

99. Lu Mingyu, "Wan Qing chuanjiaoshi yu jinhua sixiang zhi xiuzheng: Yi Renxue fanyi wei zhongxin," *Shehui kexue yanjiu* 5 (2015): 194–97. On the reception of Spencer in China, see, e.g., Ke and Li, "Spencer and Science Education."

100. N. Gist Gee, *Introduction to Biology* (Shanghai: Commercial Press, 1914), 10. On Gee, see William Haas, *China Voyager: Gist Gee's Life in Science* (New York: M.E. Sharp, 1996).

101. Huang Junjie, "Li Chunsheng dui *Tianyanlun* de pipan ji qi shixiangshi de dingwei—yi Tianyan lun shu hou wei zhongxin," *Guojia Kexue Weiyuanhui yanjiu huikan: renwen ji shehui kexue* 4, no. 2 (1994): 194–205; Wang Jiahong, "Cong Li Chunsheng dui jinhua zhuyi de fanbo kan qi zai jindai sixiangshi de diwei," *Donghai zhongwen xuebo* 19 (2007): 269–84.

102. Li, *Tianyanlun shuhou*, 38–39.

103. Li, *Tianyanlun shuhou*, 10–11.

104. Liang Qichao, "Yu zhi shengsi guan," in *Liang Qichao zhexue sixiang lunwenxuan* (Beijing: Beijing daxue chubanshe, 1984), 208–19.

105. Zhang Taiyan, "Jufen jinhua lun," *Minbao* 7 (1906); Viren Murthy, "Transfiguring Modern Temporality: Zhang Taiyan's *Yogācāra* Critique of Evolutionary History," *Modern China* 35 (2012): 483–522; Axel Schneider, "Critiques of Progress: Reflections on Chinese Conservatism," in *Chinese Visions of Progress*, ed. Thomas Fröhlich and Axel Schneider (Leiden: Brill, 2020), 263–91.

106. Huang Kewu, "Tianyan yu fofa: *Tianyanlun* dui Qingmo Minchu foxue sixiang de chongji," in *Fanyishiyanjiu 2018*, ed. Wang Hongzhi (Shanghai: Fudan Daxue chubanshe, 2020), 199–222.

107. Erik Hammerstrom, "Early Twentieth-Century Buddhist Microbiology and Shifts in Chinese Buddhism's 'Actual Canon,'" *Theology and Science* 10 (2012): 3–18.

108. Erik Hammerstrom, "Science and Buddhist Modernism in Early 20th Century China: The Life and Works of Wang Xiaoxu," *Journal of Chinese Religions* 39 (2011): 1–32; Wang Xiaoxu, *Fofa yu Kexue* (Taizhong: Xuehai shuju, 1971), esp. 28–43.

109. Xun Liu, *Daoist Modern: Innovation, Lay Practice, and the Community of Inner Alchemy in Republican Shanghai* (Cambridge, MA: Harvard East Asian Monographs, 2000).

110. Matthias Schumann, "Science and Spirit-Writing: The Shanghai Lingxuehui

and the Changing Fate of Spiritualism in Republican China," in *Text and Context in the Modern History of Chinese Religions*, ed. Philip Cart, David Ownby, and Chien-chuan Wang (Leiden: Brill 2020): 126–72; Huang Kewu, "Minguo chunian Shanghai de lingxue yanjiu: yi Shanghai Lingxuehui wei li," *Zhongyang yanjiuyuan jindaishi yanjiusuo jikan* 55 (2007): 99–136; Huang Kewu, "Lingxue yu jindai Zhonguo de zhishi zhuanxing—Minchu zhishi fenzi dui kexue, zongjiao yu mixin de zai sikao," *Sixiangshi* 2 (2014): 121–96.

111. Liu's *Daoist Modern* deals extensively with Chen's career. See esp. chaps. 2 and 3.

112. Chen Yingning, *Xianxue jiemi: Daojia yangsheng miku*, ed. Hong Jianlin (Dalian: Dalian chubanshe, 1991).

113. Chen Yingning, "Zhongmiaoju wenda xu baze," reprinted in *Shanghai daojiao* 1 (1988): 24.

114. Fa-ti Fan, "'Mr. Science,' May Fourth, and the Global History of Science," *East Asian Science, Technology, and Society: An International Journal* 16, no. 3 (2022): 279–304, https://doi.org/10.1080/18752160.2022.2095099; Fa-ti Fan, "The Controversy over Spontaneous Generation in Republican China," in *Science and Technology in Modern China, 1880s–1940s*, ed. Jing Tsu and Benjamin A. Elman (Boston: Brill, 2014), 209–44.

10: THE JŌDO SHINSHŪ EMBRACE OF SCIENCE IN LATE MEIJI AND TAISHŌ JAPAN

1. "Daruwin kinen jinseigaku kōen" [Darwin Commemorative Lectures by Jinsei Society], *Jinsei* 5, no. 12 (December 1909): 471–72.

2. Eikoh Shimao, "Darwinism in Japan, 1877–1927," *Annals of Science* 38, no. 1 (January 1981): 98, https://doi.org/10.1080/00033798100200131; G. Clinton Godart, *Darwin, Dharma, and the Divine: Evolutionary Theory and Religion in Modern Japan* (Honolulu: University of Hawai'i Press, 2017), 52.

3. "Daruwin kinen jinseigaku kōen," 472–73.

4. Godart, *Darwin, Dharma, and the Divine*. Yulia Brenina also discusses various ideological responses to the theory of evolution in her article "Japanese Responses to Evolutionary Theory, with Particular Focus on Nichiren Buddhists," in *Asian Religious Responses to Darwinism: Modern Evolutionary Theories in Middle Eastern, South Asian, and East Asian Cultural Contexts*, ed. C. Mackenzie Brown (Cham, Switzerland: Springer, 2020), 337–67.

5. Parts 2 and 3 of Fujikawa's biography, *Fujikawa Yū-sensei*, touch upon the issues of science and religion. Fujikawa-sensei kankōkai, *Fujikawa Yū-sensei* [Professor Fujikawa] (1949; Tokyo: Ōzorasha, 1988), 1. Among more recent articles, see Yūichirō Shimada, "Fujikawa Yū no shūkyōron no tenkai: 'Kagaku' to no kankeisei o megutte," [Development of Fujikawa Yū's theories of religion: In relation to "science"], *Nihon shisōshi kenkyū* 46 (Mar. 2014): 34–50.

6. Naohide Isono, "Shinkaron no nihon e no dō'nyū [Introduction of evolutionary

theory to Japan]," in *Mōsu to nihon* [E.S. Morse and Japan], ed. Takeshi Moriya (Tokyo: Shōgakukan, 1988), 303–14.

7. Frederick B. Churchill, *August Weismann: Development, Heredity, and Evolution* (Cambridge, MA: Harvard University Press, 2015), 354–56.

8. For Toyama's life and work, see Lisa Onaga, "Toyama Kametaro and Vernon Kellogg: Silkworm Inheritance Experiments in Japan, Siam, and the United States, 1900–1912," *Journal of the History of Biology* 43, no. 2 (Summer 2010): 215–64.

9. Masao Watanabe, *The Japanese and Western Science* (Philadelphia: University of Pennsylvania Press, 1990), 71.

10. Hiroyuki Katō, "Seishinteki oyobi shakaiteki shinka [Mental and social evolution]," *Jinsei* 5, no. 12 (1909): 437–45.

11. Hekkeru hakase, *Uchū no nazo*, trans. Okagami Ryō and Takahashi Masakuma (Tokyo: Yūhōkan, 1906).

12. "Daruwin kinen jinseigaku kōen," 472–73.

13. In the 1890s, 69 percent of science students who studied abroad went to Germany. In the first decade of the twentieth century this number was as high as 74 percent. James R. Bartholomew, *The Formation of Science in Japan: Building Research Tradition* (New Haven, CT: Yale University Press, 1989), 71.

14. Yū Fujikawa, "Jinsei hakkan no shushi" [Purpose of *Jinsei* publication], *Jinsei* 1, no. 1 (1905): maezuke [frontal matter], 1.

15. "Kagaku no moji ni tsuite" [About the word *kagaku*], *Jinsei* 2, no. 1 (1906): 64.

16. Quoted in Yū Fujikawa, "*Jinsei* dai ikkan no ato ni fusu" [Afterword to *Jinsei* volume 1], *Jinsei* 1, no. 9 (1905): 472.

17. Katō, "Seishinteki oyobi shakaiteki shinka," 442.

18. Watanabe, *The Japanese and Western Science*, 94–98; Godart, *Darwin, Dharma, and the Divine*, 107–17.

19. Godart, *Darwin, Dharma, and the Divine*, 92; Watanabe, *The Japanese and Western Science*, 95.

20. Asajirō Oka, "Jinrui no shōrai" [Future of humankind], in *Kindai nihon shisōtaikei*, vol. 9, *Oka Asajirō-shū*, ed. Tsukuba Hisaharu (Tokyo: Chikuma shobō, 1974), 327–28, 337.

21. Asajirō Oka, "Shinkaron kōwa" [Lectures on evolutionary theory], in *Oka Asajirō-shū*, 270–71.

22. Taisei Tamamuro, ed., *Nihon Bukkyōshi* [History of Japanese Buddhism], vol. 3 (Kyoto: Hōzōkan, 1967), 434.

23. Oka, "Shinkaron kōwa," 274–75.

24. Yōichirō Murakami, "Seibutsu shinkaron ni taisuru nihon no hannō: Meijiki no autorain" [Japanese responses to the theory of biological evolution: An outline during the Meiji period], in *Nihonjin to kindai kagaku* [The Japanese and modern science] (Tokyo: Shin'yōsha, 1980), 98–99, 137–38; Murakami, "Seibutsu shinkaron to shakai shisō" [The theory of biological evolution and social thought], in *Nihonjin*

to kindai kagaku, 147–48, 157. (These two articles were originally published in 1965 and in 1972, respectively.)

25. Katō, "Seishinteki oyobi shakaiteki shinka," 441.

26. For Ishikawa's biography, see Shingi Itō, *Ishikawa Seishō kinenshū* [Commemorative collection of Ishikawa Seishō] (Osaka: Itō byōin, 1980), 1–10.

27. There are two book-length biographies of Fujikawa. Fujikawa-sensei kankōkai, *Fujikawa Yū-sensei*, and Hideo Fujikawa, *Fujikawa Yū* (Tokyo: Ozawa shoten, 1990).

28. For a short summary of the history and teaching of Pure Land Buddhism, see Paul Williams, *Mahāyāna Buddhism: The Doctrinal Foundations* (London: Routledge, 1989), 251–76.

29. Williams, *Mahāyāna Buddhism*, 272.

30. Shinran, "Lamps for the Latter Ages," in *The Collected Works of Shinran*, vol. 1 (Kyoto: Jōdo Shinshū Hongwanji-ha, 1997), 530.

31. Masayasu Oda, "Distribution of Buddhism in Japan: An Attempt to Draw a Map of the Buddhist-Denominational Regions," *Komazawa chiri* [Komazawa geography] 39 (2003): 54.

32. Godart, *Darwin, Dharma, and the Divine*, 70–91; Brenina, "Japanese Responses," 347–49.

33. Seishō Ishikawa, *Uchū no mokushi* [Revelations of the universe] (Tokyo: Seiretsudō, 1906), https://dl.ndl.go.jp/info:ndljp/pid/898417; Seishō Ishikawa, *Shizen no myōshu* [Wonders of Nature] (Tokyo: Seiretsudō, 1907), https://dl.ndl.go.jp/info:ndljp/pid/899676; Seishō Ishikawa, *Shizen kagaku to bukkyō* [Natural science and Buddhism] (Tokyo: Seiretsudō, 1907), https://dl.ndl.go.jp/info:ndljp/pid/816525; Seishō Ishikawa, *Shizen to bukkyō* [Nature and Buddhism] (Kyoto: Bukkyō gakkai, 1916); Seishō Ishikawa, *Shizen kagaku to shūkyō* [Natural science and religion] (Kyoto: Gohōkan, 1919), https://dl.ndl.go.jp/info:ndljp/pid/958862.

34. In the back advertisement section of Ishikawa, *Shizen kagaku to bukkyō*.

35. Ishikawa, *Uchū no mokushi*, 14–15.

36. Ishikawa, *Uchū no mokushi*, 14–16.

37. Ishikawa, *Shizen kagaku to bukkyō*, 8, 13.

38. Ishikawa, *Uchū no mokushi*, 4; Ishikawa, *Shizen kagaku to bukkyō*, 58–59; Ishikawa, *Shizen kagaku to shūkyō*, 280–81.

39. Ishikawa, *Uchū no mokushi*, 11.

40. Asajirō Oka, "Jinrui no kodaikyō" [Megalomania of human race], in *Oka Asajirō-shū*, 304–7. Godart also comments on this incident and Oka's response. Godart, *Darwin, Dharma, and the Divine*, 107–8.

41. Ishikawa, *Uchū no mokushi*, 11–12.

42. Ishikawa, *Uchū no mokushi*, 4, 12–14; Ishikawa, *Shizen kagaku to bukkyō*, 19–20; Ishikawa, *Shizen to bukkyō*, 28.

43. Ishikawa, *Shizen kagaku to shūkyō*, 176–77.

44. Ishikawa, *Shizen kagaku to bukkyō*, 20–21. In the beginning of the Meiji period,

a Shinshū priest, Sata Kaiseki (1818–1882), advocated a Buddhist geocentrism with Mount Sumeru as the center of the universe.

45. Ishikawa, *Shizen kagaku to bukkyō*, 61.

46. Ishikawa, *Shizen kagaku to bukkyō*, 61.

47. Ishikawa, *Shizen kagaku to bukkyō*, 62; Ishikawa, *Shizen no myōshu*, 160–63.

48. Ishikawa, *Shizen kagaku to shūkyō*, 90–91.

49. Ishikawa, *Shizen kagaku to shūkyō*, 11.

50. Ishikawa, *Shizen kagaku to shūkyō*, 13–15.

51. Tadashi Okamoto, "'Shizen' to 'butsuri' no imi no hensen" [Transformations of the meaning of "Nature" and "Physics"], *Butsuri kyōiku* 43, no. 1 (1995): 65–66.

52. Ishikawa, *Shizen kagaku to shūkyō*, 275.

53. Ishikawa, *Shizen kagaku to shūkyō*, 279, 282. We might find a kindred spirit in Haeckel here.

54. Ishikawa, *Shizen no myōshu*, 139–49.

55. Oka, "Shinkaron kōwa," 210–11.

56. *Jinsei* 2, nos. 1–7 (1906). [This was translated in installments and can be found throughout vol. 2.]

57. Yū Fujikawa, "Ichigenteki shūkyō [Monistic religion] Part 1," *Jinsei* 10, no. 1 (1914): 6.

58. Yū Fujikawa, "Ichigenteki shūkyō Part 2," *Jinsei* 10, no. 2 (1914): 31.

59. Yū Fujikawa, "Shinran shōnin," *Chūōkōron* 30, no. 10 (1915).

60. Yū Fujikawa, *Shiran shōnin* [Holy man Shinran] (Tokyo: Muga sanbō, 1916).

61. Friedrich Schleiermacher, *On Religion: Speeches to Its Cultured Despisers*, trans. John Oman (London: Kegan Paul, 1893; first published in 1799).

62. Yū Fujikawa, *Kongōshin* [Diamond mind] (Tokyo: Rakuyōdō, 1916), 1.

63. Fujikawa, *Kongōshin*, 91.

64. Shinran, "Notes on 'Essentials of Faith Alone,'" in *Collected Works*, vol. 1, 461. Quoted by Fujikawa in *Shinran shōnin*, 38–40.

65. This word is *shinnyo* in Japanese, *tathatā* in Sanskrit, and it means "as it is." It is often translated in English as "suchness" or "truth."

66. Fujikawa, *Shinran shōnin*, 44.

67. Fujikawa, *Shinran shōnin*, 56–61.

68. Fujikawa, *Shinran shōnin*, 67.

69. Fujikawa, *Shinran shōnin*, 91.

70. Fujikawa, *Shinran shōnin*, 92–93.

71. Fujikawa, *Shinran shōnin*, 92–93.

72. Fujikawa, *Shinran shōnin*, 99–102.

73. Fujikawa, *Kongōshin*, 17–20.

74. Fujikawa, *Kongōshin*, 32–36, 42–43.

75. Fujikawa, *Kongōshin*, 23.

76. Fujikawa, *Kongōshin*, 8.

77. Fujikawa, *Kongōshin*, 85–86.

78. Yū Fujikawa, "Kagaku to shūkyō" [Science and religion], in *Dai nikai kaki daigaku kōenshū* [Lectures at second summer seminar], ed. Ejō Asai (Ishikawa: Ishikawaken Nomigun bukkyōkai, 1926), 23–24. (This text is available through NDL Digital Collections, https://dl.ndl.go.jp/ja/pid/922917.)

79. Fujikawa, "Kagaku to shūkyō," 23.

80. Godart, *Darwin, Dharma, and the Divine*, 128–30, 187.

81. Fujikawa, "Kagaku to shūkyō," 24–25; Yū Fujikawa, *Shinshū* (Tokyo: Shishin shobō, 1919), 8–9.

82. For Fujikawa's comments on Haeckel's religious views, see Yū Fujikawa, *Fujikawa Yū chojutsusen 2: Kagaku to shūkyō* [Selected writings of Fujikawa Yū, vol. 2, Science and religion] (Tokyo: Nakayama bunka kenkyūsho, 1942), 44, 240. Also see another book with the same title: Yū Fujikawa, *Kagaku to shūkyō* [Science and religion] (Tokyo: Shunjūsha, 1931), 125–26.

83. Yū Fujikawa, "Ijutsu to shūkyō" [Medicine and religion], in *Fujikawa Yū chosakushū* [Collected works of Fujikawa Yū], vol. 2 (Kyoto: Shibunkaku shuppan, 1980), 8–9, 97.

84. Fujikawa, "Ijutsu to shūkyō," 70–74.

85. Fujikawa, "Ijutsu to shūkyō," 121.

86. John William Draper's *A History of the Conflict between Religion and Science* was first translated in Japan by Kōhei Ogurusu in 1883 and was widely read.

11: COSMIC, THEISTIC EVOLUTION AND KANG YOUWEI

This chapter is an output of the project "Religious Cosmopolitanism in China, 1895–1927," funded by the Hong Kong Research Grants Council (no. 17616820). It builds on relevant portions of my doctoral thesis (Oxford University, 2019), the revised version of which has recently been published. See Wan Zhaoyuan, *Science and the Confucian Religion of Kang Youwei (1858–1927): China before the Conflict Thesis* (Leiden: Brill, 2022). Heartfelt thanks to Professor Bernard Lightman and Dr. Sarah Qidwai for their insightful comments on the various drafts of this article, and to Mr. William Hui and Ms. Barbie Halaby for their extremely helpful editorial suggestions.

1. Kang Youwei, "Yingguo youji" [An account of my travels in Great Britain (1904)], in *Kang Youwei quanji* [Complete works of Kang Youwei] (hereafter *KYWQJ*), 12 vols., ed. Jiang Yihuan and Zhang Ronghua (Beijing: Zhongguo renmin daxue chubanshe, 2007), 8:1–26. Translations in this article, unless otherwise stated, are all mine.

2. Xiaoxing Jin, "The Evolution of Evolutionism in China, 1870–1930," *Isis* 111, no. 1 (2020): 50; James R. Pusey, *China and Charles Darwin* (Cambridge, MA: Harvard University Asia Center, 1983).

3. Liang Qichao, "Yu Yan Youling shu" [A letter to Yan Fu (1897)], in *Liang Qichao quanji* [Complete works of Liang Qichao] (hereafter *LQCQJ*), ed. Zhang Pinxing (Beijing: Beijing chubanshe, 1999), 1:71–73.

4. See, for example, the view of his major biographer in the West, Kung-ch'uan Hsiao, *A Modern China and a New World: K'ang Yu-Wei, Reformer and Utopian* (Seattle: University of Washington Press, 1975), 594.

5. For a concise English introduction to Kang's life, see H.L. Boorman, ed., *Biographical Dictionary of Republican China*, vol. 2 (New York: Columbia University Press, 1968), 228–33.

6. Liang Qichao, "Nanhai Kangxiansheng zhuan" [A biography of Mr. Kang of Nanhai (1901)], in *LQCQJ*, 2:486.

7. For more on Kang's Confucian Movement and religious thought, see Hsi-yuan Chen, "Confucianism Encounters Religion: The Formation of Religious Discourse and the Confucian Movement in Modern China" (PhD diss., Harvard University, 1999); Wei Leong Tay, "Kang Youwei: The Martin Luther of Confucianism and His Vision of Confucian Modernity and Nation," in *Secularization, Religion and the State*, ed. Haneda Masashi (Tokyo: University of Tokyo Centre for Philosophy, 2010), 97–109.

8. Kang, *KYWQJ*.

9. Kang Youwei, "Woshi" [My history (1899)], in *KYWQJ*, 5:62.

10. See Sin-wai Chan, *Buddhism in Late Ch'ing Political Thought* (Hong Kong: Chinese University Press, 1985), 49.

11. Kang Youwei, "Wuxu lunzhou zhong juebishu ji Wuwu bahou" [My last note penned in a ship in 1898, with postscripts added in 1918], in *KYWQJ*, 5:4–5.

12. Kang Youwei, "Kangzi neiwai pian" [The esoteric and exoteric essays of Master Kang (1886)], in *KYWQJ*, 1:103.

13. John Fryer, "An Account of the Department for the Translation of Foreign Books at the Kiangnan Arsenal, Shanghai," *North-China Herald*, January 29, 1880.

14. Kang Youwei, "Woshi," 64; translated in "Chronological Autobiography," in *K'ang Yu-Wei: A Biography and a Symposium*, trans. and ed. Jung-pang Lo (Tucson: University of Arizona Press, 1967), 40–41.

15. For Chinese classical cosmology, see Yu-lan Fung, *A History of Chinese Philosophy*, trans. Derk Bodde, 2nd ed., 2 vols. (Princeton, NJ: Princeton University Press, 1983), 1:283–99.

16. Zhuangzi, *Zhuangzi [in Chinese-English Version]*, trans. Wang Rongpei (Beijing: Hunan People's Publishing House, 1999), 295; see also Liezi, *Liezi [in Chinese-English Version]*, trans. Liang Xiaopeng (Beijing: Zhonghua shuju, 2005), 7–8.

17. For a general discussion on these concepts, see Fung, *Chinese Philosophy*, 1:22–31; Weimin Sun, "Features of Chinese Cosmology," *Journal of East-West Thought* 2, no. 1 (March 2012): 133–44.

18. The monthly continued publication for four years in two periods of 1876–1878 and 1880–1882, and its past issues had been collected and reprinted quite a few times, made available for sale.

19. Leixiaer [Lyell], *Dixue qianshi* [Elements of geology], trans. Magaowen

[MacGowan] and Hua Hengfang, new ed., vol. 13 (Shanghai: Kiangnan Arsenal, 1873), 16a.

20. Fu Lanya [John Fryer], "Hundun shuo [On the primordial chaos]," *Gezhi huibian* [The Chinese scientific magazine] 7 (1877): 6–8.

21. Although the book does not carry the name of its author on its cover as its author, it has been identified that Robert Chambers, the publisher, wrote it. Sondra Miley Cooney, "Publishers for the People: W. & R. Chambers: The Early Years, 1832–1850" (PhD diss., Ohio State University, 1971), 201.

22. W. Chambers and R. Chambers, eds., *Introduction to the Sciences*, new ed. (London: W. Chambers and R. Chambers, 1873), preface.

23. Fu Lanya [Fryer], "Gezhi luelun" [Introduction to the sciences], *Gezhi huibian* 10 (1877): 1.

24. See Anonymous, *Vestiges of the Natural History of Creation* (London: John Churchill, 1844).

25. Kang, "Kangzi neiwai pian," 105, 110–11.

26. Kang Youwei, "Mingong pian" [On civil feats (1886)], in *KYWQJ*, 1:68.

27. Kang Youwei, "Kongzi gaizhi kao" [A treatise on Confucius as a reformer (1897)], in *KYWQJ*, 3:1–260; translated in Fung, *Chinese Philosophy*, 2:677.

28. Kang, "Yingguo youji," 23; translated in James Legge, trans., *The Chinese Classics I*, 2nd ed. (Oxford: Clarendon Press, 1893), 399.

29. Kang Youwei, *Datong shu* [The book of the great unity] (Shanghai: Zhonghua shuju, 1935), 357.

30. Kang, "Mingong pian," 68.

31. Kang, "Woshi," 81.

32. Kang, "Mingong pian," 75; see also Ding Weiliang [William A.P. Martin], *Xixue kaolue* [A brief review of Western learning] (Beijing: Zongli yamen, 1883), I, 52a.

33. Kang, "Baoguohui zhangcheng" [The Charter of the Society for Preserving China (1898)], in KYWQJ, 4:54–55; Frank Dikötter, *Imperfect Conceptions: Medical Knowledge, Birth Defects and Eugenics in China* (London: Hurst and Company, 1998), 61.

34. Kang, "Yingguo youji," 21–22.

35. Chambers, *Introduction to the Sciences*, 106.

36. Kang, "Yingguo youji," 21–22.

37. Kang Youwei, "Chunqiu bixiao dayi weiyan kao" [A Study of the profound principles and hidden meanings of the emended annals (1901)], in *KYWQJ*, 6:183.

38. Kang, "Mingong pian," 68.

39. Kang Youwei, "Yingguo Jianbuliezhu daxue huawen zongjiaoxi Zhailushi huijianji" [An interview with Herbert Allen Giles, professor of Chinese at Cambridge University of England (1904)], in *KYWQJ*, 8:34.

40. For a fuller discussion on Kang's struggle to define the Chinese concept of

jiao and the Western term "religion," see Wan, *Science and the Confucian Religion of Kang Youwei*, 140–41.

41. Kang Youwei, "Yidali youji" [An account of my travels in Italy (1904)], in *KYWQJ*, 7:374; see also Kang Youwei, "Chang'an yanjianglu" [Recorded talks (by Kang) in Xi'an (1923)], in *KYWQJ*, 11:271–99. For a discussion in detail on Kang's *shendao/rendao* categorization, see Zhaoyuan Wan, *Science and the Confucian Religion of Kang Youwei (1858–1927): China before the Conflict Thesis* (Leiden: Brill, 2022), 141–44.

42. Liang Qichao, "Lun Zhina zongjiao gaige" [On the religious reformation in China (1899)], in *LQCQJ*, 2:264–65.

43. Kang, "Chunqiu kao," 135.

44. Kang, *Datong shu*, 164; Laurence G. Thompson, *Ta T'ung Shu: The One-World Philosophy of K'ang Yu-Wei* (London: George Allen and Unwin, 1958), 126–27.

45. Kang, *Datong shu*, 452; Thompson, *Ta T'ung Shu*, 275.

46. The brief discussion here on Kang's vision of the future development of religions is based on the fuller account of the topic in Wan, *Science and the Confucian Religion*, 124–28.

47. Kang Youwei, "Zhutian jiang" [Lectures on the heavens (1927)], in *KYWQJ*, 12:11–12.

48. Benjamin Elman, *On Their Own Terms: Science in China, 1550–1900* (Cambridge, MA: Harvard University Press, 2005), 347; Jin, "Evolution of Evolutionism," 50.

49. *Gezhi huibian*, item 5, no. 1 (1876); item 31, no. 2 (1976); item 150, no. 8 (1976). See also Chambers, *Introduction to the Sciences*, 3, 63.

50. Wei Lianchen [Williamson], *Gewu tanyuan* [Natural theology and the method of divine government], 3 vols. (Shanghai: Presbyterian Mission Press, 1876).

51. Kang, "Yingguo youji," 23.

52. Kang, "Zhutian jiang," 101.

53. Kang, "Zhutian jiang," 40.

54. Kang Youwei, "Shili gongfa quanshu" [A compendium of true principles and universal laws (c. 1885)], in *KYWQJ*, 1:154. For a fuller discussion on the Term Question and Kang's deliberations on the proper names of God, see Wan, *Science and the Confucian Religion*, 63–66.

55. Kang's treatise on God has been rendered into English by me and Patrick Swire, with reference being made to the partial translation in Hsiao, *K'ang Yu-Wei*, 183–84. Kang Youwei, "Treatise on God," trans. Zhaoyuan Wan and Patrick Swire, *Renditions* 93 (Spring 2020): 43–48.

56. Kang, "Zhutian jiang," 93.

57. Kang, "Zhutian jiang," 40.

58. Kang, "Zhutian jiang," 94; Kang, "Treatise on God," 49.

59. J.E. McGuire, "Newton on Place, Time and God: An Unpublished Source," *British Journal for the History of Science* 11, no. 2 (July 1978): 121.

60. James R. Moore, "Of Love and Death: Why Darwin 'Gave Up Christianity,'" in *History, Humanity and Evolution: Essays for John C. Greene*, ed. James R. Moore (Cambridge: Cambridge University Press, 1989), 195–229.

61. See, for example, Tang Wenming, *Fujiao zaikuan: Kang Youwei Kongjiao sixiang shenlun* [Teaching with tolerance: A study of Kang's thought on Confucian religion] (Beijing: Zhongguo remin daxue chubanshe, 2012); Wei Leong Tay, "Secularization as Sacralization: Religion and the Formation of Modern Chinese Nationalism and Nation-State, 1840–1939" (PhD diss., University of Oxford, 2019); Wan, *Science and the Confucian Religion*.

62. Only a small portion of Spencer's works was translated into Chinese in the late nineteenth century, and those translations were focused on education and sociology rather than his evolutionary thought. See Zunke Ke and Bin Li, "Spencer and Science Education in China," in *Global Spencerism: The Communication and Appropriation of a British Evolutionist*, ed. Bernard Lightman (Leiden: Brill, 2016), 78–102. I doubt if Fiske had been introduced into late nineteenth-century China at all.

63. See, for example, Jürgen Osterhammel, *The Transformation of the World: A Global History of the Nineteenth Century*, trans. Patrick Camiller (Princeton, NJ: Princeton University Press, 2014), 896–97.

64. Hsiao, *K'ang Yu-Wei*, 412–14.

12: EVOLUTION AND CONSTRUCTIONS OF ISLAM IN THE OTTOMAN WORLD, 1870–1920

1. Bernard Lightman, "Creating a New Space for Debate: The Monthlies, Science, and Religion," in *Rethinking History, Science, and Religion: An Exploration of Conflict and the Complexity Principle*, ed. Bernard Lightman (Pittsburgh: University of Pittsburgh Press, 2019), 85–109; James C. Ungureanu, "Science and Religion in the Anglo-American Periodical Press, 1860–1900: A Failed Reconciliation," *Church History* 88 (2019): 120–49.

2. Ahmed Midhat, *Kâinat: Kütübhane-i Tarih*, vol. 1 (Istanbul, 1871), ز- .ك

3. Known as the "Day-Age Theory," this approach was commonly used by Christian interpreters in the nineteenth century as well. See David N. Livingstone, "The Preadamite Theory and the Marriage of Science and Religion," *Transactions of the American Philosophical Society* 3 (1992): i–78. Midhat's reference, in turn, is to the sixteenth-century Arabic-Turkish dictionary of Ahterî. See Midhat, *Kâinat*, هـ- ل.

4. Midhat, *Kâinat*, م ز— م و.

5. Midhat, *Kâinat*, ن—ا.

6. Ahmed Midhat, "Viladet," *Dağarcık* 2 (1872/1288): 49–52.

7. Ahmed Midhat, "Duvardan Bir Sada," *Dağarcık* 4 (1872/1288): 99–105.

8. This statement doesn't exist in the Qur'an in this form, but there are several verses that denote this idea, which are commonly interpreted as indicating a return to God, not a material recomposition.

9. Yusuf Imtiaz, "Discussion between Al-Ghazzālī and Ibn Rushd about the

Nature of Resurrection," *Islamic Studies* 2 (1986): 181–95; Nahyan Fancy, *Science and Religion in Mamluk Egypt: Ibn al-Nafis, Pulmonary Transit and Bodily Resurrection* (London: Routledge, 2013).

10. On this approach see Charles Kurzman, ed., *Modernist Islam, 1840–1940: A Sourcebook* (New York: Oxford University Press, 2002).

11. Auguste Debay, *Histoire naturelle de l'homme et de la femme depuis leur apparition sur le globe* (Paris: E. Dentu, 1858). Midhat quotes and paraphrases several sentences from Debay, *Histoire naturelle*, 39–41.

12. Ahmed Midhat, "Insan—Dünyada İnsanın Zuhuru," *Dağarcık* 4 (1872/1288): 116.

13. [Harputlu Ishak], "Mevâliden bir zat tarafından matbaamıza vürûd eden varakadır," *Basiret*, Muharrem 4, 1290/March 4, 1873, 1.

14. Ahmed Midhat, "Redd-i İtiraz ve İzah-i Hakikat," *Dağarcık* 8 (1873/1288): 234.

15. Midhat, "Redd-i İtiraz," 244.

16. Midhat, "Redd-i İtiraz," 240–43. The verse Midhat referred to in this context is Baqarah 28: "How can you be ungrateful to Allah Who bestowed life upon you when you were lifeless, then He will cause you to die and will again bring you back to life so that you will be returned to Him." The Baydawi commentary notes that man was but a combination of non-living substances, nutrients, amalgams, semen, and flesh, before he was granted a soul by God. See Kadı Beydâvî, *Muhtasar Beydâvî Tefsiri (Envaru't-Tenzîl ve Esrârü't-Te'vîl)*, vol.1, translated by Şadi Eren (Istanbul: Selsebil, 2011), 103. .

17. Midhat, "Redd-i İtiraz," 245.

18. Midhat was only loosely affiliated with the Young Ottoman movement, and while it is unclear whether his texts on evolution were the primary reason behind his exile, his several references to the freedom of thought in his writings and this fiery debate with a reputable theologian in which he had been accused of blasphemy placed him in the same category as the other exiles.

19. Marwa Elshakry, *Reading Darwin in Arabic, 1860–1950* (Chicago: University of Chicago Press, 2014).

20. Elshakry, *Reading Darwin*, chap. 1.

21. Elshakry, *Reading Darwin*, 99–111; Şükrü Hanioğlu, "Blueprints for a Future Society: Late Ottoman Materialists on Science, Religion, and Art," in *Late Ottoman Society: The Intellectual Legacy*, ed. Elizabeth Ozdalga (New York: Routledge, 2013), 50–138.

22. Elshakry, *Reading Darwin*, 111–13.

23. Sayyid Jamal ad-Din al-Afghani, "The Truth about the Neicheri Sect and an Explanation of the Neicheris," in *An Islamic Response to Imperialism: Political and Religious Writings of Sayyid Jamal ad-Din "al-Afghani,"* trans. Nikki R. Keddie and Hamid Algar (Berkeley: University of California Press, 1983), 130–73.

24. Nikki R. Keddie, "The Pan-Islamic Appeal: Afghani and Abdülhamid II,"

Middle Eastern Studies 3 (1966): 46–67; Azmi Özcan, *Pan-Islamism: Indian Muslims, the Ottomans and Britain, 1877–1924* (Leiden: Brill, 1997).

25. Ahmed Midhat, *Beşir Fuad* (Istanbul: Tercüman-ı Hakikat, 1887).

26. Orhan Okay, *Beşir Fuad: Ilk Turk Pozitivisti ve Materyalisti* (Istanbul: Dergah, 2008), 91.

27. The Presidential State Archives of Turkey, MF.MKT. 244-19-0, January 8, 1895/11-07-1312.

28. The Presidential State Archives of Turkey, MF.MKT. 1002-28-0, June 25, 1907/14-05-1325.

29. Elshakry, *Reading Darwin*, 131–59.

30. Rudolph Peters, "Resurrection, Revelation and Reason: Husayn al-Jisr (d. 1909) and Islamic Eschatology," in *Hidden Futures: Death and Immortality in Ancient Egypt, Anatolia, the Classical, Biblical and Arabic-Islamic World*, ed. J.M. Bremer, P. Peters, and T.P.J. van den Hout (Amsterdam: Amsterdam University Press, 1994), 221–31.

31. For examples, see Daniel A. Stolz, "'By Virtue of Your Knowledge': Scientific Materialism and the Fatwās of Rashīd Riḍā," *Bulletin of the School of Oriental and African Studies* 2 (2012): 223–47; and Alper Bilgili, "An Ottoman Response to Darwinism: Ismail Fenni on Islam and Evolution," *British Journal for the History of Science* 4 (2015): 565–82.

32. Subhi Edhem, *Darvenizm* (Manastir: Beynelmilel Ticaret Matbaasi, 1327/1910). Edhem taught natural history at the military high school in Manastir (contemporary Bitola in Macedonia). Note that despite the liberal climate of the early constitutional era, Edhem's classes were terminated in 1909, but he was able to publish his lecture notes in this format. On Edhem and Ottoman texts on Darwin in this period, see Mehmet O. Alkan, "Osmanlı Darwinizmi," *Cogito* 60 (2009): 333–58.

33. Subhi Edhem, *Darvenizm*, 184–85.

34. Subhi Edhem, *Darvenizm*, 188.

35. Osman Nuri, "Hikmet-i Maddiyeden Bir Bahis: Amal-i Terakki ve İttihad-ı Umumi," *Musavver Muhit*, April 8, 1909/Mart 26, 1325: 346.

36. Osman Nuri, "Kanun-ı Tekamül: Gaye-i Hayat ve Medeniyet-i Teselsül Ecza-yı Tabiat ve İttihad-ı Umumi-i Kainat," *Musavver Muhit*, August 12, 1909/Temmuz 10, 1325: 495.

37. Nuri, "Hikmet-i Maddiyeden Bir Bahis," 349.

38. Ernst Haeckel, *The Last Link: Our Present Knowledge of the Descent of Man* (London: Adam and Charles Black, 1898). This book is a revised version of a talk Haeckel gave at the Fourth International Congress of Zoology at Cambridge.

39. Baha Tevfik, "Bir Mukaddime: Bizde Tabiiiyyat ve Istilahlar," in Ernst Haeckel, *Insanin Menşei: Nesl-i Beşer* (Istanbul: Matbaa-i Nefaset, 1911), 14.

40. Mehmed Şemseddin, *Felsefe-i Ula: İsbat-ı Vacib ve Ruh Nazariyeleri* (Istanbul: Evkaf-ı İslamiye, 1921), 145. The preference to distinguish "Darwin the scientist" from "misguided materialists" like Haeckel was observed in the British and

Canadian contexts as well. I thank Bernard Lightman for drawing my attention to this tendency, which may be seen as a strategy for critics who did not want to be perceived as anti-science.

41. Said Cemil, "Ulum-i Hayatiyyeden Bir Yaprak—Transformizm, Alamarkizm, Darvenizm," *Resimli Kitab*, August 13, 1910/July 31, 1326: 931–32.

42. Francis Darwin, ed., *The Life and Letters of Charles Darwin, Including an Autobiographical Chapter*, vol. 1 (London: John Murray, 1887): 312–13. Since the article doesn't include a reference, it is impossible to tell if Said Cemil read this book, or its French translation, or another source quoting this section.

43. This is a very cursory description of this complex model, which was developed and interpreted in many different ways by different Sufi thinkers, both before and after Ibn Arabi. On the *insan-ı kâmil*, see Masataka Takeshita, "The Theory of the Perfect Man in Ibn 'Arabi's *Fusus Al-Hikam*," *Orient* 19 (1983): 87–102. For a broad overview of Ibn Arabi's thought, see William Chittick, "Ibn 'Arabî," *Stanford Encyclopedia of Philosophy* (Spring 2020 ed.), ed. Edward N. Zalta, https://plato.stanford.edu/archives/spr2020/entries/ibn-arabi/.

44. Among others were *istihale*, *tebeddül*, and *tahavvül*, which signify, with nuances, change from one form to another. Interestingly, while all these words have Arabic origins, Arab authors themselves commonly used the word *tatawwur* to refer to evolution. See Elshakry, *Reading Darwin*, 17, 41.

45. On the transitional forms, see, for instance, the thirteenth-century Persian mystic Aziz Nasafi's writings: *Persian Metaphysics and Mysticism: Selected Treatises of 'Azīz Nasafī*, ed. Lloyd Ridgeon (Surrey, England: Curzon Press, 2002), 234. Also see Yasien Mohamed, "The Cosmology of Ikhwān al-Ṣafā', Miskawayh and al-Iṣfahānī," *Islamic Studies* 4 (2000): 657–79. For Turkish interpretations, see Abdullah Ucman, "Devir Nazariyesi ve Osmanli Tasavvuf Edebiyatinda Devriyyeler," in *Osmanlı Toplumunda Tasavvuf ve Sufiler*, ed. Ahmet Yaşar Ocak (Ankara: TTK, 2014), 575–625.

46. For an example, see the account by Mas'udi (d. 965) in Tarif Khalidi, *Anthology of Arabic Literature from the Classical to the Modern* (Edinburgh: Edinburgh University Press, 2016), 62–63.

47. *The History of Al-Tabari*, vol. 2, *Prophets and Patriarchs*, trans. William M. Brinner (Albany: SUNY Press, 1987), 13.

48. Ismail Yakıt, "Darwin'den Önce İslam Düşünürlerinde Evrimle İlgili Fikirler," *Felsefe Arkivi* 24 (1984): 101–22.

49. Friedrich Dieterici, *Der Darwinismus im zehnten und neunzehnten Jahrhundert* (Leipzig: J.C. Hinrich, 1878); B. Placzek, "Anthropoid Mythology," *Popular Science Monthly*, September 1882, 655–63.

50. George Sarton, *Introduction to the History of Science*, vol. 3 (Baltimore: Williams and Wilkins, 1947), 211–13.

51. Alexander D. Knysh, *Ibn 'Arabi in the Later Islamic Tradition: The Making of a Polemical Image in Medieval Islam* (Albany: SUNY Press, 1999); Ahmed Zildzic,

"Friend and Foe: The Early Ottoman Reception of Ibn 'Arabi" (PhD diss., University of California, Berkeley, 2012).

52. Özkan Öztürk, "Çağdaş Türk Düşüncesinde Ibn Arabi Felsefesinin Ele Alınışı" (thesis, Marmara University, 2007).

53. Elshakry, *Reading Darwin*, 142–44.

54. See Kutluğhan Soyubol, "In Search of Perfection: Neo-spiritualism, Islamic Mysticism, and Secularism in Turkey," *Modern Intellectual History* 1 (2021): 70–94; Alireza Doostdar, "Empirical Spirits: Islam, Spiritism, and the Virtues of Science in Iran," *Comparative Studies in Society and History* 2 (2016): 322–49. On Buddhism, see Peter Harrison, "A Scientific Buddhism?," *Zygon* 4 (2010): 861–69; Donald S. Lopez Jr., *Buddhism and Science: A Guide for the Perplexed* (Chicago: University of Chicago Press, 2009). On the reinventions of Ibn Arabi as a mystic for modern times, see Gregory A. Lipton, *Rethinking Ibn 'Arabi* (New York: Oxford University Press, 2018).

55. Elshakry, *Reading Darwin*, 99.

56. On Haeckel's monism, see Niles R. Holt, "Ernst Haeckel's Monistic Religion," *Journal of the History of Ideas* 2 (1971): 265–80; and Robert J. Richards, "Ernst Haeckel and the Struggles over Evolution and Religion," *Annals of the History and Philosophy of Biology* 1 (2006): 89–116.

57. Ernst Haeckel, *Vahdet-i Mevcud: Bir tabiat âliminin dini*, trans. Ahmet Nebil and Baha Tevfik ([Istanbul]: Dersaadet Kütüphanesi, 1911), 4.

58. Haeckel, *Vahdet-i Mevcud*, 77. The original section is Ernst Haeckel, *Monism as Connecting Religion and Science* (London: Adam and Charles Black, 1894), 78.

59. Haeckel, *Vahdet-i Mevcud*, 86–88.

60. Ahmed Rıfkı, "Felsefe-i Tasavvuf: Seyh-i Ekber Muhyiddîn-i Arabî: Felsefesinden Mâ-Ba'd," *Yirminci Asırda Zekâ* 11 (July 23, 1912): 174.

61. Baha Tevfik, "Felsefiyyât: Plotin ve Muhyiddîn," *Yirminci Asırda Zekâ* 12 (August 6, 1912): 211.

62. Ahmed Avni (Konuk), *Fusûsu'l-Hikem Tercüme ve Şerhi*, vol. 1 (Istanbul: Türkiye Yazma Eserler Kurumu, 2017), 59–60.

63. Ahmed Avni (Konuk), *Fusûsu'l-Hikem Tercüme ve Şerhi*, vol. 2 (Istanbul: Türkiye Yazma Eserler Kurumu, 2017), 872.

64. Avni, *Fusûsu'l-Hikem*, 1:194. For a similar remark, see 64 as well.

65. M. Sait Özervarlı, "Alternative Approaches to Modernization in the Late Ottoman Period: İzmirli İsmail Hakki's Religious Thought against Materialist Scientism," *International Journal of Middle East Studies* 1 (2007): 84.

66. Izmirli Ismail Hakkı, *Muhtasar Felsefe-i Ula* (Istanbul: Hukuk, 1913), 184–86. The section can be found in Émile Boirac, *Cours élémentaire de philosophie* (Paris: F. Alcan, 1888), 437–40.

67. Ismail Hakkı, *Muhtasar*, 214. The word he uses in this section is *tehavvül*, which could also be translated as "transformation," but because he refers to evolutionism as *tehavvuliyyun*, I translate it as "evolution."

68. Some forty years later, Ismail Hakkı would return to the topic and discuss Muslim scholars' and Sufis' views that supported the idea of evolution, adding that the Darwinian theory was no longer endorsed by modern scientists. See his *Islam Mütefekkirleri ile Garp Mütefekkirleri Arasında Mukayese* (Ankara: Diyanet Işleri Başkanlığı, 1952), 36–41.

69. Mehmed Ali Ayni, "Şeyh İbrahim Hakkı," in *İntikad ve Mülahazalar* (Istanbul: Kütübhane-i Sudi, 1923), 123. Originally published in *Darülfünun Edebiyat Fakültesi Mecmuası* 2 (1916): 117–30.

70. Ayni mentions this in a letter he wrote on April 18, 1908, which was reproduced in the newspaper the *Vakit* on November 21, 1942, p. 8. Ayni's reference here is to Hans Kraemer, ed., *Weltall und menschheit*, 5 vols. (Berlin: Bong and Co., 1902–1904).

71. Mehmed Ali Ayni, *Şeyh Ekber'i Niçin Severim* (Istanbul: Evkaf-ı İslamiye, 1923).

72. David Kohn, "Darwin's Ambiguity: The Secularization of Biological Meaning," *British Journal for the History of Science* 2 (1989): 215–39.

73. See, among many, Robert J. Richards, "That Darwin and Haeckel Were Complicit in Nazi Biology," in *Galileo Goes to Jail and Other Myths about Science and Religion*, ed. Ronald L. Numbers (Cambridge, MA: Harvard University Press, 2009), 170–77; Tony Barta, "Mr. Darwin's Shooters: On Natural Selection and the Naturalizing of Genocide," *Patterns of Prejudice* 2 (2005): 116–37; Richard Weikart, *From Darwin to Hitler: Evolutionary Ethics, Eugenics and Racism in Germany* (New York: Palgrave Macmillan, 2016); Peter J. Bowler, *Darwin Deleted: Imagining a World without Darwin* (Chicago: University of Chicago Press, 2013).

74. Quoted in Elshakry, *Reading Darwin*, 110.

75. Abdullah Cevdet, "Fas Hükümet-i Islamiyesinin lnkırazı," *Ictihad* 5 (April 1905): 71–72; quoted in Ahmet Çapku, "Batıcılık Düşüncesi Çerçevesinde Abdullah Cevdet'te Din-Bilim-Felsefe İlişkisi," *Kutadgubilig* 24 (2013): 167–95, quotation at 176.

13: "SCIENCE IS JUSTIFIED BY WORKS, NOT BY FAITH"

1. For Haeckel's influence, see V.I. Lenin, "Ernst Haeckel and Ernst Mach," in *Materialism and Empirico-Criticism*, trans. Marxist Internet Archive (Moscow: Zveno Publishers, 1909); Robert Richards, *The Tragic Sense of Life: Ernst Haeckel and the Struggle Over Evolutionary Thought* (Chicago: University of Chicago Press, 2008), 2–3; Frank Sulloway, *Freud: Biologist of the Mind; Beyond the Psychoanalytic Legend* (Cambridge, MA: Harvard University Press, 1992), 199; Henry Adams, *The Education of Henry Adams* (New York: Houghton Mifflin, 1918), 453.

2. Erik Nordenskiöld, *The History of Biology: A Survey*, 2nd ed. (New York: Alfred A. Knopf, 1929), 506. For circulation of Haeckel's books, see Richards, *Tragic Sense*, 2.

3. Richards, *Tragic Sense*, 2.

4. Ernst Haeckel, *The History of Creation*, trans. E. Ray Lankester, 2nd ed., 2 vols. (London: Henry S. King, 1876), 1:4, 23.

5. Ernst Haeckel, *The Riddle of the Universe at the Close of the Nineteenth Century*, trans. Joseph McCabe (New York: Harper and Bros., 1900), 323.

6. Haeckel, *Riddle*, 323, 307; for Haeckel on science and culture, v; on the eternity of the cosmos, 13.

7. Haeckel, *Creation*, 1:71.

8. Haeckel, *Riddle*, 327.

9. Haeckel, *Riddle*, 312.

10. Haeckel, *Riddle*, 312.

11. "The older, naïve mythology . . . is more intelligible than the modern mystic theosophy that adores a personal God as an invisible—properly speaking, gaseous being, yet makes him think, speak, and act in human fashion; it gives us the paradoxical picture of a 'gaseous vertebrate.'" Haeckel, *Riddle*, 12.

12. Haeckel, *Riddle*, 278–79, 352.

13. Haeckel, *Riddle*, 311.

14. Haeckel, *Riddle*, 10.

15. Haeckel, *Riddle*, 309.

16. Haeckel, *Riddle*, 289, 291; emphasis in original; Ernst Haeckel, *Eternity*, trans. Thomas Seltzer (New York: Truth Seeker, 1916), 53–54.

17. Haeckel, *Creation*, 2:369; Haeckel, *Riddle*, 331; Ernst Haeckel, *Monism as Connecting Science and Religion: The Confession of Faith of a Man of Science*, trans. J. Gilchrest (London: Charles and Adam Black, 1895), 86.

18. Haeckel, *Riddle*, 336; emphasis in original.

19. Haeckel, *Creation*, 1:23; also 4, 30. Mario Di Gregorio has likewise written, "Haeckel's main concern was to build a *Weltanschauung* that was based on the scientific principles he outlined in the *GM*. [i.e., in Haeckel's first popular science book, the *Generelle Morphologie*]." Mario Di Gregorio, *From Here to Eternity: Ernst Haeckel and Scientific Faith* (Göttingen: Vandenhoeck and Ruprecht, 2005), 203.

20. Richards, *Tragic Sense*, 228–32; Gregorio, *Eternity*, 544–74; Nick Hopwood, *Haeckel's Embryos: Images, Evolution, and Fraud* (Chicago: University of Chicago Press, 2015); Daniel Gasman, *The Scientific Origins of National Socialism: Social Darwinism in Ernst Haeckel and the German Monist League* (New Brunswick: Macdonald and Co., 2004).

21. See Hopwood, *Haeckel's Embryos*; Lennart Olsson and Uwe Hossfeld, "The Haeckel Reception in Sweden," *Theory in Biosciences* 138, no. 1 (2019): 119–25; Eduard Kolchinsky and Georgy S. Levit, "The Reception of Haeckel in Pre-Revolutionary Russia and His Impact on Evolutionary Theory," *Theory in Biosciences* 138, no. 1 (2019): 73–88; Eduard I. Kolchinsky, "Russian Editions of E. Haeckel's Works and the Evolution of their Perception," *Theory in Biosciences* 138, no. 1 (2019): 49–71.

22. Haeckel, *Riddle*, 321.

23. William Keith Brooks, "The Tyranny of the Monistic Creed, a Review," *Science*, n.s. 1, no. 14 (Apr. 5, 1895): 382.

24. Louis Agassiz, "Evolution and Permanence of Type," *Atlantic Monthly* 33 (Jan. 1874): 101. Agassiz criticized Haeckel at length in an appendix to the French translation of his *Essay on Classification*, but as this did not appear in English it exerted little influence on the English-language discussion.

25. Agassiz, "Evolution and Permanence," 97.

26. Stephen Jay Gould, "Agassiz's Later, Private Thoughts on Evolution: His Marginalia in Haeckel's *Natürliche Schöpfungsgeschichte*," in *Two Hundred Years of Geology in America*, ed. Cecil J. Schneer (Hanover, NH: University Press of New England, 1979), 280, 281.

27. Haeckel, *Creation*, 1:66, 70.

28. Haeckel, *The Evolution of Man* (London: C. Kegan Paul and Co., 1879), I:116.

29. Quoted in Christoph Irmscher, *Louis Agassiz: Creator of American Science* (Boston: Houghton Mifflin Harcourt, 2013), 12. According to Irmscher, "Haeckel had nothing good to say about Agassiz" (162).

30. Ernst Haeckel, *The Wonders of Life* (London: Watts, 1904), 352.

31. Edward Lurie, *Louis Agassiz: A Life in Science* (Chicago: University of Chicago, 1960), 169, 179–84.

32. "Agassiz, Jean Louis Rodolphe," in *Complete Dictionary of Scientific Biography*, 26 vols. (Detroit, MI: Charles Scribner's Sons, 2008), 1:72; Irmscher, *Louis Agassiz*, 2, 111–20; Lurie, *Louis Agassiz*, 333; Irmscher, *Louis Agassiz*, 3, 38–40; "Agassiz, Jean Louis Rodolphe," 74.

33. "Gray, Asa," in *Complete Dictionary of Scientific Biography*, 5:513.

34. T. Russell Hunter, "Making a Theist Out of Darwin: Asa Gray's Post-Darwinian Natural Theology," *Science & Education* 21, no. 7 (July 2012): 967.

35. Hunter, "Making a Theist," 973.

36. Hunter, "Making a Theist," 961, 967–68.

37. Joseph LeConte, *Evolution* (New York: D. Appleton, 1891), 288.

38. LeConte, *Evolution*, 295–96. LeConte denied that evolution had materialistic implications: 285–87.

39. LeConte, *Evolution*, 282–83.

40. Charles Sedgwick Minot, "Criticisms of Haeckel," *American Naturalist: An Illustrated Magazine of Natural History*, no. 11 (1877): 368.

41. Minot, "Criticisms," 369.

42. Minot, "Criticisms," 369.

43. The *American Naturalist* aspired to a general as well as a technical audience in the 1870s. For the changing editorial posture of the publication, see Betty Smocovitis et al., "Trends and Transitions in 150 Years of the American Naturalist," *American Naturalist* 196, no. 6 (Dec. 2020): 663–78, especially 665. Perhaps Packard took less offense at Haeckel's abuse of Agassiz than Minot, on account of a quarrel between him and Louis Agassiz in the 1860s.

44. Charles Sedgwick Minot, "Sketch of Comparative Embryology," *American Naturalist* 14 (1880): 479–80.

45. Samuel Irenaeus Prime, "Made Without a Maker," in *Irenaeus Letters: Originally Published in the New York Observer* (New York: New York Observer, 1881), 335.

46. For a fuller discussion of these illustrations, see Hopwood, *Haeckel's Embryos*, 73–84; Richards defends Haeckel in *Tragic Sense*, 234–44.

47. Richards, *Tragic Sense*, 242–44.

48. "Notes and Literature: General Biology," *American Naturalist* 41 (1907): 589–92. Haeckel has also been charged with fraud by contemporary biologists. See Elizabeth Pennisi, "Haeckel's Embryos: Fraud Rediscovered," *Science*, n.s., 277, no. 5331 (Sep. 5, 1997): 1435; Michael Richardson and Gerhard Keuck, "A Question of Intent: When Is 'Schematic' Illustration a Fraud?" *Nature* 410, no. 144 (2001): 144. Richards rebuts in "Haeckel's Embryos: Fraud Not Proven," *Biology & Philosophy* 24, no. 1 (2009): 147–54; and *Tragic Sense*, 305–12.

49. John W. Judd, "Alexander Agassiz, For. MEM. R. S.," *Nature* 83 (Apr. 1910): 163. Judd refers to Alexander Agassiz as an oceanographer but also "a great authority on the *Echinodermata*" (a large phylum of marine invertebrates, including starfish, sea urchins, and sea cucumbers) and lauds his extensive research on coral (also invertebrates). "Agassiz, Alexander," in *Complete Dictionary of Scientific Biography*, 1:71; George Lincoln Goodale, "Biographical Memoir of Alexander Agassiz, 1835–1910," *Biographical Memoirs*, vol. 7 (Washington, DC: National Academy of Sciences, 1912), 297.

50. Goodale, "Biographical Memoir," 298.

51. Alexander Agassiz, "Scientific Intelligence III: Botany and Zoology," *American Journal of Science and Arts*, ser. 3, v. 11 (original ser. v. 111) (Jan.–Jun. 1876): 74.

52. Alexander Agassiz, *Revision of the Echini* (Cambridge, MA: University Press, 1874), 754.

53. Agassiz, *Revision*, 762.

54. "Brooks, William Keith," in *Complete Dictionary of Scientific Biography*, 2:501.

55. William Keith Brooks, "The Tyranny of the Monistic Creed, a Review," *Science*, n.s. 1, no. 14 (Apr. 5, 1895): 382.

56. Brooks, "Tyranny," 382; emphasis in original.

57. Brooks, "Tyranny," 382.

58. Brooks, "Tyranny," 384.

59. William Keith Brooks, *Foundations of Zoölogy* (New York: Columbia University Press, 1899), 136, 132–33.

60. David Starr Jordan, "A Sage in Science," *Science*, n.s. 9, no. 224 (Apr. 14, 1899): 529–32.

61. "Jordan, David Starr," in *Complete Dictionary of Scientific Biography*, 7:169–70.

62. "Jordan, David Starr," 126–27; David Starr Jordan, *Footnotes to Evolution* (New York: D. Appleton, 1907), 368, 375.

63. David Starr Jordan, *The Stability of Truth* (New York: H. Holt, 1911), 66. The American philosopher Frank Thilly (1865–1934), of Missouri University, argued that Haeckel's monism did not even make sense on its own terms, independent of

Jordan's insistence on empirical demonstrability. "The World-View of a Scientist," *Popular Science Monthly* 61 (Nov. 1901–Apr. 1902): 407–25.

64. Jordan, *Stability*, 72.

65. Jordan, *Stability*, 73–77.

66. Jordan, *Footnotes to Evolution*, 368.

67. Jordan, *Footnotes to Evolution*, 376. Jordan refers to the liberal politician William Ewart Gladstone (1809–1898), who was prime minister of Great Britain several times in the latter half of the nineteenth century. Between 1885 and 1891 he and Huxley carried on a debate over the merits of evolution and theology in the pages of the *Nineteenth Century*. See Jonathan Conlin, "Gladstone, Development, and the Discipline of History, 1840–96," *Historical Journal* 63, no. 4 (2020): 911–34.

68. Vernon L. Kellogg, *Darwinism To-Day* (New York: H. Holt, 1907), 130–31.

69. Kellogg, *Darwinism*, 21.

70. M. Elizabeth Barnes, "Ernst Haeckel's Biogenetic Law (1866)," in *The Embryo Project Encyclopedia*, May 3, 2014, http://embryo.asu.edu/handle/10776/7825.

71. Vernon L. Kellogg, "Review: Haeckel's *Evolution of Man*," *Science*, n.s. 32, no. 827 (Nov. 4, 1910): 629.

72. Weismann had achieved fame by snipping the tails off of several generations of mice and watching for signs that their tails had grown shorter as a result (1888). According to Weismann, this should have been observed if Lamarckian "use-inheritance" really conditioned heredity. No such shortening was observed, however, and this was regarded as an important point in the debate between Darwinians and neo-Lamarckians, c. 1900. See Peter J. Bowler, *Evolution* (Berkeley: University of California Press, 2009), 255. Thus Weismann, like Haeckel, was one of Germany's leading Darwinists during the war years.

73. Haeckel, *Eternity*, 149–50; Di Gregorio, *Eternity*, 18.

74. Vernon L. Kellogg, *Headquarters Nights* (Boston: Atlantic Monthly Press, 1917), 28.

75. Kellogg, *Headquarters Nights*, 21–22.

76. Richards, *Tragic Sense*, 432.

77. Minot, "Criticisms," 368.

78. Kellogg, "Review," 629. "Der Haeckelismus in der Zoologie" was the title of a polemical pamphlet written against Haeckel by the German zoologist Carl Semper (1832–1893). To the best of my knowledge, it has not been translated into English.

79. Richards, *Tragic Sense*, 3, 24.

80. Ichnologists study the fossilized footprints of extinct animals.

81. Eric Schocket, "Proletarian Paperbacks: The Little Blue Books and Working-Class Culture," *College Literature* 29, no. 4 (Fall 2002): 67.

82. Carroll Lane Fenton, *Haeckel's Monistic Philosophy* (Girard, KS: Haldeman-Julius Co., 1924), 6–7, 29, 32.

83. Fenton, *Haeckel's Monistic Philosophy*, 6–7.

84. Bernard Lightman, "Fiske, Youmans, and the Appropriation of the System," in *Global Spencerism*, ed. Bernard Lightman (Boston: Brill, 2016), 136.

85. See Bernard Lightman, ed., *Victorian Scientific Naturalism* (Chicago: University of Chicago Press, 2014); Frederick Gregory, *Scientific Materialism in Nineteenth Century Germany* (Boston: D. Reidel, 1977).

86. Bowler, *Evolution*, 97.

87. Gregorio, *Eternity*, 21.

88. Lenin, "Ernst Haeckel and Ernst Mach."

89. Kolchinsky and Levit, "Reception of Haeckel," 74.

90. Kolchinsky, "Russian Editions," 60.

91. James Moore, "Deconstructing Darwinism: The Politics of Evolution in the 1860's," *Journal of the History of Biology* 24, no. 3 (Autumn 1991): 353–408; see especially 383–89.

92. It is reasonable to suppose that as the X-Club gained influence, as evolution became less controversial, and as "progressive" ideas such as social Darwinism and eugenics entered the mainstream, attitudes toward Haeckel may have softened, but no extensive study of Haeckel's influence in Britain has yet been undertaken. Undoubtedly he had at least one ally in the British scientific establishment—the Darwinian zoologist Edwin Ray Lankester, who facilitated the translation of Haeckel's *Natural History of Creation*, his first book to appear in English.

93. Olsson and Hossfeld, "Haeckel Reception," 119–25.

94. Gabriel Finkelstein, "Haeckel and du-Bois Reymond: Rival German Darwinists," *Theory in Biosciences* 138, no. 1 (2019): 105–12; Frederick C. Beiser, *After Hegel: German Philosophy, 1840–1900* (Princeton, NJ: Princeton University Press, 2014), 128–32.

95. Richards, *Tragic Sense*, chaps. 8, 9.

96. Far too much has been made of Hodge's opposition in *What Is Darwinism?* For a fuller picture, see Bradley J. Gundlach, *Process and Providence: The Evolution Question at Princeton, 1845–1929* (Grand Rapids, MI: William B. Eerdmans, 2013).

97. Lightman, "Fiske, Youmans," 137–41.

98. "Significant Books of Science," *Atlantic Monthly: A Magazine of Literature, Science, Art, and Politics* 96 (Jul.–Dec. 1905): 682.

99. K. Liu, "Yan Fu's Daoist Reinterpretation of Evolutionism," in *Asian Religious Responses to Darwinism: Evolutionary Theories in Middle Eastern, South Asian, and East Asian Cultural Contexts*, ed. C. Mackenzie Brown (Cham, Switzerland: Springer, 2020), 298.

100. Martin Reixinger, "South Asian Muslim Responses to Darwinism," in Brown, *Asian Religious Responses*, 43.

101. D. Killingley, "The Hindu Heritage and Hindu Criticism of Darwinism," in Brown, *Asian Religious Responses*, 149.

102. Richards, *Tragic Sense*, 166, 205.

103. Haeckel, *Creation*, 1:1–2.

104. Haeckel, *Riddle*, 336; Gregorio, *Eternity*, 1, 368–69, 574.

105. Gregorio, *Eternity*, 18, 26–27, 44–46.

106. Richards, *Tragic Sense*, 278.

107. Haeckel, *Riddle*, 382.

108. Haeckel, *Riddle*, 291.

109. Haeckel, *Riddle*, 336.

110. Jordan, *Stability of Truth*, 90–91.

CODA: FROM COMPLEXITY TO THE TRANSNATIONAL

1. Ronald Numbers, "Simplifying Complexity: Patterns in the History of Science and Religion," in *Science and Religion: New Historical Perspectives*, ed. Thomas Dixon, Geoffrey Cantor, and Stephen Pumfrey (Cambridge: Cambridge University Press, 2010), 264.

2. Numbers, "Simplifying Complexity," 272.

3. Peter Harrison, *The Territories of Science and Religion* (Chicago: University of Chicago Press, 2015).

4. Harrison, *Territories*, x.

5. Myrna Perez Sheldon, Ahmed Ragab, and Terence Keel, eds., *Critical Approaches to Science and Religion* (New York: Columbia University Press, 2023).

SELECT BIBLIOGRAPHY

This bibliography includes key scholarly works that discuss modern science, evolution, and religion, and can serve as a guide to further reading. The bibliography is divided into general studies, and local and regional studies.

GENERAL STUDIES OF SCIENCE, EVOLUTION, AND RELIGION

Brooke, John Hedley, and Ronald L. Numbers, eds. *Science and Religion around the World*. Oxford: Oxford University Press, 2011.

Dixon, Thomas, Geoffrey Cantor, and Stephen Pumphrey, eds. *Science and Religion: New Historical Perspectives*. Cambridge: Cambridge University Press, 2010.

Fehige, Yiftach, ed. *Science and Religion: East and West*. London: Routledge, 2016.

Ferngren, Gary B., ed. *Science and Religion: A Historical Introduction*. 2nd ed. Baltimore: Johns Hopkins University Press, 2017.

Glick, Thomas F., ed. *The Comparative Reception of Darwinism*. Chicago: University of Chicago Press, 1988.

Harrison, Peter. *The Territories of Science and Religion*. Chicago: University of Chicago Press, 2015.

Lightman, Bernard, ed. *Global Spencerism: The Communication and Appropriation of a British Evolutionist*. Leiden: Brill, 2016.

Lightman, Bernard, ed. *Rethinking History, Science and Religion: An Exploration of Conflict and the Complexity Principle*. Pittsburgh: University of Pittsburgh Press, 2019.

Livingstone, David N. *Dealing with Darwin: Place, Politics, and Rhetoric in Religious Engagements with Evolution*. Baltimore: Johns Hopkins University Press, 2014.

Numbers, Ronald L., and John Stenhouse, eds. *Disseminating Darwinism: The Role of Place, Race, Religion, and Gender*. Cambridge: Cambridge University Press, 1999.

Richards, Robert J. *The Tragic Sense of Life: Ernst Haeckel and the Struggle over Evolutionary Thought*. Chicago: University of Chicago Press, 2008.

Ruse, Michael, ed. *The Cambridge Encyclopedia of Darwin and Evolutionary Thought*. Cambridge: Cambridge University Press, 2013.

Sivasundaram, Sujit. "A Global History of Science and Religion." In *Science and Religion: New Historical Perspectives*, edited by Thomas Dixon, G.N. Cantor, and Stephen Pumfrey, 177–97. Cambridge: Cambridge University Press, 2010.

Slotten, Hugh Richard, Ronald L. Numbers, and David N. Livingstone, eds. *Modern Science in National, Transnational and Global Context*. Vol. 8 of *The Cambridge History of Science*, edited by David C. Lindberg and Ronald L. Numbers. Cambridge: Cambridge University Press, 2020.

LOCAL AND REGIONAL STUDIES

ANGLO-AMERICAN

Dawson, Gowan, and Bernard Lightman, eds. *Victorian Scientific Naturalism: Community, Identity, Continuity*. Chicago: University of Chicago Press, 2014.

Gundlach, Bradley J. *Process and Providence: The Evolution Question at Princeton, 1845–1929*. Grand Rapids, MI: William B. Eerdmans, 2013.

Larson, Edward J. *Summer for the Gods: The Scopes Trial and America's Continuing Debate over Science and Religion*. New York: Basic Books, 2006.

Larson, Edward J. *Trial and Error: The American Controversy over Creation and Evolution*. Oxford: Oxford University Press, 2003.

Livingstone, David. *Adam's Ancestors: Race, Religion, and the Politics of Human Origins*. Baltimore: Johns Hopkins University Press, 2008.

Moore, James R. *The Post-Darwinian Controversies: A Study of the Protestant Struggle to Come to Terms with Darwin in Great Britain and America, 1870–1900*. Cambridge: Cambridge University Press, 1981.

Newell, Catherine L. *Destined for the Stars: Faith, the Future, and America's Final Frontier*. Pittsburgh: University of Pittsburgh Press, 2019.

Numbers, Ronald L. *The Creationists: From Scientific Creationism to Intelligent Design*. Cambridge, MA: Harvard University Press, 2008.

Roberts, John H. *Darwinism and the Divine in America: Protestant Intellectuals and Organic Evolution, 1859–1900*. Madison: University of Wisconsin Press, 1988.

Secord, James A. *Victorian Sensation: The Extraordinary Publication, Reception, and Secret Authorship of "Vestiges of the Natural History of Creation."* Chicago: University of Chicago Press, 2000.

Sera-Shriar, Efram, ed. *Historicizing Humans: Deep Time, Evolution, and Race in Nineteenth-Century British Sciences*. Pittsburgh: University of Pittsburgh Press, 2018.

Stanley, Matthew. *Huxley's Church and Maxwell's Demon: From Theistic Science to Naturalistic Science*. Chicago: University of Chicago Press, 2015.

Ungureanu, James C. *Science, Religion, and the Protestant Tradition: Retracing the Origins of Conflict*. Pittsburgh: University of Pittsburgh Press, 2019.

ASIA

Andresen, Jensine. "Barbour, Buddhism, Bohm, and Beyond." In *Fifty Years in Science and Religion: Ian G. Barbour and His Legacy*, edited by Robert John Russell, 329–40. Aldershot, UK: Ashgate, 2004.

Bayly, C.A. *Empire and Information: Intelligence Gathering and Social Communication in India 1780–1870*. Cambridge: Cambridge University Press, 1996.

Brown, C. Mackenzie, ed. *Asian Religious Responses to Darwinism: Evolutionary Theories in Middle Eastern, South Asian, and East Asian Cultural Contexts*. Cham, Switzerland: Springer, 2020.

Brown, C. Mackenzie. *Hindu Perspectives on Evolution: Darwin, Dharma, and Design*. London: Routledge, 2012.

Chakrabarti, Pratik. *Inscriptions of Nature: Geology and the Naturalization of Antiquity*. Baltimore: Johns Hopkins University Press, 2020.

Davis, Winston. *The Moral and Political Naturalism of Baron Katō Hiroyuki*. Berkeley, CA: Institute of East Asian Studies, 1996.

Diagne, Souleymane Bachir. *Postcolonial Bergson*. New York: Fordham University Press, 2019.

Dodson, Michael S. *Orientalism, Empire and National Culture*. Basingstoke, UK: Palgrave Macmillan, 2007.

Elman, Benjamin. *On Their Own Terms: Science in China, 1550–1900*. Cambridge, MA: Harvard University Press, 2005.

Godart, G. Clinton. *Darwin, Dharma, and the Divine: Evolutionary Theory and Religion on Modern Japan*. Honolulu: University of Hawai'i Press, 2017.

Hammerstrom, Erik. *The Science of Chinese Buddhism: Early Twentieth-Century Engagements*. New York: Columbia University Press, 2015.

Howland, Douglas R. *Translating the West: Languages and Political Reason in Nineteenth-Century Japan*. Honolulu: University of Hawai'i Press, 2002.

Huang, Kewu. "Tianyan yu Fofa: *Tianyanlun* dui Qingmo Minchu Foxue sixiang de chongji" [Evolution and Buddhism: The impact of *Tianyanlun* on Buddhist thinking in late Qing and early Republic of China], in *Fanyishi yanjiu 2018*, edited by Wang Hongzhi, 199–222. Shanghai: Fudan daxue chubanshe, 2020.

Jin, Xiaoxing. "The Evolution of Evolutionism in China, 1870–1930." *Isis* 111, no. 1 (2020): 46–66.

Jin, Xiaoxing. "Translation and Transmutation: The *Origin of Species* in China." *British Journal for the History of Science* 52, no. 1 (2019): 117–41.

Kahandagama, Anushka. "Buddhist Revivalist Movement under British Rule in the Nineteenth Century Sri Lanka: 'Sciences' of Enlightenment and Buddhism." MPhil diss., South Asian University, Delhi, 2019.

Kern, Emily. "Out of Asia: A Global History of the Scientific Search for the Origins of Humankind, 1800–1965." PhD thesis, Princeton University, 2018.

Kwok, Danny W.Y. *Scientism in Chinese Thought: 1900–1950*. New Haven, CT: Yale University Press, 1965.

Li, Wenyu. *Tianyanlun boyi* [Refutations of evolutionism]. 2nd ed. Shanghai: Xujiahui tushanwan yinshuguan, 1932.

Lopez, Donald S., Jr. *The Scientific Buddha: His Short and Happy Life*. New Haven, CT: Yale University Press, 2012.

McMahan, David L. *The Making of Buddhist Modernism*. Oxford: Oxford University Press, 2009.

Mukharji, Projit. "Vernacularizing the Body: Informational Egalitarianism, Hindu Divine Design and Race in Physiology Schools, Bengal 1859–1877." *Bulletin of the History of Medicine* 91, no. 3 (2017): 554–85.

Pusey, James R. *China and Charles Darwin*. Cambridge, MA: Harvard University Asia Center, 1983.

Quack, Johannes. *Disenchanting India: Organized Rationalism and Criticism of Religion in India*. New York: Oxford University Press, 2012.

Raina, Dhruv, and S. Irfan Habib. "The Moral Legitimation of Modern Science: Bhadralok Reflections on Theories of Evolution." *Social Studies of Science* 26 (1996): 9–42.

Raj, Kapil. *Relocating Modern Science: Circulation and the Construction of Scientific Knowledge in South Asia and Europe, Seventeenth to Nineteenth Centuries*. Delhi: Permanent Black, 2006.

Ramaswamy, Sumathi. *Terrestrial Lessons: The Conquest of the World as Globe*. Chicago: University of Chicago Press, 2017.

Savary, Luzia. *Evolution, Race and Public Spheres in India*. New York: Routledge, 2019.

Schwartz, Benjamin. *In Search of Wealth and Power: Yan Fu and the West*. Cambridge, MA: Harvard University Press, 1964.

Sen, Joydeep. *Astronomy in India 1784–1876*. Abingdon, UK: Routledge, 2016.

Sivasundaram, Sujit. *Islanded: Britain, Sri Lanka, and the Bounds of an Indian Ocean Colony*. Chicago: University of Chicago Press, 2013.

Sponsel, Leslie E., and Poranee Natadecha-Sponsel. "Environment and Nature in Buddhism." In *Encyclopaedia of the History of Science, Technology, and Medicine in Non-Western Cultures*, edited by Helaine Selin, 775–76. Dordrecht: Springer Netherlands, 2008.

Thomas, Renny. *Science and Religion in India: Beyond Disenchantment*. Abingdon, UK: Routledge, 2022.

Wallace, B. Alan, ed. *Buddhism and Science: Breaking New Ground*. New York: Columbia University Press, 2003.

Wan, Zhaoyuan. *Science and the Confucian Religion of Kang Youwei (1858–1927): China before the Conflict Thesis*. Leiden: Brill, 2022.

Wang, Tiangen. *Tianyanlun chuanbo yu Qingmo Minchu de shehui dongyuan* [The spread of *Tianyanlun* and the social mobilization in late Qing and early Republic of China]. Hefei: Hefei gongye daxue chubanshe, 2006.

Watanabe, Masao. *The Japanese and Western Science*. Philadelphia: University of Pennsylvania Press, 1990.

Xue, Yuqin, and Xu Huabo. "Qingmo Minchu zhishijie dui Yan yi *Tianyanlun* de pipan yu fansi-yi Li Wenyu de *Tianyanlun boyi* wei zhongxin" [The intellectual refutations of and reflections on Yan's translation of *Evolution and Ethics* in late Qing and early Republic of China]. *Shijie zongjiao yanjiu* 1 (2016): 109–25.

Yang, Haiyan. "Knowledge Across Borders: The Early Communication of Evolution in China." In *The Circulation of Knowledge between Britain, India and China*, edited by Bernard Lightman, Gordon McOuat, and Larry Stewart, 179–208. Leiden: Brill, 2013.

AUSTRALIA AND NEW ZEALAND

Butcher, Barry W. "Darwinism and Australia, 1836–1914." PhD diss., University of Melbourne, 1992.

Butcher, Barry W. "Darwin's Australian Correspondents: Deference and Collaboration in Colonial Science." In *Nature in Its Greatest Extent: Western Science in the Pacific*, edited by Roy MacLeod and Philip F. Rehbock, 139–57. Honolulu: University of Hawai'i Press, 1988.

Butcher, Barry W. "Gorilla Warfare in Melbourne: Halford, Huxley and 'Man's Place in Nature.'" In *Australian Science in the Making*, edited by R.W. Home, 153–69. Cambridge: Cambridge University Press, 1988.

Finney, Colin. *Paradise Revealed: Natural History in Nineteenth-Century Australia*. Melbourne: Museum of Victoria, 1993.

Frame, Tom. *Evolution in the Antipodes: Charles Darwin and Australia*. Sydney: University of New South Wales Press, 2009.

MacLeod, Roy, and Philip F. Rehbock, eds. *Darwin's Laboratory: Evolutionary Theory and Natural History in the Pacific*. Honolulu: University of Hawai'i Press, 1994.

Moyal, Ann, with Robert E. Marks. "The Scientists and Darwin's *The Origin of Species* in Nineteenth Century Australia: A Re-evaluation." *Journal and Proceedings of the Royal Society of New South Wales* 152 (2019): 5–26.

Mozley, Ann. "Evolution and the Climate of Opinion in Australia, 1840–76." *Victorian Studies* 10 (1967): 411–30.

Mozley Moyal, Ann. "Sir Richard Owen and His Influence on Australian Zoological and Palaeontological Science." *Records of the Australian Academy of Science* 3, no. 2 (1975): 41–56.

LATIN AMERICA

Català, José Sala. *Ideología y Ciencia Biológica en España entre 1860 y 1881: La Difusión de un Paradigma*. Madrid: CSIC, 1987.

Domingues, H.M. Bertol, M. Romer Sá, and Thomas Glick, eds. *A recepção do Darwinismo no Brasil*. Rio de Janeiro: Editora Fiocruz, 2003.

Glick, Thomas. *Darwin en España*. Barcelona: Ediciones Península, 1982.

Glick, Thomas, M. Puig-Samper, and R. Ruiz Gutiérrez, eds. *The Reception of Darwinism in the Iberian World: Spain, Spanish America and Brazil*. Boston: Springer, 2001.

Pelayo, Francisco. *Ciencia y Creencia en España durante el Siglo XIX*. Madrid: CSIC, 1999.

MIDDLE EAST

Buchwald, Jed Z., and Diane Greco Josefowicz. *The Zodiac of Paris: How an Improbable Controversy over an Ancient Egyptian Artifact Provoked a Modern Debate between Religion and Science*. Princeton, NJ: Princeton University Press, 2010.

Elshakry, Marwa. "Knowledge in Motion: The Cultural Politics of Modern Science Translations in Arabic." *Isis* 99, no. 4 (2008): 701–30.

Elshakry, Marwa. "Muslim Hermeneutics and Arabic Views of Evolution." *Zygon* 46, no. 2 (2011): 330–44.

Elshakry, Marwa. *Reading Darwin in Arabic, 1860–1950*. Chicago: University of Chicago Press, 2013.

Gold, Meira. "Ancient Egypt and the Geological Antiquity of Man, 1847–1863." *History of Science* 57, no. 2 (2019): 194–230.

Hanioğlu, Şükrü. "Blueprints for a Future Society: Late Ottoman Materialists on Science, Religion, and Art." In *Late Ottoman Society: The Intellectual Legacy*, edited by Elizabeth Özdalga, 50–138. New York: Routledge, 2013.

Howard, Damian. *Being Human in Islam: The Impact of the Evolutionary Worldview*. New York: Routledge, 2011.

Keddie, Nikki R., and Jamāl ad-Dīn Afghānī. *An Islamic Response to Imperialism: Political and Religious Writings of Sayyid Jamāl Ad-Din "Al-Afghānī."* Berkeley: University of California Press, 1968.

Mestyan, Adam. *Primordial History, Print Capitalism, and Egyptology in Nineteenth-Century Cairo*. Cairo: Institut Français D'Archéologie Orientale, 2021.

Schaffer, Simon. "Oriental Metrology and the Politics of Antiquity in Nineteenth-Century Survey Sciences." *Science in Context* 30, no. 2 (2017): 173–212.

Sedra, Paul. *From Mission to Modernity: Evangelicals, Reformers and Education in Nineteenth-Century Egypt*. New York: Palgrave Macmillan, 2011.

Stolz, Daniel A. *The Lighthouse and the Observatory: Islam, Science, and Empire in Late Ottoman Egypt*. Cambridge: Cambridge University Press, 2018.

RUSSIA

Siddiqi, Asif A. *The Red Rockets' Glare: Spaceflight and the Russian Imagination, 1857–1957*. Cambridge: Cambridge University Press, 2010.

Stites, Richard. "Fantasy and Revolution: Alexander Bogdanov and the Origins of Bolshevik Science Fiction." In *Red Star: The First Bolshevik Utopia*, edited by Loren R. Graham and Richard Stites, 1–16. Bloomington: Indiana University Press, 1984.

Vucinich, Alexander. *Darwin in Russian Thought*. Berkeley: University of California Press, 1988.

CONTRIBUTORS

JOEL BARNES is a historian based at the University of Queensland, Australia. He was previously a postdoctoral research fellow on the Science and Religion: Exploring the Spectrum of Global Perspectives project. In 2018–2020 he was a research associate in the Australian Centre for Public History at the University of Technology Sydney, working on an Australian Research Council–funded project on the history of humanities institutions in Australia since 1945. His published work has appeared in journals including *History Australia*, *History of Education*, *History of Education Review*, *History of Humanities*, *History of the Human Sciences*, and *Historical Research*.

FA-TI FAN teaches history at Binghamton University, State University of New York. He is the author of *British Naturalists in Qing China: Science, Empire, and Cultural Encounter* and several dozen essays on science in East Asia and the global history of science. He is currently president of the History of Science Society.

TAMARA FERNANDO is an assistant professor in history at SUNY Stony Brook, New York. Her research has three main focal points: the construction of scientific knowledge about the ocean; the connected intellectual, scientific, and labor histories of the Indian Ocean between the Middle East, South Asia, and Southeast Asia; and the question of how the scientific archive addresses histories of the non-human. Her research has been published in *Past & Present* and *Comparative Studies in Society and History*. Her present book project is a multisited history of pearling in the Indian Ocean.

MEIRA GOLD is an assistant professor/faculty fellow in history of science at the Gallatin School, New York University. After earning her PhD from the University of Cambridge in 2020, she held postdoctoral fellowships at the European University Institute and York University. Her research explores colonial Egyptology, human antiquity, and the politics of fieldwork in the long nineteenth century. Gold received the Alexander Prize from the Royal Historical Society and the Reingold Prize from the History of Science Society for her articles on these topics. She is currently completing her first monograph, *Archaeology from Ruins: Victorian Egyptology and the Making of a Colonial Field Science.*

EDWARD GUIMONT is assistant professor of world history at Bristol Community College in Fall River, Massachusetts. He is coauthor of the book *When the Stars Are Right: H.P. Lovecraft and Astronomy* and is currently writing a monograph on the history of the Flat Earth movement in the British empire.

DANIEL HALVERSON is a PhD student at University of Toronto, Institute for the History and Philosophy of Science and Technology, where he studies history of science and religion: especially evolution, freethinking, and Christianity in the nineteenth century. He earned his BA, with honors, from Florida Atlantic University, and his MA from Case Western Reserve University. His master's thesis, "Monographs on the Universe: Americans Respond to Ernst Haeckel's Evolutionary Science and Theology, 1866–1883," won the Melvin Kranzberg Prize for best master of arts thesis written that year and was later published in *Past Tense*. He is a past editor of *Spontaneous Generations: A Journal for the History and Philosophy of Science.*

BERNARD LIGHTMAN is Distinguished Research Professor in the Humanities Department at York University and past president of the History of Science Society. Lightman's research focuses on the cultural history of Victorian science. Among his most recent publications are the edited collections *Rethinking History, Science and Religion*, *Science Periodicals in Nineteenth Century Britain* (coedited with Gowan Dawson, Sally Shuttleworth, and Jonathan Topham), and *Identity in a Secular Age* (coedited with Fern Elsdon-Baker). He is one of the general editors of the John Tyndall Correspondence Project, an international collaborative effort to obtain, digitalize, transcribe, and publish all surviving letters to and from Tyndall.

JAUME NAVARRO is Ikerbasque Research Professor at the University of the Basque Country (Spain), with a long career in the history of science, mostly the history of physics, the philosophy of science (with a focus in historical epistemology), and the historiography of science and religion. He is author of *A History of the Electron: J.J. and G.P. Thomson* (Cambridge University Press, 2012) and *Ciencia-Religión y sus Tradiciones Inventadas* (Tecnos, 2022) and editor, among others, of *Science and Faith Within Reason: Reality, Creation, Life and Design* (Ashgate, 2011), *Research and Pedagogy: A History of Quantum Physics Through Its Early Textbooks* (Editions Open Access, 2013), and *Ether and Modernity: The Recalcitrance of an Epistemic Object in the Early Twentieth Century* (Oxford University Press, 2018).

SARAH QIDWAI is a postdoctoral researcher at the University of Regensburg. She is a historian of science and empire in the nineteenth century. Broadly speaking, she works on transnational and local perspectives of various scientific disciplines during the long nineteenth century.

SIMON SCHAFFER is an emeritus fellow of Darwin College Cambridge. Recent publications include "Late Enlightenment Crises of Facts" (*Configurations*), "The Ark and the Archive" (*Studies in Romanticism*), "Instruments and Ingenuity between Britain and India" (*Bulletin of the Scientific Instruments Society*), and *A Book of the Body Politic* (with Bruno Latour and Pasquale Gagliardi). He is a member of the Leverhulme Trust research project "Making Climate History," based at the University of Cambridge. In 2021 he was supported by the Translitterae PSL program at the Ecole Normale Supérieure, Paris.

IGNACIO SILVA is professor of science and theology at the Universidad Austral, Argentina. Formerly a research fellow at the Ian Ramsey Centre for Science and Religion, University of Oxford, Silva focuses on the relations between science and religion. He is the editor of *Latin American Perspectives on Science and Religion* (Pickering and Chatto, 2014) and of *Human and Divine Providence* (Routledge, 2020) and the author of *Science and Providence in a World of Contingency* (Routledge, 2021). His papers are featured in the *Journal of Religion*, the *European Journal for Philosophy of Religion*, *Zygon: Journal of Religion and Science*, and the *Heythrop Journal*.

JOHN STENHOUSE is associate professor of history at the University of Otago, where he teaches the history of science, intellectual history, European history, and New Zealand history. Recent publications include *Eugenics at the Edges of Empire: New Zealand, Australia, Canada and South Africa* (Palgrave Macmillan, 2018), coedited with Diane B. Paul and Hamish G. Spencer; and "Missionary Science" in *The Cambridge History of Science*, vol. 8, *Modern Science in National, Transnational and Global Context*, ed. Hugh R. Slotten, Ronald L. Numbers, and David N. Livingstone (Cambridge University Press, 2020).

STEPHEN P. WELDON is professor and chair of the Department of History of Science, Technology, and Medicine at the University of Oklahoma. His historical focus is on science and society, especially science and religion in the modern era. His book, *The Scientific Spirit of American Humanism* (Johns Hopkins University Press, 2020), explores how scientists and science-minded intellectuals contributed to liberal and radical religion in twentieth-century America. Weldon also edits the *Isis Bibliography of the History of Science*, the definitive bibliographical resource for the discipline of history of science, and he developed and manages an online open-access service called IsisCB Explore based on that bibliography.

WAN ZHAOYUAN, DPhil (Oxon), is associate professor in the history of science at the Centre for Studies of Values and Culture and the School of Philosophy, Beijing Normal University, China. His research areas include science and religion, Newton studies, and the history of scientific thought. He has published *Science and the Confucian Religion of Kang Youwei (1858–1927): China Before the Conflict Thesis* (Brill, 2022), as well as a dozen translations into Chinese.

M. ALPER YALÇINKAYA is assistant professor of sociology at TED University in Ankara, Türkiye. He previously taught at Ohio Wesleyan University in the United States and was a research fellow at the Carlos III University of Madrid. Yalçınkaya's research and publications focus on the cultural history of science in Muslim contexts, particularly the nineteenth-century Ottoman Empire, and Muslim contributions to the global debate on science and religion.

TOMOKO YOSHIDA's scholarship focuses on Japanese religious history and on the intersection between science and religion in Asia. She has published and translated material related to the historiography of medieval Buddhism in Japan, medieval Japanese court culture, and Shinto. She is the recipient of the 38th Akegarasu Haya Prize for her essay on Fujikawa Yū and his scientific interpretation of Buddhist doctrine, and her coauthored article "Asian Traditions" appeared in *Science and Religion: A Historical Introduction* (2017). She has taught Chinese and Japanese history, language, and literature.

INDEX